Hurricane Climatology

Hurricane Climatology

A Modern Statistical Guide Using R

James B. Elsner and Thomas H. Jagger

OXFORD
UNIVERSITY PRESS

OXFORD
UNIVERSITY PRESS

Oxford University Press is a department of the University of Oxford.
It furthers the University's objective of excellence in research, scholarship,
and education by publishing worldwide.

Oxford New York
Auckland Cape Town Dar es Salaam Hong Kong Karachi
Kuala Lumpur Madrid Melbourne Mexico City Nairobi
New Delhi Shanghai Taipei Toronto

With offices in
Argentina Austria Brazil Chile Czech Republic France Greece
Guatemala Hungary Italy Japan Poland Portugal Singapore
South Korea Switzerland Thailand Turkey Ukraine Vietnam

Published in the United States of America by
Oxford University Press
198 Madison Avenue, New York, NY 10016

Library of Congress Cataloging-in-Publication Data
Elsner, James B., author.
Hurricane climatology : a modern statistical guide using R / James B. Elsner and Thomas H. Jagger.
pages ; cm
Includes bibliographical references and index.
ISBN 978-0-19-982763-3 (hardcover : alk. paper)
1. Hurricanes–Forecasting–Statistical methods.
2. R (Computer program language) I. Jagger, Thomas H., author. II. Title.
QC944.E47 2013
551.55′20285555—dc23
2012019317

CONTENTS

Part Two: Models and Methods

PREFACE

"The goal is to provide analytical tools that will last students a lifetime."
—Edward Tufte

A hurricane is nature's most destructive storm. Violent wind, flooding rain, and powerful surge pose hazards to life and property. Changes in hurricane activity could have significant societal consequences. Hurricanes are the leading source of insured losses from natural catastrophes worldwide and global warming could make the strongest hurricanes even stronger.

Most of what we know about hurricanes comes from past storms. Hurricane climatology is the study of hurricanes as a collection of past events. It answers questions about when, where, and how often. This book is an argument that there is much more to learn. The goal is to show you how to analyze, model, and predict hurricane climate using data. It shows you how to create statistical models from hurricane data that are accessible, reproducible, and explanatory.

The book is didactic. It teaches you how to learn about hurricane climate. It uses statistics. Statistics is the science of organizing and interpreting data. Statistics is best done using R. The text is written around code that when copied to an R session reproduces the graphs, tables, and maps presented. The approach is different from other books that use R. Rather than showing you how to use R for a variety of purposes, this book focuses on a single topic and shows you how to better understand the topic by using R.

The first five chapters provide background material on R and doing statistics. This material is appropriate for an undergraduate course on statistical methods in the environmental sciences. Chapter 6 presents details on the data sets that are used in the later chapters. Chapters 7, 8, and 9 lay out the building blocks of models for hurricane climate research. This material is appropriate for graduate-level courses in climatology. Chapters 10–13 give examples from our more recent research that could be used in a seminar on methods and models for hurricane climate analysis and prediction.

The book benefited from the research conducted with our students including Robert Hodges, Jill Trepanier, and Kelsey Scheitlin. Editorial assistance came from Laura Michaels with additional help from Sarah Strazzo. We thank Ian Elsner for the book's Web site. Our sincere thanks go to the R Core Development Team for maintaining the software and to the many package authors who enrich the software environment. We thank Richard Murnane and Tony Knap of the Risk Prediction Initiative for their unwavering support over the years.

Gratitude goes to Svetla for her wry humor.

James B. Elsner and Thomas H. Jagger
Tallahassee, Florida
August 2012

Hurricane Climatology

PART ONE

Data, Statistics, and Software

1

HURRICANES, CLIMATE, AND STATISTICS

"Chaos was the law of nature; Order was the dream of man."
—Henry Adams

This book is about hurricanes, climate, and statistics. These topics may not seem related. Hurricanes are violent winds and flooding rains, climate is about weather conditions from the past, and statistics is about numbers. But what if you wanted to estimate the probability of winds exceeding 60 m s^{-1} in Florida next year. The answer involves all three, hurricanes (fastest winds), climate (weather of the past), and statistics (probability). This book teaches you how to answer these questions in a rigorous and scientific way. We begin here with a short description of the topics and a few notes on what this book is about.

1.1 HURRICANES

A hurricane is an area of low air pressure over the warm tropical ocean. The low pressure creates showers and thunderstorms that start the winds rotating. The rotation helps to develop new thunderstorms. A tropical storm forms when the rotating winds exceed 17 m s^{-1} and a hurricane when they exceed 33 m s^{-1}.[1] Once formed, the winds continue to blow despite friction by an in-up-and-out circulation that imports heat at high temperature from the ocean and exports heat at lower temperature in the upper troposphere (near 16 km), which is similar to the way a steam engine converts thermal energy to mechanical motion. In short, a hurricane is powered by moisture and heat.

Strong winds are a hurricane's defining characteristic. Wind is caused by the change in air pressure between two locations. In the center of a hurricane, the air

[1] The winds are estimated over a one-minute duration at 10 m above the ocean surface.

pressure, which is the weight of a column of air from the surface to the top of the atmosphere, is quite low compared with the air pressure outside the hurricane. This difference causes the air to move from the outside inward toward the center. By a combination of friction as the air rubs on the ocean below and the spin of the earth as it rotates on its axis, the air does not move directly inward but rather spirals in a counterclockwise direction toward the region of lowest pressure.

Pressure differences between the cyclone center and the surrounding air determine the speed of the wind. Since the pressure outside most hurricanes is the same, a hurricane's central pressure is a good measure of a hurricane's intensity. The lower the central pressure, the more intense the hurricane. Pressures inside the most intense hurricanes are among the lowest that occur anywhere on the earth's surface at sea level. In the largest and most intense hurricanes, the strongest winds are located in the wall of thunderstorms that surrounds the nearly calm central eye. If the hurricane is stationary (spinning, but with no forward motion), the field of winds is shaped like a bagel, with a calm center and the fastest winds forming a ring around the center.

The distance from the center of the hurricane to the location of the hurricane's strongest winds is called the radius to maximum winds. The strongest winds are in the eye-wall and the radius to maximum winds ranges from several kilometers in the smallest hurricanes to several hundred or more kilometers in the largest. While the wind just above the ocean surface spirals counterclockwise toward the center (in the Northern Hemisphere), the air at high altitudes blows outward in a clockwise spiral. This outward flowing air produces thin cirrus clouds that extend distances of thousands of kilometers from the center of circulation. The presence of these clouds may be the first visible sign that a hurricane is approaching.

Landfall occurs when the hurricane center crosses a coastline. Because the fastest winds are located in the eye-wall, it is possible for a hurricane's fastest winds to be over land even if landfall does not occur. Similarly, it is possible for a hurricane to make landfall and have its fastest winds remain out at sea. Fortunately, the winds slacken quickly after the hurricane moves over land. High winds destroy poorly constructed buildings and mobile homes. Flying debris such as signs and roofing material add to the destructive power.

Hurricanes also cause damage as a result of flooding rain and storm surge. Rainfall is the amount of water that falls over an area during a given time interval. Hurricanes derive energy from the ocean by evaporating water into the air, which then gets converted back to liquid through condensation inside the thunderstorm clouds. The liquid falls from the clouds as rain. The stronger the thunderstorms, the more the rain and the potential for flooding.

At a given location, the amount of rain that falls depends on factors including hurricane intensity, forward speed, and the underlying topography. Forward speed refers to how fast the center moves. Rainfall can be enhanced by winds blowing up a mountain. Antecedent moisture conditions also play a role. Freshwater flooding can be a serious danger even hundreds of kilometers from the point of landfall. Bands of showers and thunderstorms that spiral inward toward the center are the first weather experienced as a hurricane approaches. High wind gusts and heavy downpours occur in the individual rain bands, with relatively calm weather occurring between the bands. Brief tornadoes sometimes form in the rain bands especially as the cyclone moves inland.

The force of the winds moving around the storm creates the storm surge by pushing the ocean water toward the shore. Over the open ocean, the water can flow in all directions away from the storm. Strong winds blowing across the ocean surface create a stress that forces the water levels to increase downwind and decrease upwind. This wind setup is inversely proportional to ocean depth, so over the deep ocean away from land the water level rises are minimal. As the hurricane approaches shallow water, there is no room for the water to flow underneath so it rises and gets pushed by the wind as surge. Slope of the continental shelf and low atmospheric pressure also play a role. A shallow slope allows for greater surge. The ocean level rise due to the low air pressure adds to the surge, but its magnitude (about 1 cm per hectopascal drop in pressure) is less than that caused by the wind. The advancing surge may increase the water level 5 m or more above the sea level. In addition, wind-driven waves are superimposed on the storm surge. The total water level can cause severe surge impacts in coastal areas, particularly when the storm surge coincides with astronomical tides caused by the moon and sun.

1.2 CLIMATE

Climate is a set of weather patterns. One description of hurricane climate is the number of hurricanes over a given region and time. Other descriptions include the average hurricane intensity, the percentage of hurricanes that make it to land, or metrics that combine several attributes into one. Elsner and Kara (1999) present a wealth of climatological information on North Atlantic hurricanes.

On average, 50 hurricanes occur worldwide each year. Hurricanes develop during the time of the year when the ocean temperatures are the hottest. Over the North Atlantic, this includes the months of June through November, with a sharp peak from late August through the middle of September when the direct rays of the summer sun have had the largest impact on sea temperature. Worldwide, May is the least active month while September is the most active. During the twentieth century, hurricanes made landfall in the United States at an average rate of five every 3 years.

Hurricanes vary widely in intensity as measured by their fastest moving winds. Hurricane intensities are grouped into five categories (Saffir–Simpson scale) with the weakest category-one winds blowing at most 42 mps and the strongest category-five winds exceeding speeds of 69 mps. Three category-five hurricanes hit the United States during the twentieth century, including the Florida Keys Hurricane in 1935, Hurricane Camille in 1969, and Hurricane Andrew in 1992. Hurricanes also vary considerably in size (spatial extent), with the smallest hurricanes measuring only a few hundred kilometers in radius (measured from the eye center to the outermost closed line of constant surface pressure) and the largest exceeding a thousand kilometers or more.

Hurricanes are steered by large-scale wind patterns in the atmosphere above the surface and by the earth's spin. In the deep tropics, these forces push a hurricane slightly north of due west (in the Northern Hemisphere). Once north of about 23 degrees of latitude, a hurricane tends to take a more northwestward track and then eventually a northeastward at still higher latitudes. Local fluctuations in the magnitude and direction of steering can result in tracks that deviate significantly from this pattern.

Hurricane climate is linked to what is happening to the earth's global climate. Greater ocean heat, for example, enhances the potential for stronger hurricanes. This can occur on time scales of weeks to years. El Niño is a good example. It is the name given to climate changes over the tropical Pacific region caused by warming of the ocean surface by a few degrees. This leads to a shift in the tropical thunderstorms normally over the western Pacific near Indonesia eastward toward the central and eastern Pacific, closer to the Atlantic. The thunderstorms create upper air winds that inhibit hurricane formation. Another example is the position and strength of the subtropical high-pressure ridge, which on average is situated at a latitude of 30° N. A hurricane forming on the equatorward side of the ridge gets steered toward the northwest. During some years, the subtropical ridge is farther south and west, forcing the hurricanes westward through the Caribbean Sea and toward Mexico and the United States.

1.3 STATISTICS

Statistics is the collection, organization, analysis, and interpretation of data. Here you will use them to describe, model, and predict hurricane climate. Like people, all hurricanes are unique. A person's behavior might be difficult to anticipate, but the average behavior of a group of people is quite predictable. A particular hurricane may move eastward through the Caribbean Sea, but hurricanes typically travel westward in this part of the world. Statistical models quantify this typicalness. Furthermore, statistical models provide a syntax for describing uncertainty.

Statistics is not the same as arithmetic. A strong hurricane might be described as a 1-in-100-year event. If you do the arithmetic, you might conclude that after the event there is plenty of time ($100 - 1 = 99$ years) before the next one. This is wrong. Statistics tells you that you can expect a hurricane of this magnitude or larger to occur once in 100 years *on average*, but it could occur next year with a probability of 1 percent. Moreover, hurricanes a bit weaker only will be more frequent and perhaps a lot more so.

The adage is that you can use statistics to say anything. But this assumes deceit. Without statistics anything you say is mere opinion. Recent upward trends in hurricane activity have spurred a debate as to the cause. Statistics helps clarify the issues. A good reference textbook for statistical methods in weather and climate is available from Wilks (2006). Statistics can be divided into descriptive and inferential. Descriptive statistics is used to summarize data. Data are hurricane records. Inferential statistics is used to draw conclusions about the processes that regulate hurricane climate. They help you understand how climate operates.

Inference is used to test competing scientific hypotheses. Inference from data requires attention to randomness, uncertainty, and data biases. It also requires a model. A model is a concise (as few parameters as possible) description of your data. A set of counts representing the number of hurricanes occurring in a region each year, can be described using a Poisson distribution (model) and a rate (parameter). Given the rate, the modeled counts should match the observed counts. This allows you to understand hurricane frequency and how it varies with climate through variations in the rate.

Statistical models are needed because proof is impossible. The primary goal is to determine what is *likely* to be true. When you toss a frisbee, you know it will come down and where. This is physics. By contrast, if you smoke you might get lung cancer. If the seas get hotter, hurricanes might get stronger. You, know that your odds of getting cancer are higher if you smoke. And you know that the odds of stronger hurricanes are higher if the seas are warmer. You can work out these odds with a statistical model, but you cannot know with certainty whether you as an individual smoker will get cancer, or whether the next hot year will bring stronger hurricanes.

The relationship between hurricanes and climate can be subtle. Differences in the spatial and temporal scales are large. Hurricanes form in many different ways, are impacted by various environmental conditions, and dissipate under all sorts of influences. Environmental factors, genesis mechanisms, and feedbacks interact in complex and multifaceted ways, so establishing proof is impossible. Instead, you observe, compare, and weigh the evidence for or against various hypotheses.

Despite limitations to predicting individual behavior of hurricanes beyond a few days imposed by thus complexity, it is possible to predict average behavior at lead times of weeks to decades. Hurricane climate predictability arises from the slow evolution of ocean heating and due to processes associated with monthly to seasonal dynamics of the atmosphere–ocean system. Progress has been made, much of it from using statistical models. Most of what we know about hurricanes derives from observation. Statistical models are built from observational data. Observations become more precise over time leading to heterogeneous data sets, but a good modeling strategy is to use as much information as possible.

Inferential statistics come in two flavors: frequentist and Bayesian. Frequentist (classical) inference is what you learn in school. It involves methods of hypothesis testing and confidence intervals.

Bayesian inference is a way to calculate a probability that your hypothesis is true by combining data with prior belief. The term "Bayesian" comes from Bayes theorem, which provides the formula for the calculation. For example, what is the probability that you are correct in claiming that the strongest hurricanes are getting stronger? With Bayesian inference, probability is your degree of belief rather than an intrinsic characteristic of the world.

The underlying Bayesian principle is the accumulation of evidence. Evidence comes from historical and geologic data that, by their very nature, are incomplete and fragmentary. Much has been written about the differences in philosophy between classical and Bayesian approaches. Here we are pragmatic and use both. Frequentist methods are used except in cases where we think Bayesian methods provide an advantage.

1.4 R

Science demands openness and reproducibility. It is essential that you explain exactly what you did and how you did it. All computations should be reproducible. The R language makes this easy to do. Redoing all ecological fieldwork is asking too much. But with climate research, reproducibility is perfectly possible.

R contains a number of built-in functions for organizing, graphing, and modeling your data. Directions for obtaining R, accompanying packages, and other sources of documentation are provided at `http://www.r-project.org/`. R is an open-source project, which means that it depends on a community of active developers to grow and evolve. No person or company owns it. R is maintained and supported by thousands of individuals worldwide who use it and who contribute to its ongoing development. Base R is mostly written in C (with R and some Fortran); R packages are mostly written in R and C.

The book is interlaced with R code, so you can read it as a workbook. It was written in the Sweave format of Leisch (2003). Sweave is an implementation of the literate programming style advocated by Knuth (1992) which permits an interplay between code written in R, the output of that code, and commentary on the code. Sweave documents are preprocessed by R to produce a LaTeX document. The book also shows you how to use R to construct graphs and maps for presenting your data, including modern tools for data inspection and visualization. The text uses Arno Pro font, with italics for data set names outside of R. A bold face is used for package names and a typewriter font for the R code.

1.5 ORGANIZATION

The book has two parts. The first part (Chapters 1–6) concerns the background material focusing on software, statistics, and data. The second part (Chapters 7–13) concerns methods and models used in hurricane climate research. Chapter 2 is a tutorial on using R. If you have never used R before, start here. The material is presented using examples from hurricane climatology. Chapter 3 provides a review of introductory statistics. Topics include distributions, two-sample tests, correlation, and regression. Chapter 4 is an introduction to Bayesian statistics. Chapter 5 is how to make graphs and maps from your data. Chapter 6 gives information about the various data sets used in the later chapters.

Chapter 7 is on models for hurricane occurrence. Emphasis is on the Poisson distribution for count data. Chapter 8 is on models for hurricane intensity including quantile regression and models from extreme value theory. Chapter 9 is on spatial models for analyzing point patterns, area averages, and field data. Chapter 10 looks at various types of cluster models including temporal, spatial, and feature clusters. Chapter 11 examines time series models including a look at the novel application of networks. Chapter 12 provides examples of Bayesian models including those that make use of Markov chain Monte Carlo methods. Chapter 13 considers a few impact models.

Chapters 2–6 provide introductory material and a tutorial in using R. The material can be presented as part of a course on statistical methods in the environmental sciences at the undergraduate level. Chapter 6 provides details on the data sets that are used in the later chapters and can be skipped on first reading. Chapters 7, 8, and 9 provide the basic building blocks of models for hurricane climate research. This material is appropriate for advanced undergraduate- and graduate-level courses in climatology. Chapters 10–13 show examples from more recent research that can be used in a graduate seminar on methods and models for hurricane analysis and prediction.

The examples come from our research on hurricanes. All the data sets and code used in the book are available at www.hurricaneclimate.com, and readers are encouraged to reproduce the figures, graphs, and tables. We provide references to the relevant literature on the methods, but do not focus on this. We use standard statistical notation and a voice that makes the reader the subject.

This book does not cover global climate models (GCMs). Interest in GCMs is predicated on the assumption that progress in forecasting hurricanes will come only through improvements in dynamical models. If the hurricane response to variations in climate is highly nonlinear or involves many intervening variables then dynamical model skill will outpace skill from statistical models. On the other hand, if the response is near linear and associated with only a handful of variables, improvements to dynamical models will result in forecasts that largely duplicate the skill available using statistical models but at substantially greater expense.

Consider that a dynamical model of all the cells in your body and the associated chemistry and physics is not likely to provide a more accurate prediction of when you will die (from "natural" causes) than a statistical model based on your cohorts (age, sex, etc.) that takes into account a few important variables related to diet and exercise. The analogy is not perfect as there is not a cohort of earth climates from which to build a statistical model. But under the assumption of homogeneity and stationarity, you have pseudoreplicates that can be used as cohorts. If we are right, then greater attention needs to be paid to statistics.

And there is no better way to do statistics than with R. Chapter 2 provides a tutorial, on how to get started.

2

R TUTORIAL

"I think it is important for software to avoid imposing a cognitive style on workers and their work."
—Edward Tufte

This chapter is a tutorial on using R. To get the most out of it, you should open an R session and type the commands into the console as you read the text. You should be able to use copy-and-paste if you have access to an electronic version of the book. All code is available on the book's Web site.

2.1 INTRODUCTION

Science requires transparency and reproducibility. The R language for statistical modeling makes this easy. Developing, maintaining, and documenting your R code is simple. R contains numerous functions for organizing, graphing, and modeling your data. Directions for obtaining R, accompanying packages, and other sources of documentation are available at `http://www.r-project.org/`. Anyone serious about applying statistics to climate data should learn R.

The book is self-contained. It presents R code and data (or links to data) that can be copied to reproduce the graphs and tables. This reproducibility provides you with an enhanced learning opportunity. Here we present a tutorial to help you get started. This can be skipped if you already know how to work with R.

2.1.1 What Is R?

R is the 'lingua franca' of data analysis and statistical computing. It helps you perform a variety of computing tasks by giving you access to commands. This is similar to other programming languages such as Python and C++. R is particularly useful to researchers because it contains a number of built-in functions for organizing data, performing calculations, and creating graphics.

R is an open-source statistical environment modeled after S. The S language was developed in the late 1980s at AT&T labs. The R project was started by Robert Gentleman and Ross Ihaka of the Statistics Department of the University of Auckland in 1995. It now has a large audience. It is currently maintained by the R core-development team, an international group of volunteer developers. To get to the R project Web site, open a browser and, in the search window, type the keywords "R project" or directly link to the Web page using `http://www.r-project.org/`. Directions for obtaining the software, accompanying packages, and other sources of documentation are provided at the site.

Why use R? It is open source, free, runs on the major computing platforms, has built-in help and excellent graphing capabilities, and it is powerful, extensible, and contains thousands of functions. A drawback for some is the limited graphical user interface (GUI). This means that it is harder to learn at the outset and you need to appreciate syntax. Modern ways of working with computers include browsers, music players, and spreadsheets. R is nothing like these. Infrequent users forget commands. There are no visual cues; a blank screen is intimidating. At first, working with R might seem daunting. However, with a little effort, you will quickly learn the basic commands and then realize how it can help you do much, much more.

R is really a library of modern statistical tools. It is unmatched in its breadth and scope. A climate scientist whose research requires customized scripting, extensive simulation analysis, or state-of-the-art statistical analysis will find R to be a solid foundation. R is also a language. It has rules and syntax. R can be run interactively, and analysis can be done on the fly. It is not limited by a finite set of functions. You can download packages to perform specialized analysis and graph the results. R is object oriented, allowing you to define new objects and create your own methods for them.

Many people use spreadsheets. This is good for tasks such as data storage and manipulation. Unfortunately, spreadsheets are unsuitable for serious research. This is because it is hard to make publication-quality graphs, there is limited community support for new methods, and it can be a challenge to share your calculations with others. If you are serious about reproducible research, you should not use a spreadsheet for statistics.

2.1.2 Get R

At the R project Web site, click on CRAN (Comprehensive R Archive Network) and select a nearby mirror site. To download the base distribution of R, follow the instructions appropriate for your operating system. On your computer, click on the download icon and follow the install instructions. Click on the icon to start R. If you are using the Linux operating system, type the letter R from a command window.

R is most easily used in an interactive manner. You ask R a question and it gives you an answer. Questions are asked and answered on the command line. The > (greater than symbol) is used as the prompt. Throughout this book, it is the character that is

NOT typed or copied into your R session. If a command is too long, a + (plus symbol) is used as a continuation prompt. If you get lost on a command, you can use Ctrl-c or Esc to get the prompt back and start over. Most commands are functions, and most functions require parentheses.

2.1.3 Packages

A package is a set of functions for doing specific things. As of early 2012, there were over 3,200 of them. Indeed, this is one of the main attractions, the availability of literally hundreds of packages for performing statistical analysis and modeling. And the range of packages is enormous. The **BiodiversityR** offers a graphical interface for calculations of environmental trends, the package **Emu** analyzes speech patterns, and the package **GenABEL** is used to study the human genome.

To install and load the package called **UsingR** type

```
> install.packages("UsingR")
```

Note that the syntax is case sensitive. `UsingR` is not the same as `usingR` and `Install.Packages` is not the same as `install.packages`. After the package is downloaded to your computer, you make it available to your R session by typing

```
> library(UsingR)
```

Or

```
> require(UsingR)
```

Note again that the syntax is case sensitive. When installing the package, the package name needs to be in quotes (either signal or double, but not directional). Quotes are not needed when making the package available to your working session. Each time you start a new R session, the package needs to be made available, but it does not need to be installed. If a package is not available from a CRAN site, you can try another. To change your CRAN site, type

```
> chooseCRANmirror()
```

Then select a different location.

2.1.4 Calculator

R evaluates commands typed at the prompt and returns the result of the computation to the screen or saves it to an object. For example, to find the sum of the square root of 25 and 2, type

```
> sqrt(25) + 2
[1] 7
```

The `[1]` says "first requested element will follow." Here there is only one element, the answer 7. The > prompt that follows indicates R is ready for another command. For example, type

```
> 12/3 - 5
[1] -1
```

R uses order of operations so that, for instance, multiplication comes before addition.
How would you calculate the 5th power of 2? Type

```
> 2^5
[1] 32
```

How about the amount of interest on $1,000, compounded annually at 4.5 percent (annual percentage) for 5 years? Type

```
> 1000 * (1 + .045)^5 - 1000
[1] 246
```

2.1.5 Functions

There are numerous mathematical and statistical functions available in R. They are all used in a similar manner. A function has a name, which is typed, followed by a pair of parentheses (required). Arguments are added inside the parentheses as needed. For example, the square root of two is given as

```
> sqrt(2)     # the square root
[1] 1.41
```

The # is the comment character. Any text in the line following this character is treated as a comment and is not evaluated. Some other examples include

```
> sin(pi)     # the sine function
[1] 1.22e-16
> log(42)    # log of 42 base e
[1] 3.74
```

Most functions have arguments that allow you to change the default behavior. For example, to use base 10 for the logarithm, you can use either of the following:

```
> log(42, 10)
[1] 1.62
> log(42, base=10)
[1] 1.62
```

To understand the first function, `log(42,10)`, you need to know that R expects the base to be the second argument (after the comma) of the function. The second example uses a named argument of the type `base=10` to explicitly set the base value.

The first style contains less typing, but the second style is easier to remember and is good coding practice.

2.1.6 Warnings and Errors

When R does not understand your function, it responds with an error message. For example

```
> srt(2)

Error in try(srt(2)) : could not find function "srt"
```

If you get the function correct, but your input is not acceptable, then

```
> sqrt(-2)

NaN
```

The output `NaN` is used to indicate "not a number."

As mentioned, if R encounters a line that is not complete, a continuation prompt, `+`, is printed, indicating more input is expected. You can complete the line after the continuation prompt.

2.1.7 Assignments

It is convenient to name an object so that you can use it later. Doing so is called an assignment. Assignments are straightforward. You put a name on the left-hand side of the equals sign and a value, function, object, and so on, on the right. Assignments do not usually produce printed output.

```
> x = 2     # assignments return a prompt only
> x + 3     # x is now 2
[1] 5
```

Remember, the pound symbol (#) is used as a comment character. Anything after the it is ignored. Adding comments to your code is a way of recalling what you did and why.

You are free to make object names out of letters, numbers, and the dot or underline characters. A name starts with a letter or a dot (a leading dot may not be followed by a number). You are not allowed to use math operators, such as `+`, `-`, `*`, and `/`. The help page for `make.names` describes this in more detail (`?make.names`).

Note that case is also important in names; `x` is different than `X`. It is good practice to use conventional names for data types. For instance, `n` is for the number of data records or the length of a vector, `x` and `y` are for spatial coordinates, and `i` and `j` are for integers and indices for vectors and matrices. These conventions are not forced, but consistency makes it easier for others to understand your code.

Variables that begin with the dot character are reserved for advanced programmers. Unlike many programming languages, the period in R is used only as a punctuation and can be included in an object name (e.g., `my.object`).

2.1.8 Help

Using R to do statistics requires knowing many functions—more than you can likely keep in your head. R has built-in help for information about what is returned by the function, for details on additional arguments, and for examples. If you know the name of a function, type

```
> help(var)
```

This brings up a help page for the function named inside the parentheses. (`?var` works the same way). The function and package names are given in the preamble. This is followed by a brief description of the function and how it is used. Arguments are explained along with function and argument details. Examples are given toward the bottom of the page. A good way to understand what a function does is to run the examples.

Most help pages provide examples. The examples help you understand what the function does. You can try the examples individually by copying and pasting them into your R session. You can also try them all at once by using the `example` function. For instance, type

```
> example(mean)
```

Help pages work great if you know the name of the function. If not, the function `help.search("mean")` searches each entry in the help system and returns matches (often many) of functions that mention the word `"mean"`. The function `apropos` searches through only function names and variables for matches. Type

```
> apropos("mean")
```

To end your R session, type

```
> q(save="no")
```

Like most R functions, `q` needs an open (left) and a close (right) parentheses. The argument `save="no"` says do not save the workspace. Otherwise the workspace and session history are saved to a file in your working directory. By default, the file is called `.RData`. The workspace is your current R working environment and includes all your objects, like vectors, matrices, data frames, lists, functions and so on.

2.2 DATA

2.2.1 Small Amounts

To begin, you need to get your data into R. For small amounts, you use the `c` function, which combines (concatenates) items. Consider a set of hypothetical hurricane counts, where in the first year there were two landfalls, in the second there were three, landfalls and so on. To enter these values, type

```
> h = c(2, 3, 0, 3, 1, 0, 0, 1, 2, 1)
```

The 10 values are stored in a vector object of class numeric called `h`.

```
> class(h)
[1] "numeric"
```

To show the values, type the name of the object.

```
> h
 [1] 2 3 0 3 1 0 0 1 2 1
```

Take note. You assigned the values to an object called `h`. The assignment operator is an equal sign (=). Another assignment operator used frequently is `<-`, a left-pointing arrow that consists of two keystrokes (the less-than sign and the hyphen). This is the more common of the assignment operators. The equal sign is used in declaring argument values, so confusion can arise if it is also used as the assignment operator.

With most assignments, only the prompt is returned to the screen with nothing printed. The object to the left of the assignment operator is given the values of whatever is to the right of the operator. Objects are printed by typing the object name as you just did. Finally, the values when printed are prefaced with a `[1]`. This indicates that the first entry in the object has a value of 2 (the number immediately to the right of `[1]`). More on this later.

The arrow keys are used to retrieve previous functions. This saves typing. Each command is stored in a history file and the up arrow key moves backward through the history file and the down arrow moves forward. The left and right arrow keys work as expected. Changes can be made to a mistyped function followed by a return without the need to go to the end of the line.

You can also enter small amounts of data with the `scan` function. You enter data values one at a time line by line. When finished, type enter (return). R has a wide range of functions for inputting data. We will look at a few of them as needed.

2.2.2 Functions

Once the data are stored as an object, you call functions to do various things. For example, you find the total number of landfalls occurring over the set of years by typing

```
> sum(h)
[1] 13
```

The function adds up the values of the vector elements. The number of years is found by typing

```
> length(h)
[1] 10
```

The average number of hurricanes over this 10 year period is found by typing

```
> sum(h)/length(h)
[1] 1.3
```

or

```
> mean(h)
[1] 1.3
```

Other useful functions include `sort`, `min`, `max`, `range`, `diff`, and `cumsum`. Try them on the object `h` of landfall counts. For example, what does the function `diff` do?

Most functions have a name followed by a left parenthesis, then a set of arguments separated by commas followed by a right parenthesis. Arguments have names. Some are required, but many are optional with R providing default values. Throughout this book, function name referenced in line are left without arguments and without parentheses.

In summary, consider the code

```
> x = log(42, base=10)
```

Here `x` is the object name, = is the assignment operator, `log` is the function, 42 is the value for which the logarithm is being computed, and 10 is the argument corresponding to the logarithm base. Note here the equal sign is used in two different ways: as an assignment operator and to specify a value for an argument.

2.2.3 Vectors

Your earlier data object `h` is stored as a vector. This means that R keeps track of the order you entered the data. The vector contains a first element, a second element, and so on. This is convenient.

Your data of landfall counts has a natural order—year 1, year 2, and so on—so you want to keep this order. You would like to be able to make changes to the data item by item instead of reentering the entire set of values. R lets you do this. Also vectors are math objects so math operations can be performed on them.

Let us see how these concepts apply to your data. Suppose `h` contains the annual landfall counts from the first decade of a longer record. You want to keep track of counts over a second decade. This can be done as follows:

```
> d1 = h        # make a copy
> d2 = c(0, 5, 4, 2, 3, 0, 3, 3, 2, 1)
```

Most functions will operate on each vector component (element) all at once.

```
> d1 + d2
 [1] 2 8 4 5 4 0 3 4 4 2
> d1 - d2
```

```
 [1]  2 -2 -4  1 -2  0 -3 -2  0  0
> d1 - mean(d1)
 [1]  0.7  1.7 -1.3  1.7 -0.3 -1.3 -1.3 -0.3  0.7
[10] -0.3
```

In the first two cases, the first-year count of the first decade is added (and subtracted) from the first-year count of the second decade and so on. In the third case, a constant (the average of the first decade) is subtracted from each count of the first decade. This is an example of recycling. R repeats values from one vector so as to match the length of the other vector. Here the mean value is computed and then repeated 10 times. The subtraction then follows on each component one at a time.

Suppose you are interested in the variability of hurricane counts from one year to the next. An estimate of this variability is the variance. The sample mean of a set of numbers y is

$$\bar{y} = \frac{1}{n}\sum_{i=1}^{n} y_i, \tag{2.1}$$

where n is the sample size. And the sample variance is

$$s^2 = \frac{1}{n-1}\sum_{i=1}^{n} (y_i - \bar{y})^2. \tag{2.2}$$

Although the function `var` will compute the sample variance, to see how vectorization works in R, you write few lines of code.

```
> dbar = mean(d1)
> dif = d1 - dbar
> ss = sum(dif^2)
> n = length(d1)
> ss/(n - 1)
[1] 1.34
```

Note how the different parts of the equation for the variance (2.2) match what you type in R. To verify your code, type

```
> var(d1)
[1] 1.344444
```

To change the number of significant digits printed to the screen from the default of 7, type

```
> options(digits=3)
> var(d1)
[1] 1.34
```

The standard deviation, which is the square root of the variance, is obtained by typing

```
> sd(d1)
[1] 1.16
```

One restriction on vectors is that all the components must have the same type. You cannot create a vector with the first component a numeric value and the second component a character text. A character vector can be a set of text strings as in

```
> Elsners = c("Jim", "Svetla", "Ian", "Diana")
> Elsners
[1] "Jim"     "Svetla" "Ian"     "Diana"
```

Note that character strings are made by matching quotes on both sides of the string, either double, ", or single, '. Caution: The quotes must not be directional. If you copy your code from a word processor (such as MS Word) the program will insert directional quotes. It is better to copy from a text editor such as Notepad.

You add another component to the vector `Elsners` by using the `c` function.

```
> c(Elsners, 1.5)
[1] "Jim"     "Svetla" "Ian"     "Diana"   "1.5"
```

The component 1.5 gets coerced to a character string. Coercion occurs for mixed types where the components get changed to the lowest common type, which is usually a character. You cannot perform arithmetic on a character vector.

Elements of a vector can have names. The names will appear when you print the vector. You use the `names` function to retrieve and set names as character strings. For instance, you type

```
> names(Elsners) = c("Dad", "Mom", "Son", "Daughter")
> Elsners
     Dad      Mom      Son Daughter
   "Jim" "Svetla"    "Ian"  "Diana"
```

Unlike most functions, `names` appears on the left side of the assignment operator. The function adds the names attribute to the vector. Names can be used on vectors of any type.

Returning to your hurricane example, suppose the National Hurricane Center (NHC) finds a previously undocumented hurricane in the sixth year of the second decade. In this case, you type

```
> d2[6] = 1
```

This changes the sixth element (component) of vector `d2` to 1, leaving the other components alone. Note the use of square brackets (`[]`). Square brackets are used to subset components of vectors (and arrays, lists, etc.), whereas parentheses are used with functions to enclose the set of arguments.

You list all values in vector d2 by typing

```
> d2
 [1] 0 5 4 2 3 1 3 3 2 1
```

To print the number of hurricanes during the third year only, type

```
> d2[3]
[1] 4
```

To print all the hurricane counts except from the fourth year, type

```
> d2[-4]
[1] 0 5 4 3 1 3 3 2 1
```

To print the hurricane counts for the odd years only, type

```
> d2[c(1, 3, 5, 7, 9)]
[1] 0 4 3 3 2
```

Here you use the c function inside the subset operator [.

Since here you want a regular sequence of years, the expression c(1,3,5,7,9) can be simplified using structured data.

2.2.4 Structured Data

Sometimes a set of values has a pattern. The integers from 1 through 10, for example. To enter these one by one using the c function is tedious. Instead, the colon function is used to create sequences. For example, to sequence the first 10 positive integers, you type

```
> 1:10
 [1]  1  2  3  4  5  6  7  8  9 10
```

Or to reverse the sequence, you type

```
> 10:1
 [1] 10  9  8  7  6  5  4  3  2  1
```

You create the same reversed sequence of integers using the rev function together with the colon function as

```
> rev(1:10)
```

The seq function is more general than the colon function. It allows for not only start and end values, but also a step size or sequence length. Some examples include

```
> seq(1, 9, by=2)
[1] 1 3 5 7 9
> seq(1, 10, by=2)
[1] 1 3 5 7 9
```

```
> seq(1, 9, length=5)
[1] 1 3 5 7 9
```

Use the `rep` function to create a vector with elements having repeat values. The simplest usage of the function is to replicate the value of the first argument the number of times specified by the value of the second argument.

```
> rep(1, times=10)
 [1] 1 1 1 1 1 1 1 1 1 1
```

or

```
> rep(1:3, times=3)
[1] 1 2 3 1 2 3 1 2 3
```

You create more complicated patterns by specifying pairs of equal-sized vectors. In this case, each component of the first vector is replicated the corresponding number of times as specified in the second vector.

```
> rep(c("cold", "warm"), c(1, 2))
[1] "cold" "warm" "warm"
```

Here the vectors are implicitly defined using the `c` function and the name of the second argument (`times`) is left off. Again it is good coding practice to name the arguments. If you name the arguments, then the order in which they appear in the function is not important.

Suppose you want to repeat the sequence of cold, warm, warm three times. You nest the above sequence generator in another repeat function as follows:

```
> rep(rep(c("cold", "warm"), c(1, 2)), 3)
[1] "cold" "warm" "warm" "cold" "warm" "warm" "cold" [8]
"warm" "warm"
```

Function nesting gives you a lot of flexibility.

2.2.5 Logic

As you have seen, there are functions like `mean` and `var` that when applied to a vector of data output a statistic. Another example is the `max` function. To find the maximum number of hurricanes in a single year during the first decade, type

```
> max(d1)
[1] 3
```

This tells you that the worst year had three hurricanes.
To determine which years had this many hurricanes, type

```
> d1 == 3
 [1] FALSE  TRUE FALSE  TRUE FALSE FALSE FALSE FALSE
 [9] FALSE FALSE
```

Note the double equal sign. Recall a single equal sign assigns `d1` the value 3. With the double equal sign you are performing a logical operation on the components of the vector. Each component is matched in value with the value 3, and a true or false is returned. That is, is component one equal to 3? No, so return `FALSE`; is component two equal to 3? Yes, so return `TRUE`, and so on. The length of the output will match the length of the vector.

Now how can you get the years corresponding to the `TRUE` values? To rephrase, which years have three hurricanes? If you guessed the function `which`, you are on your way to mastering R.

```
> which(d1 == 3)
[1] 2 4
```

You might be interested in the number of years in each decade without a hurricane.

```
> sum(d1 == 0); sum(d2 == 0)
[1] 3
[1] 1
```

Here we apply two functions on a single line by separating the functions with a semicolon.

Or how about the ratio of the number of hurricanes over the two decades.

```
> mean(d2)/mean(d1)
[1] 1.85
```

So there is 85 percent more landfalls during the second decade. The statistical question is, is this difference significant?

Before moving on, it is recommended that you remove objects from your workspace that are no longer needed. This helps you recycle names and keeps your workspace clean. First, to see the objects in your workspace, type

```
> objects()
 [1] "Elsners"              "RweaveLatex2"
 [3] "RweaveLatexRuncode2"  "Sweave2"
 [5] "d1"                   "d2"
 [7] "dbar"                 "dif"
 [9] "h"                    "n"
[11] "ss"                   "tikz.Swd"
[13] "try_out"              "x"
```

Then, to remove only selected objects, type

```
> rm(d1, d2, Elsners)
```

To remove all objects, type

```
> rm(list=objects())
```

This will clean your workspace completely. To avoid name conflicts it is good practice to start a session with a clean workspace. However do not include this command in code you give to others.

2.2.6 Imports

Most of what you do in R involves data. To get data into R, first you need to know your working directory. You do this with the `getwd` function by typing

```
> getwd()
[1] "/Users/jelsner/Dropbox/book/Chap02"
```

The output is a character string in quotes that indicates the full path of your working directory on your computer. It is the directory where R will look for data. To change your working directory, you use the `setwd` function and specify the path name within quotes. Alternatively, you should be able to use one of the menu options in the R console. To list the files in your working directory, you type `dir()`.

Second, you need to know the type of data file. This will determine the read function. For example, the data set *US.txt* contains a list of tropical cyclone counts by year making landfall in the United States (excluding Hawaii) at hurricane intensity. The file is a space-delimited text file. In this case, you use the `read.table` function to import the data.

Third, you need to know whether your data file has column names. These are given in the first line of your file, usually as a series of character strings. The line is called a "header", and if your data file has one, you need to specify `header=TRUE`.

Assuming the text file *US.txt* is in your working directory, type

```
> H = read.table("US.txt", header=TRUE)
```

If R returns a prompt without an error message, the data have been imported. Here your file contains a header so the argument `header` is used.

At this stage, the most common mistake is that your data file is not in your working directory. This will result in an error message along the lines of "cannot open the connection" or "cannot open file."

The function has options for specifying the separator character or characters between columns in the file. For example, if your file has commas between columns, then you use the argument `sep=","` in the `read.table` function. If the file has tabs, then you use `sep="\t"`. Note that R makes no changes to your original file.

You can also change the missing value character. By default, it is NA. If the missing value character in your file is coded as 99, specify `na.strings="99"`, and it will be changed to NA in your R data object.

There are several variants of `read.table` that differ only in the default argument settings. Note in particular `read.csv`, which has settings that are suitable for comma delimited (csv) files that have been exported from a spreadsheet. Thus, the typical work flow is to export your data from a spreadsheet to a csv file, then import it to R using the `read.csv` function.

You can also import data directly from the web by specifying the URL instead of the local file name.

```
> loc = "http://myweb.fsu.edu/jelsner/US.txt"
> H = read.table(loc, header=TRUE)
```

The object `H` is a data frame and the function `read.table` and variants return data frames. A data frame is similar to a spreadsheet. The values are arranged in rows and columns. The rows are the cases and the columns are the variables. To check the dimensions of your data frame, type

```
> dim(H)
[1] 160   6
```

This tells you that there are 160 rows and 6 columns in your data frame.

To list the first six lines of the data object, type

```
> head(H)
  Year All MUS G FL E
1 1851   1   1 0  1 0
2 1852   3   1 1  2 0
3 1853   0   0 0  0 0
4 1854   2   1 1  0 1
5 1855   1   1 1  0 0
6 1856   2   1 1  1 0
```

The columns include year, number of U.S. hurricanes, number of major U.S. hurricanes, number of Gulf coast hurricanes, number of Florida hurricanes, and number of East coast hurricanes in that order. Note that the column names are given as well. The last six lines of your data frame are listed similarly using the `tail` function. The number of lines listed is changed using the argument `n` (for example, n=3).

If your data reside in a directory other than your working directory, you can use the `file.choose` function. This will open a dialog box allowing you to scroll and choose the file. Note that the default for this function has no arguments: `file.choose()`.

To make the individual columns available by column name, type

```
> attach(H)
    All, E, FL, G, MUS, Year
```

The total number of years in the record is obtained and saved in `n`, and the average number of U.S. hurricanes is saved in `rate` using the following two lines of code:

```
> n = length(All)
> rate = mean(All)
```

By typing the names of the saved objects, the values are printed.

```
> n
[1] 160
```

```
> rate
[1] 1.69
```

Thus over the 160 years of data, the average number of U.S. hurricanes per year is 1.69.

If you want to change the names of the data frame, type

```
> names(H)[4] = "GC"
> names(H)
[1] "Year" "All"  "MUS"  "GC"   "FL"   "E"
```

This changes the fourth column name from G to GC. Note that this is done to the data frame in R and not to your original data file.

While attaching a data frame is convenient, it is not a good strategy as name conflicts can easily arise. If you do attach your data frame, make sure you use the function `detach` after you are finished.

2.3 TABLES AND PLOTS

Now that you know a bit about using R, you are ready for some data analysis. R has a wide variety of data structures including scalars, vectors, matrices, data frames, and lists.

2.3.1 Tables and Summaries

Vectors and matrices must have a single class. For example, the vectors `A`, `B`, and `C` below are constructed as numeric, logical, and character respectively.

```
> A = c(1, 2.2, 3.6, -2.8) #numeric vector
> B = c(TRUE, TRUE, FALSE, TRUE)  #logical vector
> C = c("Cat 1", "Cat 2", "Cat 3") #character vector
```

To view the data class, type

```
> class(A); class(B); class(C)
[1] "numeric"
[1] "logical"
[1] "character"
```

Let the vector `wx` denote the weather conditions for five forecast periods as character data.

```
> wx = c("sunny", "clear", "cloudy", "cloudy", "rain")
> class(wx)
[1] "character"
```

Character data are summarized using the `table` function. To summarize the forecast weather conditions over the 6 days, type

```
> table(wx)
wx
 clear cloudy   rain  sunny
     1      2      1      1
```

The output is a list of the unique character strings and the corresponding number of occurrences of each string.

As another example, let the object `ss` denote the Saffir–Simpson category for a set of five hurricanes.

```
> ss = c("Cat 3", "Cat 2", "Cat 1", "Cat 3", "Cat 3")
> table(ss)
ss Cat 1 Cat 2 Cat 3
       1     1     3
```

Here the character strings correspond to intensity levels as ordered categories with `Cat 1` < `Cat 2` < `Cat 3`. In this case, it is better to convert the character vector `ss` to an ordered factor with levels. This is done using the `factor` function.

```
> ss = factor(ss, order=TRUE)
> class(ss)
[1] "ordered" "factor"
> ss
[1] Cat 3 Cat 2 Cat 1 Cat 3 Cat 3
Levels: Cat 1 < Cat 2 < Cat 3
```

The class of `ss` gets changed to an ordered factor. A print of the object results in a list of the elements in the vector and a list of the levels in order. Note, if you do the same for the `wx` object, the order is alphabetical by default. Try it.

You can also use the `table` function on discrete numeric data. For example,

```
> table(All)
All
 0  1  2  3  4  5  6  7
34 48 38 27  6  1  5  1
```

The table tells you that your data has 34 zeros, 48 ones, and so on. Since these are annual U.S. hurricane counts, you know, for instance, that there are 6 years with four hurricanes and so on.

Recall from the previous section that you attached the data frame `H` so you can use the column names as vectors. The data frame remains attached for the entire session. Remember that you detach it with the function `detach`.

The summary function is used to a get a description of your object. The form of the value(s) returned depends on the class of the object being summarized. If your object is a data frame consisting of numeric data the output are six summary statistics (mean, median, minimum, maximum, first quartile, and third quartile) for each of the columns.

```
> summary(H)
      Year            All             MUS
 Min.   :1851   Min.   :0.00   Min.   :0.0
 1st Qu.:1891   1st Qu.:1.00   1st Qu.:0.0
 Median :1930   Median :1.00   Median :0.0
 Mean   :1930   Mean   :1.69   Mean   :0.6
 3rd Qu.:1970   3rd Qu.:2.25   3rd Qu.:1.0
 Max.   :2010   Max.   :7.00   Max.   :4.0
       GC              FL               E
 Min.   :0.000   Min.   :0.000   Min.   :0.000
 1st Qu.:0.000   1st Qu.:0.000   1st Qu.:0.000
 Median :0.500   Median :0.000   Median :0.000
 Mean   :0.688   Mean   :0.681   Mean   :0.469
 3rd Qu.:1.000   3rd Qu.:1.000   3rd Qu.:1.000
 Max.   :4.000   Max.   :4.000   Max.   :3.000
```

Each column of your data frame `H` is labeled and summarized by six numbers including the minimum, the maximum, the mean, the median, and the first (lower) and third (upper) quartiles. For example, you see that the maximum number of major U.S. hurricanes (`MUS`) in a single season is 4. Since the first column is the year, the summary is not particularly meaningful.

2.3.2 Quantiles

The quartiles from the `summary` function are examples of quantiles. Sample quantiles cut a set of ordered data into equal-sized bins. The ordering comes from rearranging the data from lowest to highest. The first, or lower, quartile corresponding to the .25 quantile (25th percentile) indicates that 25 percent of the data have a value less than this quartile value. The third, or upper, quartile corresponding to the .75 quantile (75th percentile) indicates that 75 percent of the data have a smaller value than this quartile value.

The `quantile` function calculates sample quantiles. For example, consider the set of North Atlantic Oscillation (NAO) index values for the month of June from the period 1851 to 2010. The NAO is a variation in the climate over the North Atlantic Ocean featuring fluctuations in the difference of atmospheric pressure at sea level between the Iceland and the Azores. The index is computed as the difference in standardized sea-level pressures. The standardization is done by subtracting the mean and dividing by the standard deviation. The unit on the index is standard deviation. See Chapter 6 for more details on these data.

First read the data consisting of monthly NAO values, then apply the `quantile` function to the June values.

```
> NAO = read.table("NAO.txt", header=TRUE)
> quantile(NAO$Jun, probs=c(.25, .5))
    25%    50%
-1.405 -0.325
```

Note the use of the $ sign to point to a particular column in the data frame. Recall that to list the column names of the data frame object called NAO, type names(NAO).

Of the 160 values, 25 percent of them are less than -1.4 standard deviations (s.d.), 50 percent are less than -0.32 s.d. Thus there are an equal number of years with June NAO values between -1.4 and -0.32 s.d.

The third quartile value corresponding to the .75 quantile (75th percentile) indicates that 75% of the data have a value less than this. The difference between the first and third quartile values is called the interquartile range (IQR). Fifty percent of all values lie within the IQR. The IQR of a data vector can be found by using the IQR function.

2.3.3 Plots

R has a wide range of plotting capabilities. It takes time to master, but a little effort goes a long way. You will create a lot of plots as you work through this book. Here are a few examples to get you started. Chapter 5 provides more details.

Bar Plots

The bar plot (or bar chart) is a way to compare categorical or discrete data. Levels of the variable are arranged in some order along the horizontal axis and the frequency of values at each level is plotted as a bar with the bar height proportional to the frequency.

To make a bar plot of your U.S. hurricane counts, type

```
> barplot(table(All), ylab="Number of Years",
+           xlab="Number of Hurricanes")
```

The plot in Figure 2.1 is a concise summary of the number of hurricanes. The bar heights are proportional to the number years with that many hurricanes. The plot conveys the same information as the table. The purpose of the bar plot is to illustrate the difference between data values. Readers expect the plot to start at zero, so you should try to draw it that way. Also usually there is little scientific reason to make the bars appear three dimensional.

Note that the axis labels are set using the ylab and xlab arguments with the label as a character string in quotation. Be careful to avoid the directional quotes that appear in word-processing type.

Although many of the plotting commands are simple and somewhat intuitive, to get a publication-quality figure requires tweaking the default settings. You will see some of these tweaks as you work through the book.

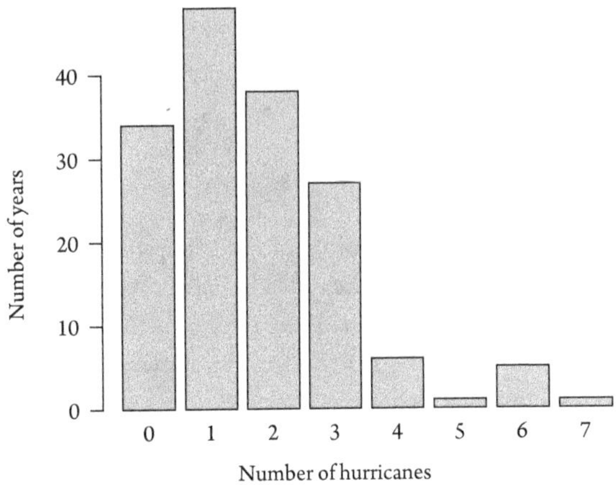

Figure 2.1 Bar plot of U.S. hurricanes.

When a function like `barplot` is called, the output is sent to the graphics device (Windows, Quartz, or X11) for your computer screen. There are also devices for creating postscript, pdf, png, and jpeg output and sending them to a file outside of R. For publication-quality graphics, the postscript and pdf devices are preferred because they produce scalable images. For drafts use the bitmap device.

The sequence is to first specify a graphics device, then call your graphics functions, and finally close the device. For example, to create an encapsulated postscript file (eps) of your bar plot placed in your working directory, type

```
> postscript(file="MyFirstRPlot.eps")
> barplot(table(All), ylab="Number of Years",
+         xlab="Number of Hurricanes")
> dev.off()  #close the graphics device
```

The file containing the bar plot is placed in your working directory. Note that the `postscript` function opens the device and `dev.off()` closes it. Make sure you close the device. To list the files in your working directory type `dir()`.

The pie chart is used to display relative frequencies (`?pie`). It represents this information with wedges of a circle or pie. Since your eye has difficulty judging relative areas (Cleveland, 1985), it is better to use a dot chart. To find out more type `?dotchart`.

Scatter Plots

Perhaps the most useful graph is the scatter plot. You use it to represent the relationship between two continuous variables. It is a graph of the values of one variable against the values of the other as points (x_i, y_i) in a plane.

You use the `plot` function to make a scatter plot. The syntax is `plot(x, y)`, where `x` and `y` are vectors containing the paired data. Values of the variable named in the first argument (here `x`) are plotted along the horizontal axis.

For example, to graph the relationship between the February and March values of the NAO, type

```
> plot(NAO$Feb, NAO$Mar, xlab="February NAO",
+           ylab="March NAO")
```

The plot is shown in Figure 2.2. It is a summary of the relationship between the NAO values in February and March. Low values of the index in February tend to be followed by low values in March and high values in February tend to be followed by high values in March. There is a direct (or positive) relationship between the two variables although the points are scattered widely indicating the relationship is not tight.

The relationship between two variables can be visualized with a scatter plot. You can change the point symbol with the argument pch. If your goal is to model the relationship, you should plot the dependent variable (the variable you are interested in modeling) on the vertical axis. Here it makes sense to put the March values on the

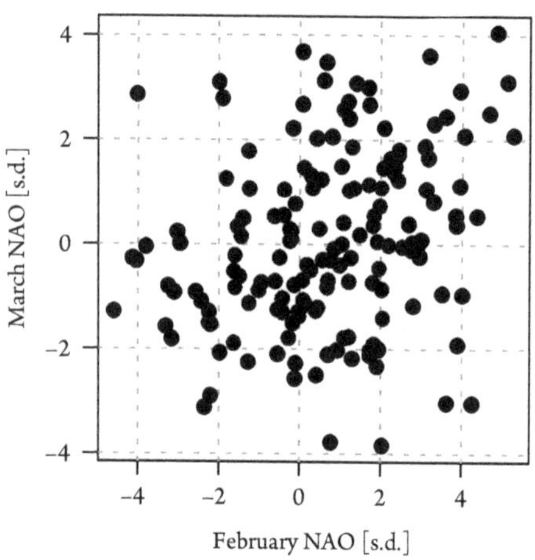

Figure 2.2 Scatter plot of February and March NAO values.

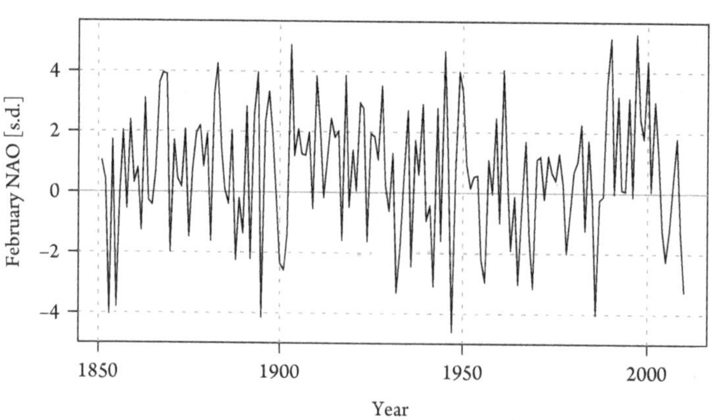

Figure 2.3 Time series of February NAO.

vertical axis since a predictive model would use February values to forecast March values.

The plot produces points as a default. This is changed using the argument `type` where the letter ℓ is placed in quotation. For example, to plot the February NAO values as a time series, type

```
> plot(NAO$Year, NAO$Feb, ylab="February NAO",
+   xlab="Year", type="l")
```

Table 2.1 R functions used in this chapter.

Function	*Description*
Numeric Functions	
`sqrt(x)`	Square root of x
`log(x)`	Natural logarithm of x
`length(v)`	Number of elements in vector v
`summary(d)`	Statistical summary of columns in data frame d
Statistical Functions	
`sum(v)`	Summation of the elements in v
`max(v)`	Maximum value in v
`mean(v)`	Average of the elements in v
`var(v)`	Variance of the elements in v
`sd(v)`	Standard deviation of the elements in v
`quantile(x, prob)`	Prob quantile of the elements in x
Structured Data Functions	
`c(x, y, z)`	Concatenate the objects x, y, and z
`seq(from, to, by)`	Generate a sequence of values
`rep(x, n)`	Replicate x n times
Table and Plot Functions	
`table(a)`	Tabulate the characters or factors in a
`barplot(h)`	Bar plot with heights h
`plot(x, y)`	Scatter plot of the values in x and y
Input, Package, and Help Functions	
`read.table("file")`	Input the data from connection file
`head(d)`	List the first six rows of data frame d
`objects()`	List all objects in the workspace
`help(fun)`	Open help documentation for function fun
`install.packages("pk")`	Install the package pk on your computer
`require(pk)`	Make functions in package pk available

The plot is shown in Figure 2.3. The values fluctuate about zero and do not appear to have a long-term trend. With time series data, it is better to connect the values with lines. More details on how to make time series plots and other graphs are given throughout the book.

This concludes your introduction to R. We showed you where to get R, how to install it, and how to obtain packages. We showed you how to use R as a calculator, how to work with functions, make assignments, and get help. We also showed you how to work with small amounts of data and how to import data from a file. We concluded with how to tabulate, summarize, and make some simple plots. There is much more ahead, but you have made a good start. Table 2.1 lists most of the functions used in the chapter. A complete list of the functions used in the book is given in Appendix A.

In the next chapter we provide an introduction to statistics. If you have taken a course in statistics, this will be a review, but we encourage you to follow along anyway as you will learn new things about using R.

3

CLASSICAL STATISTICS

"The difference between 'significant' and 'not significant' is not itself statistically significant."
—Andrew Gelman

All hurricanes are different. Statistics helps you characterize hurricanes from the typical to the extreme. In this chapter, we provide an introduction to classical (or frequentist) statistics. To get the most out of it, we again encourage you to open an R session and type in the code as you read along.

3.1 DESCRIPTIVE STATISTICS

Descriptive statistics are used to summarize your data. The mean and the variance are good examples. So is correlation. Data can be a set of weather records or output from a global climate model. Descriptive statistics provide answers to questions like does Jamaica experience more hurricanes than Puerto Rico?

In Chapter 2, you learned some functions for summarizing your data, let us review. Recall that the data set *H.txt* is a list of hurricane counts by year making landfall in the United States (excluding Hawaii). To input the data and save them as a data object, type

```
> H = read.table("H.txt", header=TRUE)
```

Make sure the data file is located in your working directory. To check your working directory, type `getwd()`.

3.1.1 Mean, Median, and Maximum

Sometimes all you need are a few summary statistics from your data. You can obtain the mean and variance by typing

```
> mean(H$All); var(H$All)
[1] 1.69375
[1] 2.10059
```

Recall that the semicolon acts as a return so you can place multiple functions on the same text line. The sample mean is a measure of the central tendency and the sample variance is a measure of the spread. These are called the first- and second- moment statistics. Like all statistics, they are random variables. A random variable can be thought of as a quantity whose value is not fixed; it changes depending on the values in your sample.

If you consider the number of hurricanes over a different sample of years, the sample mean will almost certainly be different. Same with the variance. The sample mean provides an estimate of the population mean (the mean over all past and future years). Different samples drawn from the same population will result in different mean values, but as the sample size increases, the values will get closer to the population mean.

The values printed on your screen have too many digits. Since there are only 160 years, you want to report the mean using only 2 or 3 significant digits. This is done with the `signif` function.

```
> signif(mean(H$All), digits=3)
[1] 1.69
```

Notice how the functions are nested. Here the `mean` function is nested in the `signif` function. You can also set the number of digits globally using the `options` function as mentioned in Chapter 2.

The median, standard deviation, maximum, and minimum are obtained by typing

```
> median(H$All); sd(H$All); max(H$All); min(H$All)
[1] 1
[1] 1.45
[1] 7
[1] 0
```

At least one year had seven hurricanes hit the U.S. coast. Higher-order moments like skewness and kurtosis are available in the `moments` package.

To determine which year has the maximum, you first test each year's count against the maximum using the logical operator ==. This returns a vector of TRUEs and FALSEs. You then subset the hurricane year according to that vector. For example, type

```
> maxyr = H$All == max(H$All)
> H$Year[maxyr]
[1] 1886
```

This tells you that 1886 is the single year with the most hurricanes.

The `which.max` function is similar but returns the row number index corresponding only to the *first* occurrence of the maximum. You can then subset the hurricane year on the index.

```
> H$Year[which.max(H$All)]
[1] 1886
```

Similarly, to find the years with no hurricanes, type

```
> H$Year[H$All == min(H$All)]
 [1] 1853 1862 1863 1864 1868 1872 1884 1889 1890
[10] 1892 1895 1902 1905 1907 1914 1922 1927 1930
[19] 1931 1937 1951 1958 1962 1973 1978 1981 1982
[28] 1990 1994 2000 2001 2006 2009 2010
```

And to determine how many years have no hurricanes, type

```
> sum(H$All == 0)
[1] 34
```

The `sum` function counts a TRUE as 1 and a FALSE as 0, so the result tells you how many years have a count of zero hurricanes.

You also might be interested in streaks. For instance, what is the longest streak of years without a hurricane? To answer this, first you create an ordered vector of years with at least one hurricane. Next you use the `diff` function to take differences between sequential years given in the ordered vector. Finally, you find the maximum of these differences minus one.

```
> st = H$Year[H$All > 0]
> max(diff(st) - 1)
[1] 3
```

Thus the longest streak without hurricanes is only 3 years.

Alternatively, you can use the `rle` function to compute the length and value of runs in a vector and then table the results.

```
> st = H$All == 0
> table(rle(st))
         values
lengths FALSE TRUE
      1     2   21
      2     6    5
      3     4    1
      4     4    0
      5     2    0
      6     3    0
      7     2    0
```

```
8       1    0
10      1    0
11      1    0
13      1    0
```

The results show five sets of two consecutive years without a hurricane and one set of three consecutive years without a hurricane.

3.1.2 Quantiles

Percentiles also help you describe your data. The pth percentile ($p/100$ quantile) is the value that cuts off the first p percent of the data when the data values are sorted in ascending order. The `quantile` function is used to obtain these values.

First import the NAO data file. Refer to Chapter 6 for a description of these data.

```
> NAO = read.table("NAO.txt", header=TRUE)
```

To obtain quantiles of the June NAO values, type

```
> nao = NAO$Jun
> quantile(nao)
     0%     25%     50%     75%    100%
 -4.050 -1.405 -0.325  0.760  2.990
```

By default, you get the minimum, the maximum, and the three quartiles (the 0.25, 0.5, and 0.75 quantiles), so named because they correspond to a division of the values into four equal parts. The difference between the first and third quartiles is called the IQR, which is used as an alternative to the standard deviation because it is less affected by extremes.

To obtain other quantiles, you include the argument `prob` in the function. For example, to find the 19th, 58th, and 92nd percentiles of the June NAO values, type

```
> quantile(nao, prob=c(.19, .58, .92))
    19%     58%     92%
 -1.620 -0.099   1.624
```

Be aware that there is more than one way to compute quantiles. Details can be found in the documentation (type `help(quantile)`).

3.1.3 Missing Values

Things become a bit more complicated if your data contain missing values. R handles missing values in different ways depending on the context. With a vector of values, some of which are missing and marked with `NA`, the `summary` function computes statistics and returns the number of missing values. For example, read in the monthly sea-surface temperatures (*SST.txt*), create a vector of August values, and summarize them.

```
> SST = read.table("SST.txt", header=TRUE)
> sst = SST$Aug
> summary(sst)
   Min. 1st Qu.  Median    Mean 3rd Qu.    Max.
   22.6    23.0    23.1    23.1    23.3    23.9
   NA's
    5.0
```

Here you see the summary statistics and note that there are 5 years with missing values during August. The summary statistics are computed by removing the missing values.

However, a function that computes a statistic, like the `mean`, applied to a vector with missing values returns an NA.

```
> mean(sst)
[1] NA
```

In this case, R does not remove the missing values unless requested to do so. The mean of a vector with an unknown value is unknown. If you wish to have the missing values removed, you need to include the argument `na.rm=TRUE`,

```
> mean(sst, na.rm=TRUE)
[1] 23.1
```

An exception is the `length` function, which does not understand the argument `na.rm`, so you cannot use it to count the number of missing values. Instead, you use the `is.na` function, which returns a TRUE for a missing value and FALSE otherwise. You then use the `sum` function to count the number of TRUEs. For your August SST data, type

```
> sum(is.na(sst))
[1] 5
```

The number of nonmissing data values is obtained by using the logical negation operator `!` (read as 'not'). For example, type

```
> sum(!is.na(sst))
[1] 155
```

This tells you there are 155 years with August SST values.

3.2 PROBABILITY AND DISTRIBUTIONS

Climate is the set of all weather patterns. You can think of weather as a data-generating machine. For example, with each season the number of hurricanes is recorded as a count. This count is a data value from the weather machine that gets collected alongside counts from other years. Additional data are available like the highest wind speed, and so on. This view of data arising from a generating process gives statistics a prominent role in understanding climate.

3.2.1 Random Samples

Statistics is an application of probability theory. Probability theory arose from studying games of chance like rolling dice and picking cards at random. Randomness and probability are central to statistics.

You can simulate games of chance with the `sample` function. For instance, to pick 4 years at random from the 1990s, you type

```
> sample(1990:1999, size=4)
[1] 1997 1995 1998 1996
```

The sequence function : is used to create a vector of 10 values representing the years in a decade. The `sample` function is then used to pick, at random without replacement, a set of four (`size=4`) values. This is called a *random sample.*

Notice that in deciphering R code, it is helpful to read from right to left and from inside to outside. That is, you start by noting a size of four from a sequence of numbers from 1990 through 1999 and then taking a sample from these numbers.

The default is "sample without replacement." The sample will not contain a repeat year. If you want to sample with replacement, use the argument `replace=TRUE`. Sampling with replacement is suitable for modeling the occurrence of El Niño (E) and La Niña (L) events. An El Niño event is characterized by a warm ocean along the equator near the coast of Peru. A La Niña event is characterized by a cool ocean. The fluctuation between El Niño and La Niña events coincides with the fluctuation in hurricane activity.

To model the occurrence over 10 random seasons, type

```
> sample(c("E", "L"), size=10, replace=TRUE)
 [1] "L" "E" "L" "E" "E" "E" "L" "L" "E" "E"
```

Historically, the probability of an El Niño is about the same as the probability of a La Niña, but the idea of random events is not restricted to equal probabilities. For instance, suppose you are interested in the occurrence of hurricanes hitting Florida. Let the probability be 12 percent that a hurricane picked at random hits Florida. You simulate 24 random hurricanes by typing

```
> sample(c("hit", "miss"), size=24, replace=TRUE,
+   prob=c(.12, .88))
 [1] "miss" "miss" "miss" "miss" "hit"  "miss" "miss"
 [8] "miss" "hit"  "miss" "miss" "miss" "miss" "miss"
[15] "miss" "miss" "miss" "miss" "hit"  "miss" "miss"
[22] "miss" "miss" "miss"
```

The simulated frequency of hits will not be exactly 12 percent, but the variation about this percentage decreases as the sample size increases according to the law of large numbers.

3.2.2 Combinatorics

Returning to your set of years from a decade. Common sense tells you that the probability of getting each of the 10 years in a sample of size 10 with each year picked at random without replacement is one. But what is the probability of randomly drawing a set of three particular years?

This is worked out as follows. The probability of getting a particular year (say 1992) as the first one in the sample is $1/10$, the next one is $1/9$, and the next one is $1/8$. Thus the probability of a given sample is $1/(10 \times 9 \times 8)$. The `prod` function calculates the product of a set of numbers, so you get the probability by typing

```
> 1/prod(10:8)
[1] 0.00139
```

Note that this is the probability of getting a set of 3 years in a particular order (i.e., 1992, 1993, 1994). If you are not interested in the arrangement of years, then you need to include the cases that give the 3 years in a different order. The probability will be the same if the years are in any order, so you need to know the number of combinations and then multiply this number by the previous probability.

Given a sample of three numbers, there are three possibilities for the first number, two possibilities for the second, and only one possibility for the third. Thus the number of combinations of three numbers in any order is $3 \times 2 \times 1$ or $3!$. You can use the `factorial` function and multiply this number by the above probability to get

```
> factorial(3)/prod(10:8)
[1] 0.00833
```

Thus the probability of a given sample of 3 years in any order is 0.83 percent.

You get the same result using the `choose` function. For any set containing n elements, the number of distinct x-element samples of it that can be formed (the x combinations of its elements) is given by

$$\binom{n}{x} = \frac{n!}{x!(n-x)!}, \tag{3.1}$$

which is read as "n choose x." The multiplicative inverse of this number is the probability. You can check the equality of these two ways of working out the probability by typing

```
> factorial(3)/prod(10:8) == 1/choose(10,3)
[1] TRUE
```

Recall that the double equal signs indicate a logical operator with two possible outcomes (TRUE and FALSE), so a return of TRUE indicates equality.

3.2.3 Discrete distributions

It is likely that you are more interested in some calculated value from a random sample. Instead of a set of hits and misses on Florida from a sample of hurricanes, you might want to know the number of hits. Since the set of hits is random so is the total number over all hits.

The number of hits is another example of a *random variable*. In this case, it is a nonnegative integer that can take on values in $0, 1, 2, \ldots, n$, where n is the total number of hurricanes. Said another way, given a set of n North Atlantic hurricanes, the number that hit Florida is a discrete random variable H.

A random variable H has a *probability distribution* that is described using $f(h) = P(H = h)$. The set of all possible Florida counts is the random variable denoted with a large H, while a particular count is denoted with a small h. This is standard statistical notation. Thus $f(h)$ is a function that assigns a probability to each possible count. It is written as

$$f(h|p,n) = \binom{n}{h} p^h (1-p)^{n-h}, \tag{3.2}$$

where the parameter p is the probability of a Florida hit given a North Atlantic hurricane. This is known as the *binomial distribution* and $\binom{n}{h}$ is known as the binomial coefficient.

Returning to your example above, in a set of 24 hurricanes with a probability $p = 12\%$ that a hurricane picked at random hits Florida (hit rate), the probability that exactly three of them will strike Florida $[P(H = 3)]$ is found by typing

```
> choose(24, 3) * .12^3 * (1 - .12)^(24 - 3)
[1] 0.239
```

Thus there is a 23.9 percent chance of having 3 of the 24 hurricanes hit Florida. Note, there is a probability associated with all nonnegative integers from 0 to 24. For counts 0 to 15, the probabilities are plotted in Figure 3.1. The distribution is discrete with probabilities assigned only to counts. The distribution peaks at three and four hurricanes, and there are small but nonzero probabilities for the largest counts.

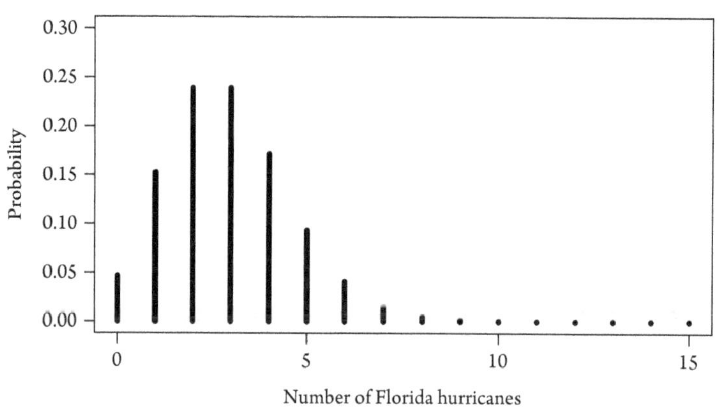

Figure 3.1 Probability of h Florida hurricanes, given a random sample of 24 North Atlantic hurricanes using a binomial model with a hit rate of 12%.

3.2.4 Continuous Distributions

Most climate data are available as values on a continuous scale. For instance, the NAO index used in Chapter 2 is given in units of standard deviation. Values are recorded to a finite precision, but in practice this is generally not relevant. What is relevant is the fact that the values tend to cluster near a central value; values far from this central value are more rare than values closer to it.

A continuous value has no probability associated with it. This is because there are infinitely many values between any two values, so the probability at any particular one is zero. Instead we have the concept of *density*. The density of a continuous random variable is a function that describes the relative likelihood that the variable will have a particular value. The density is the probability associated with a small region around the value divided by the size of the region. This probability density function is nonnegative and its sum over the set of all possible values is equal to one.

The *cumulative distribution function* for continuous random variables is given by

$$F(x) = \int_{-\infty}^{x} f(u)du \tag{3.3}$$

If f is continuous at x, then

$$f(x) = \frac{\mathrm{d}}{\mathrm{d}x}F(x) \tag{3.4}$$

You can think of $f(x)\mathrm{d}x$ as the probability of X falling within the small interval $[x, x+\mathrm{d}x]$.

The most common continuous distribution is the normal (or Gaussian distribution). It has a density given by

$$f(x|\mu,\sigma^2) = \frac{1}{\sqrt{2\pi\sigma^2}} e^{-\frac{(x-\mu)^2}{2\sigma^2}} \tag{3.5}$$

where the parameter μ is the mean and the parameter σ^2 is the variance. We write $N(\mu,\sigma^2)$ as a shorthand for this distribution. The normal distribution has the characteristic bell shape with the mean located at the peak of the distribution.

Changing μ shifts the distribution left or right changing σ^2 widens and narrows the distribution, while the values remain symmetric about the peak (Fig. 3.2). The distance between the two inflection points, where the curves change from opening downward to opening upward, is two standard deviations. The normal distribution is a family of distributions, where the family members have different parameter values. The family member with $\mu = 0$ and $\sigma^2 = 1$ is called the standard normal distribution.

The normal distribution is the foundation for many statistical models. It is analytically tractable and arises as the outcome of the *central limit theorem*, which states that the sum of a large number of random variables, regardless of their distribution, is distributed approximately normally. Thus the normal distribution is commonly encountered in practice as a model for complex phenomena. In climatology, it is used as a model for observational error and for uncertainty.

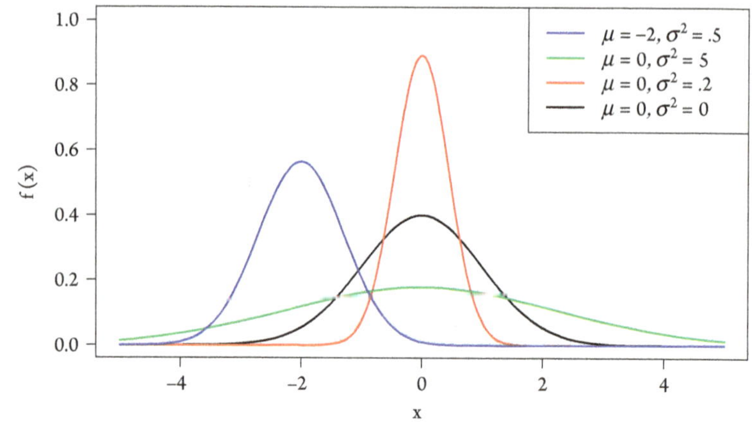

Figure 3.2 Probability density functions for a normal distribution.

3.2.5 Distributions

In R distributions are functions. This eliminates the need for lookup tables. Distributions come in families. Each family is described by a function with parameters. The normal distribution is a family of distributions with each member having a different mean and variance.

The *uniform distribution* is a family of continuous distributions on the interval $[a, b]$ that assigns equal probability to equal-sized areas in the interval, where the parameters a and b are the endpoints of the interval. Here we look at the normal and Poisson distributions, but the others follow the same pattern.

Four items can be calculated for a statistical distribution:

- Density or point probability
- Cumulative distribution function
- Quantiles
- Random numbers

For each distribution, there is a function corresponding to each of the four items. The function has a prefix letter indicating which item. The prefix letter `d` is used for the probability density function, `p` is used for the cumulative distribution function, `q` is used for the quantiles, and `r` is used for random samples. The root name for the normal distribution in R is `norm`, so the probability density function for the normal distribution is `dnorm`, the cumulative distribution function is `pnorm`, the quantile function is `qnorm`, and the random sample function is `rnorm`.

3.2.6 Density

The density for a continuous distribution is a measure of the relative probability of getting a value close to x. With "close" defined as a small interval, this probability is the area under the curve between the endpoints of the interval. The density for a discrete distribution is the probability of getting exactly the value x. It is a density with respect to a counting measure.

You can use the density function to draw a picture of the distribution. First create a sequence of x values, then plot the sequence along with the corresponding densities from a distribution with values for the parameters.

As an example, the Weibull continuous distribution provides a good description for wind speeds. To draw its density, type

```
> w = seq(0, 80, .1)
> plot(w, dweibull(w, shape=2.5, scale=40), type="l",
+   ylab="Probability density")
```

The `seq` function generates equidistant values in the range from 0 to 80 in steps of 0.1. The distribution has two parameters: the shape and scale. Along with the vector of w values, you must specify the shape parameter. The default for the scale parameter is 1. The result is shown in Figure 3.3. Here the parameters are set to values that describe tropical cyclone wind speeds. The family of Weibull distributions includes members that are not symmetric. In this case, the left tail of the distribution is shorter than the left tail of a normal distribution and the right tail is longer.

You create a similar plot using the `curve` function by typing

```
> curve(dweibull(x, shape=2.5, scale=40), from=0,
+   to=80)
```

Note that the first argument (here `dweibull`) must be a function or expression that contains `x`.

For discrete distributions, it is better to use pins as discrete symbols rather than a continuous curve. As an example, the Poisson distribution is a good description for

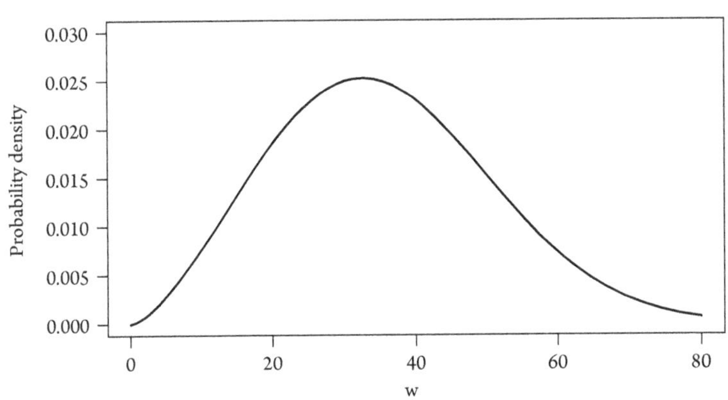

Figure 3.3 Probability density function for a Weibull distribution.

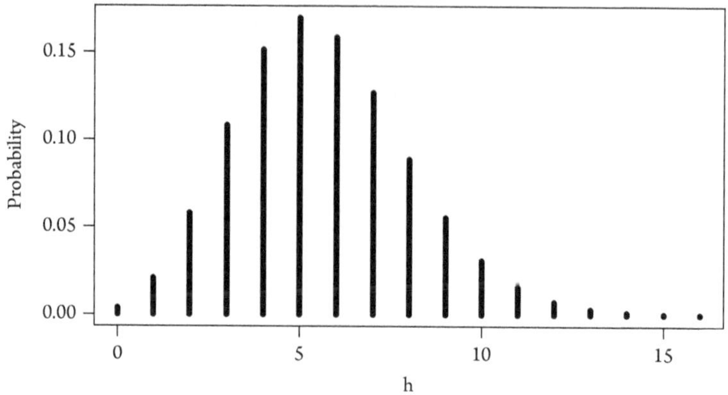

Figure 3.4 Probability mass function for a Poisson distribution.

hurricane counts. To draw its distribution, type

```
> h = 0:16
> plot(h, dpois(h, lambda=5.6), type="h", lwd=3,
+   ylab="Probability distribution")
```

The result is shown in Figure 3.4. The distribution corresponds to the probability of observing h number of hurricanes, given a value for the rate parameter. The argument `type="h"` causes pins to be drawn. The Poisson distribution is a limiting form of the binomial distribution with no upper bound on the number of occurrences. The rate parameter λ characterizes this process. For small values of λ, the distribution is positively skewed.

3.2.7 Cumulative Distribution Functions

The cumulative distribution function describes the probability less than or equal to a value x. It can be plotted, but it is often necessary to get probabilities for distinct values. The function pnorm returns the probability of getting a value equal to or less than the value specified in the first argument assuming a normal distribution with a given mean and standard deviation.

For example, consider again the NAO data for June. Assuming these values are described by a normal distribution with a mean of -0.38 and standard deviation of 1.43, the chance that a June value is less than -1.5 is gotten by typing

```
> pnorm(-1.5, mean=mean(nao), sd=sd(nao))
[1] 0.217
```

or approximately 22 percent. That is, only about 22 percent of the June NAO values are less than or equal to -1.5.

Consider the hurricane data as another example. The annual rate of East Coast hurricanes is obtained by typing `mean(H$E)`. This is the rate (λ) parameter value for the

Poisson distribution. So the probability of next year having no East Coast hurricane is obtained by typing

```
> ppois(0, lambda=mean(H$E))
[1] 0.626
```

You express the probability as a percentage and round to three significant figures by typing

```
> round(ppois(0, lambda=mean(H$E)), digits=3) * 100
[1] 62.6
```

3.2.8 Quantile Functions

The quantile function is the inverse of the cumulative distribution function. The p-quantile is the value such that there is a p probability of getting a value less than or equal to it. The median value is, by definition, the .5 quantile.

In tests of statistical significance (see the next section), the p-quantile is usually set at $p = 0.05$ and is called the $\alpha = p \times 100\%$ significance level. You are interested in knowing the threshold that a test statistic must cross in order to be considered significant at that level. The p-value is the probability of obtaining a value as large, or larger, than the p-quantile.

Theoretical quantiles are also used to calculate confidence intervals. If you have n normally distributed observations from a population with mean μ and standard deviation σ, then the average $\bar{x}$ is described by a normal distribution with mean μ and standard deviation $\sigma/\sqrt{n}$. A 95 percent confidence interval for μ is obtained as

$$\bar{x} + \sigma/\sqrt{n} \times N_{.025} \leq \mu \leq \bar{x} + \sigma/\sqrt{n} \times N_{.975} \tag{3.6}$$

where $N_{0.025}$ and $N_{0.975}$ are the 2.5 and 97.5 percentiles of the standard normal distribution, respectively.

You compute a 95 percent confidence interval about the population mean using the sample of June NAO values by typing

```
> xbar = mean(nao)
> sigma = sd(nao)
> n = length(nao)
> sem = sigma/sqrt(n)
> xbar + sem * qnorm(.025)
[1] -0.604
> xbar + sem * qnorm(.975)
[1] -0.161
```

This produces a 95% confidence interval for the population mean of $[-0.6, -0.16]$.

The normal distribution is symmetric so $-N_{0.025} = N_{0.975}$. You can verify this by typing

```
> -qnorm(.025); qnorm(.975)
[1] 1.96
[1] 1.96
```

Note that `qnorm(.025)` gives the left tail value of the normal distribution and that `qnorm(.975)` gives the right tail value.

Also the quantile for the standard normal is often written as $\Phi^{-1}(.975)$, where Φ is the notation for the cumulative distribution function of the standard normal. Thus it is common to write the confidence interval for the population mean as

$$\bar{x} \pm \sigma/\sqrt{n} \times \Phi^{-1}(.975) \tag{3.7}$$

Another application of the quantile function is to help assess the assumption of normality for a set of observed data. This is done by matching the empirical quantiles with the quantiles from the standard normal distribution. We give an example of this in Chapter 5.

3.2.9 Random Numbers

Random numbers are important in examining climate variation. They are also important in simulating synthetic data that have the same statistical properties as your observations. This is useful when you want to know what effect your assumptions and approximations have on your results.

Random numbers are generated from algorithms. But the sequence of values appear as if they were drawn randomly. That is why they are sometimes referred to as "pseudo-random" numbers.

The distribution functions can generate random numbers (deviates) for you. The first argument specifies the number of deviates to generate, and the subsequent arguments are the parameters of the distribution. For instance, to generate 10 random numbers from a standard normal distribution, type

```
> rnorm(10)
 [1] -0.1950 -1.0521 -0.5509  1.2252 -0.3236  1.6867
 [7]  0.3956 -0.2206 -0.0383 -1.3363
```

Your deviates will be different than those printed here since they are generated randomly. In an experimental setting it is a good strategy to generate the same set of deviates each time. You do this by specifying a random number generator (RNG) and a seed value. If the RNG is the same and the seed value is the same, the set of random numbers will always be the same.

```
> set.seed(3042)
> rnorm(10)
 [1]  0.644 -0.461  1.400  1.123  0.908  0.320 -1.014
 [8] -0.241  0.523 -1.694
```

Here the `set.seed` function uses the default Mersenne Twister RNG with a seed value of 3042. Note that the commands must by typed in the same order; first set the seed, then generate the random numbers.

Specifying an RNG and a seed value allows your results to be replicated exactly. This is important when your results depend on a method that exploits random values as is the case with some of the Bayesian models you will consider in Chapters 4 and 12.

To simulate the next 20 years of Florida hurricane counts based on the counts over the historical record, you type

```
> rpois(20, lambda=mean(H$FL))
 [1] 1 1 1 1 3 1 0 0 0 0 1 1 0 1 0 1 0 0 0
```

To simulate the maximum wind speed (m s^{-1}) from the next 10 tropical cyclones occurring over the North Atlantic, you type

```
> rweibull(10, shape=2.5, scale=50)
 [1]  51.7  38.7  18.2  30.5  41.2  82.7  47.9 100.4
 [9]  26.1  21.3
```

Note that if your numbers are the same as those here, you continued with the same sequence of RNG initiated with the seed value provided earlier.

3.3 ONE-SAMPLE TEST

Inferential statistics refers to using your data to draw conclusions about the population. The simplest case involves testing whether your data support a particular mean value. The population mean is a model for your data and your interest is whether your single sample of values is consistent with this simple mean model.

The one-sample t (Student's t) test is based on the assumption that your data values $(x_i,\ldots,x_n)$ are independent and come from a normal distribution with a mean μ and variance σ^2. The shorthand notation is

$$x_i \sim \text{iid } N(\mu,\sigma^2) \tag{3.8}$$

where the abbreviation iid indicates "independent and identically distributed." You wish to test the *null hypothesis* that $\mu = \mu_0$.

You estimate the parameters of the normal distribution from your sample. The average $\bar{x}$ is an estimate of μ and the sample variance s^2 is an estimate of σ^2. It is important to keep in mind that you can never know the true parameter values. In statistical parlance, they are said to be fixed, but unknowable.

The key concept is that of the *standard error*. The standard error of the mean (or s.e.$(\bar{x})$) describes the variation in your average calculated from your n values. That is, suppose you had access to another set of n values (from the same set of observations or from the same experiment) and you again compute $\bar{x}$ from these values. This average will almost surely be different from the average calculated from the first set.

Statistics from different samples taken from the same population will vary. You can demonstrate this. First generate a population from a distribution with a fixed mean (3) and standard deviation (4).

```
> X = rnorm(2000, mean=3, sd=4)
```

Next take five samples each with a sample size of six and compute the average. This can be done using a `for` loop. The loop structure is `for(i in 1:n){cmds}`, where `i` is the loop index and `n` is the number of times the loop will execute the commands (`cmds`). Here your command is print the mean of a sample of six from the vector `X`.

```
> for(i in 1:5) print(mean(sample(X, size=6)))
[1] 1.68
[1] 1.34
[1] 3.05
[1] 4.35
[1] 1.01
```

The list of sample means are not all the same and not equal to three.

What happens to the variability in the list of means when you increase your sample size from 6 to 60? What happens when you increase the population standard deviation from 4 to 14? Try it.

The standard error of the mean is

$$\text{s.e.}(\bar{x}) = \frac{\sigma}{\sqrt{n}} \tag{3.9}$$

where σ is the population standard deviation and n is the sample size. Even with a single sample, you can estimate s.e.($\bar{x}$) by substituting the sample standard deviation s for σ.

The s.e.($\bar{x}$) tells you how far the sample average may reasonably be from the population mean. With data that are normally distributed there is a 95 percent probability of the sample average staying within $\mu \pm 1.96\sigma$. Note how this is worded. It implies that if you take many samples from the population, computing the average for each sample, you will find that 95 percent of the samples have an average that falls within about two s.e.($\bar{x}$)s of the population mean.

The value of 1.96 comes from the fact that the difference in cumulative probability distribution from the standard normal between ± 1.96 is .95. To verify, type

```
> pnorm(1.96) - pnorm(-1.96)
[1] 0.95
```

With a bit more code, you can verify this for your sample of data saved in the object `X`. This time instead of printing five sample averages, you save 1,000 of them in an object called `Xb`. You then sum the number of TRUEs when logical operators are used to define the boundaries of the interval.

```
> set.seed(3042)
> X = rnorm(2000, mean=3, sd=4)
```

```
> Xb = numeric()
> for(i in 1:1000) Xb[i] = mean(sample(X, size=6))
> p = sum(Xb > 3 - 2 * 4/sqrt(6) &
+     Xb < 3 + 2 * 4/sqrt(6))/1000
> p
[1] 0.952
```

That is, for a given sample, there is a 95 percent chance that the interval defined by the s.e.($\bar{x}$) will cover the true (population) mean.

In the case of a one-sample test, you postulate a population mean and you have a single sample of data. For example, let μ_0 be a guess at the true mean, then you calculate the t statistic as

$$t = \frac{\bar{x} - \mu_0}{\text{s.e.}(\bar{x})} \tag{3.10}$$

With $(\bar{x} - \mu_0) = 2 \times \text{s.e.}(\bar{x})$, your t statistic is two. Your sample mean could be larger or smaller than μ_0 so t can be between -2 and $+2$ with $\bar{x}$ within 2 s.e.($\bar{x}$)s of μ_0.

If you have few data (less than about 30 cases), you need to correct for the fact that your estimate of s.e.($\bar{x}$) uses the sample standard deviation rather than σ. By using s instead of σ in Eq. 3.9, your chance of being farther from the population mean is larger. The correction is made by substituting the t-distribution (Student's t-distribution) for the standard normal distribution.

Like the standard normal, the t-distribution is continuous, symmetric about the origin, and bell-shaped. It has one parameter called the degrees of freedom (ν) that controls the relative "heaviness" of the tails. The degrees of freedom (d.f.) parameter $\nu = n - 1$, where n is the sample size. For small samples, the tails of the t-distribution are heavier than the tails of a standard normal distribution (see Fig. 3.5), meaning that it is more likely to produce values that fall far from the mean.

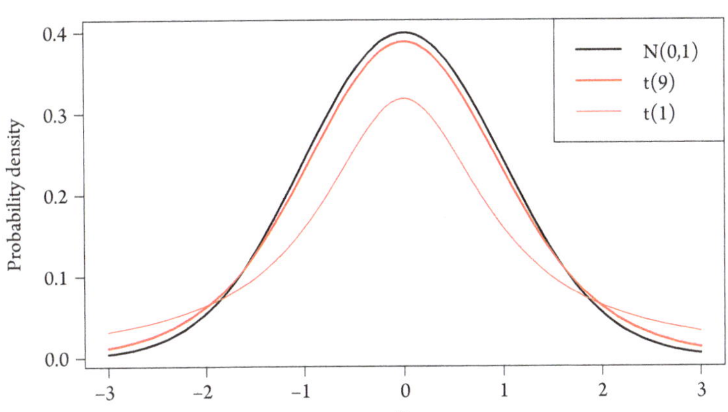

Figure 3.5 Probability density functions. Standard normal distribution [N(0, 1)] and t-distributions with 9 and 1 d.f.

For instance, the difference in cumulative probabilities at the ± 1.96 quantile values from a t-distribution with 9 d.f. is given by

```
> pt(q=1.96, df=9) - pt(q=-1.96, df=9)
[1] 0.918
```

This probability is smaller than that for the standard normal distribution. This indicates that only about 92 percent of the time the interval between ± 1.96 will cover the true mean. Note in the figure how the t-distribution approximates the normal distribution as the sample size increases. For sample sizes of 30 or more, there is little difference.

Given a hypothesized population mean (μ_0), the t statistic is computed from your sample of data using Eq. 3.10. Next the cumulative probability of that value is determined from the t-distribution with d.f. equal to sample size minus one. Finally, the probability is multiplied by two to obtain the p-value. A small p-value leads to a rejection of the null hypothesis and a large p-value leads to a failure to reject the null hypothesis.

The p-value is an estimate of the probability that a particular result, or a result more extreme than the result observed, could have occurred by chance if the null hypothesis is true. In the present case, if the true mean is μ_0, what is the probability that your sample mean is as far or farther from μ_0 as it is? In short, the p-value is a measure of the credibility of the null hypothesis. The higher the p-value, the more credible the null hypothesis appears given your sample of data.

But the p-value is best interpreted as evidence *against* the null hypothesis, thus a small value indicates evidence to reject the null. The interpretation is not black and white. A convenient way to express the evidence is given in Table 3.1.

To illustrate, consider the area-averaged North Atlantic SST values each August in units of °C as an example. Input the monthly data and save the values for August in a separate vector by typing

```
> SST = read.table("SST.txt", header=TRUE)
> sst = SST$Aug
```

Begin with a look at a summary table of these values.

Table 3.1 The p-Value as evidence against the null hypothesis.

p-value Range	*Evidence Against Null Hypothesis*
0–0.01	Convincing
0.01–0.05	Moderate
0.05–0.15	Suggestive, but inconclusive
>0.15	No

```
> summary(sst)
   Min. 1st Qu.  Median    Mean 3rd Qu.    Max.
   22.6    23.0    23.1    23.1    23.3    23.9
   NA's
    5.0
```

The median temperature is 23.1°C with a maximum of maximum of 23.9°C over the 155 years. Note that the object `sst` has a length of 160 years, but the first 5 years have missing values.

You might be interested in the hypothesis that the values deviate significantly from 23.2°C. Although we spent some effort to explain the test statistic and procedure, the application is rather straightforward. The task for you is to test whether this distribution has a mean $\mu = 23.2$. Assuming the data come from a normal distribution, this is done using the `t.test` function as follows:

```
> t.test(sst, mu=23.2)
    One Sample t-test

data:  sst
t = -2.94, df = 154, p-value = 0.003759
alternative hypothesis: true mean is not equal to 23.2
95 percent confidence interval:
 23.1 23.2
sample estimates:
mean of x
     23.1
```

Here there are several lines of output. The output begins with a description of the test you asked for followed by the name of the data object used (here `sst`). The next line contains the value of the *t*-statistic (`t`) as defined in Eq. 3.10, degrees of freedom (`df`), and the *p*-value (`p-value`).

The degrees of freedom is the number of values in calculating the *t*-statistic that are free to vary. Here it is equal to the number of years (number of independent pieces of information) that go into calculating the *t*-statistic minus the number of statistics used in the intermediate steps of the calculation. Equation 3.10 shows only one statistic (the mean) is used, so the number of degrees of freedom is 154.

You do not need a table of the *t*-distribution to look at which quantiles the *t*-statistic belongs to. You can see that the *p*-value is 0.004 indicating convincing evidence against the null hypothesis that the mean is 23.2. When the argument `mu=` is left off, the default is `mu=0`.

A sentence regarding the alternative hypothesis is printed next. It has two pieces of information: the value corresponding to your hypothesis and whether your test is one- or two-sided. Here it states `not equal to`, indicating a two-sided test.

You specify a one-sided test against the alternative of a larger μ by using the `alternative="greater"` argument. For instance, you might hypothesize the temperature to exceed a certain threshold value. Note that abbreviated argument names often work. For example, here it is okay to write `alt="g"` to get the one-sided, greater than, alternative.

The next output is the 95 percent confidence interval for the true mean. You can think of it as defining a set of hypothetical mean values, such that if they were used as values for your null hypothesis (instead of 23.2), they would lead to a p-value of 0.05 or greater (failure to reject the null). You can specify a different confidence level with the `conf.level` argument. For example, `conf.level=.99` will give you a 99 percent interval.

The final bit of output is the mean value from your sample. It is the best estimate for the true mean. Note that you are not testing the data. You are testing a hypothesis about some hypothetical value using your data.

To summarize, in classical statistical inference, you state a hypothesis that, for example, the population mean has a value equal to μ_0. You then use your data to see if there is evidence to reject it. The evidence is summarized as a p-value. A p-value less than 0.15 is taken as suggestive but inconclusive evidence that your hypothesis is wrong, while a p-value less than 0.01 is convincing evidence you are wrong. The larger the value of $|t|$, the smaller the p-value.

3.4 WILCOXON SIGNED-RANK TEST

Even if you cannot assume your data are sampled from a normal distribution, the t-test will provide a robust inference concerning the population mean. By robust we mean that the test results are not overly sensitive to departures from normality, especially in large samples. Keep in mind that the assumption of normality is about the distribution of the population of values, not just the sample you have.

The Wilcoxon signed-rank test does not require you to make the normality assumption. First, the hypothesized value (μ_0) is subtracted from each observation. Next, the absolute value of each difference is taken and sorted from smallest to largest. Ranks are assigned to each difference according to the sorting with the smallest difference given a rank of one. Then, the set of absolute magnitudes of the differences are ranked. Ranking is done by ordering from lowest to highest all magnitudes and counting the number of magnitudes with this value or lower. The lowest magnitude gets a rank of one.

The function `rank` is used to obtain ranks from a set of values. For example, type

```
> rank(c(2.1, 5.3, 1.7, 1.9))
[1] 3 4 1 2
```

The function returns the ranks with each value assigned a ranking from lowest to highest. Here the value of 2.1 in the first position of the data vector is ranked third and the value of 1.7 in the fourth position is ranked one.

Returning to your SST data, to see the ranks of the first 18 differences, type

```
> x = sst - 23.2
> r = rank(abs(x))
> r[1:18]
```

```
 [1] 156.0 157.0 158.0 159.0 160.0  34.0 112.0 124.0
 [9]  91.0 102.0  11.0 147.0 150.0  37.0  96.0   3.5
[17]  77.0  20.0
```

This says there are 156 years that have a difference (absolute value) less than or equal to the first year's SST value. By default ties are handled by averaging the ranks, so for an even number of ties, the rank are expressed as a fractional half, otherwise they are a whole number.

The test statistic (V) is the sum of the ranks corresponding to the values that are above the hypothesized mean

```
> sum(r[x > 0], na.rm=TRUE)
[1] 4224
```

Assuming only that the distribution is symmetric around μ_0, the test statistic corresponds to selecting each rank from 1 to n with a probability of 0.5 and calculating the sum. The distribution of the test statistic can be calculated exactly, but becomes computationally prohibitive for large samples. For large samples, the distribution is approximately normal.

The application of the nonparametric Wilcoxon signed-rank test in R is done in the same way as the t-test. You specify the data values in the first argument and the hypothesized population mean in the second argument.

```
> wilcox.test(sst, mu=23.2)
    Wilcoxon signed rank test with continuity
    correction

data:   sst
V = 4170, p-value = 0.001189
alternative hypothesis: true location is not equal
   to 23.2
```

The p-value of 0.0012 indicates convincing evidence against the null hypothesis, which is somewhat greater evidence than provided by the t-test.

There is less output as there is no parameter estimate and no confidence limits, although it is possible under some assumptions to define a location measure and confidence intervals for it (Dalgaard, 2002). The continuity correction refers to a small adjustment to the test statistic when approximating the discrete ranks with a continuous (normal) distribution. Note that the test statistic is slightly different than what you calculated preiviously. See the help file for additional details.

Although a nonparametric alternative to the t-test can be valuable, caution is advised. If the assumptions are met, then the t- test will be more efficient by about 5 percent relative to the nonparametric Wilcoxon test. That is, for a given sample size, the t-test better maximizes the probability that the test will reject the null hypothesis when it is false. That is, the t-test has more power than the Wilcoxon test. However, in the presence of outliers, the nonparametric Wilcoxon test will less likely indicate spurious significance compared with the parametric t-test.

The Wilcoxon test has problems when there are ties in the ranks for small samples. By default, an exact p-value is computed if the samples contain less than 50 values and there are no ties. Otherwise a normal approximation is used to obtain the p-value.

3.5 TWO-SAMPLE TEST

It is often useful to compare two samples of climate data. For instance, you might be interested in examining whether El Niño influences hurricane rainfall. Here you would create two samples with one sample containing hurricane rainfall during years denoted as El Niño, and the other sample containing hurricane rainfall during all other years.

The two-sample t-test is used to test the hypothesis that two samples come from distributions with the same population mean. The theory is the same as that employed in the one-sample test. Vectors are now doubly indexed ($x_{1,1},\ldots,x_{1,n_1}$ and $x_{2,1},\ldots,x_{2,n_2}$). The first index identifies the sample and the second identifies the case. The assumption is that the values follow normal distributions $N(\mu_1,\sigma_1^2)$ and $N(\mu_2,\sigma_2^2)$, and your interest is to test the null hypothesis $\mu_1=\mu_2$. You calculate the t-statistic as

$$t=\frac{\bar{x}_1-\bar{x}_2}{\sqrt{\text{s.e.}(\bar{x}_1)^2+\text{s.e.}(\bar{x}_2)^2}} \tag{3.11}$$

where the denominator is the standard error of the difference in means and $\text{s.e.}(\bar{x}_i)=s_i/\sqrt{n_i}$. If you assume the two samples have the same variance ($s_1^2=s_2^2$), then you calculate the s.e.($\bar{x}$)s using a single value for s based on the standard deviation of all values over both samples. Under the null hypothesis that the population means are the same, the t-statistic will follow a t-distribution with n_1+n_2-2 degrees of freedom.

If you do not assume equal variance, the t-statistic is approximated by a t-distribution after adjusting the degrees of freedom by the Welch procedure. By default, the function uses the Welch procedure resulting in a noninteger degrees of freedom. Regardless of the adjustment, the two-sample test will usually give about the same result unless the sample sizes and the standard deviations are quite different.

As an example, suppose you are interested in whether the June NAO values have mean values that are different depending on hurricane activity along the Gulf coast later in the year. First create two samples of NAO values. The first sample contains June NAO values in years with no Gulf hurricanes and the second sample contains June NAO values in years with at least two Gulf hurricanes. This is done using the `subset` function.

```
> nao.s1 = subset(nao, H$G == 0)
> nao.s2 = subset(nao, H$G >= 2)
```

You then summarize the two sets of values with the mean, standard deviation, and sample size.

```
> mean(nao.s1); sd(nao.s1); length(nao.s1)
[1] -0.277
[1] 1.51
[1] 80
> mean(nao.s2); sd(nao.s2); length(nao.s2)
[1] -1.06
[1] 1.03
[1] 22
```

The mean NAO value is larger during inactive years, but is it significantly larger? The standard deviation is also larger and so is the sample size.

Your null hypothesis is that the population mean of June NAO values during active Gulf years is equal to the population mean of June NAO values during inactive years. The test is performed with the `t.test` function with the two data vectors as the two arguments. By default, the function uses the Welch procedure.

```
> t.test(nao.s1, nao.s2)
    Welch Two Sample t-test

data:  nao.s1 and nao.s2
t = 2.82, df = 48.7, p-value = 0.007015
alternative hypothesis: true difference in means is not
   equal to 0
95 percent confidence interval:
 0.223 1.337
sample estimates:
mean of x mean of y
   -0.277    -1.057
```

The output is similar to that from the one-sample test provided earlier. The type of test and the data objects are given in the preamble. The value of the t-statistic, the degrees of freedom, and the p-value follow.

Here you find a t-statistic of 2.815. Assuming that the null hypothesis of no difference in population means is correct, this t value (or a value larger in magnitude) has a probability 0.007 of occurring by chance given a t-distribution with 48.67 degrees of freedom. Thus, there is convincing evidence that June NAO values are different between the two samples.

As with the one-sample test, the alternative hypothesis, which is that the true difference in means is not equal to zero, is stated as part of the output. This is the most common alternative in these situations.

The confidence interval (CI) refers to the difference in sample means (mean from sample 1 minus mean from sample 2). So you state that the difference in sample means is 0.78 [(0.22, 1.34), 95% CI]. The interval does not include zero consistent with the conclusion from the test statistic and the corresponding *p*-value indicating convincing evidence against the null hypothesis (less than the 5% significance level).

If you are willing to assume the variances are equal (for example, both samples come from the same population), you can specify the argument `var.equal=T`. In this case, the number of degrees of freedom is a whole number, the *p*-value is larger, and the confidence interval is wider.

3.6 STATISTICAL FORMULA

Instead of creating subsets of the object `nao` based on values in the object `H`, you can create a data frame with two parallel columns. Include all values for the NAO in one column and the result of a logical operation on Gulf hurricane activity in a separate column.

```
> gulf = H$G > 1
> nao.df = data.frame(nao, gulf)
> tail(nao.df)
      nao  gulf
155 -1.00  TRUE
156 -0.41 FALSE
157 -3.34 FALSE
158 -2.05  TRUE
159 -3.05 FALSE
160 -2.40 FALSE
```

This displays the NAO values and whether or not there was two or more Gulf hurricanes in corresponding years.

The goal is to see whether there is a shift in the level of the NAO between the two groups of hurricane activity years (`TRUE` and `FALSE`). Here the groups are years with two or more Gulf hurricanes (TRUE) and years with one or fewer hurricanes (FALSE).

With this setup, you specify a two-sample *t*-test using the tilde (~) operator as

```
> t.test(nao ~ gulf, data=nao.df)
    Welch Two Sample t-test

data:  nao by gulf
t = 3.1, df = 36, p-value = 0.00373
alternative hypothesis: true difference in means is not
   equal to 0
95 percent confidence interval:
 0.271 1.293
sample estimates:
mean in group FALSE  mean in group TRUE
             -0.275              -1.057
```

The object to the left of the tilde is the variable you want to test and the object to the right is the variable used for testing. The tilde is read as "described by" or "conditioned on." That is, the June NAO values are described by Gulf coast hurricane activity. This is how statistical models are specified in R. You will see this model structure throughout the book. Note that by using the `data=nao.df,` you can refer to the column vectors in the data frame by name in the model formula.

The conclusion is the same. Years of high and low Gulf hurricane activity appear to be presaged by June NAO values that are significantly different. The output is essentially the same although the group names are taken from the output of the logical operation. Here `FALSE` refers to inactive years.

3.7 TWO-SAMPLE WILCOXON TEST

If the normality assumption is suspect or sample sizes are small, you might prefer a nonparametric test for differences in the mean. As with the one-sample Wilcoxon test, the two-sample counterpart is based on replacing your data values by their corresponding rank. This is done without regard to group. The test statistic W is then computed as the sum of the ranks in one group.

The function is applied using the model structure as

```
> wilcox.test(nao ~ gulf, data=nao.df)
    Wilcoxon rank sum test with continuity
    correction

data:  nao by gulf
W = 2064, p-value = 0.006874
alternative hypothesis: true location shift is not
     equal to 0
```

The results are similar to those found using the t-test and are interpreted as convincing evidence of a relationship between late spring NAO index values and hurricane activity along the Gulf coast of the United States.

3.8 COMPARE VARIANCES

It is not necessary to assume equal variances when testing for differences in means. Indeed, this is the default option with the `t.test` function. Yet your interest could be whether the variability is changing. For instance, you might speculate that the variability in hurricane activity will increase with global warming.

Note that the variance is strictly positive. Given two samples of data, the ratio of variances will be unity if the variances are equal. Under the assumption of equal population variance, the F-statistic, as the ratio of the sample variances, has an F-distribution with two parameters. The parameters are the two sample sizes minus one.

The F-distribution is positively skewed meaning the tail on the right is longer than the tail on the left. Figure 3.6 shows the probability density for two F-distributions. A larger sample size results in a density centered on one and more symmetric.

The function `var.test` is called in the same way as the function `t.test`, but performs an F test on the equality of the variances.

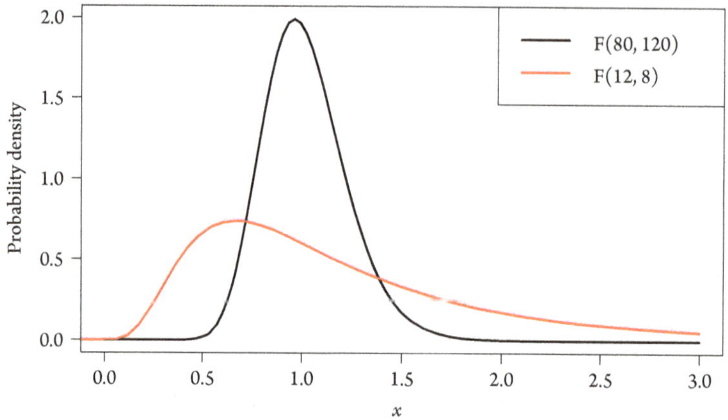

Figure 3.6 Probability density functions for an F-distribution.

```
> var.test(nao ~ gulf, data=nao.df)
    F test to compare two variances

data:  nao by gulf
F = 2, num df = 137, denom df = 21, p-value =
0.06711
alternative hypothesis: true ratio of variances is not
     equal to 1
95 percent confidence interval:
 0.95 3.58
sample estimates:
ratio of variances
                 2
```

Results show an F-statistic of 2 with degrees of freedom equal to 137 and 21 resulting in a p-value of 0.067 under the null hypothesis of equal variance. The magnitude of the p-value provides suggestive but inconclusive evidence of a difference in population variance. Note that the 95 percent confidence interval on the F-statistic includes the value of one as you would expect given the p-value.

The F-test is sensitive to departures from normality. Also, for small data sets, the confidence interval will be quite wide (see Fig. 3.6) often requiring you to take the assumption of equal variance as a matter of belief.

3.9 CORRELATION

Correlation extends the idea of comparing one variable in relation to another. Correlation indicates the amount and the direction of association between two variables. If hurricanes occur more often when the ocean is warmer, then you say that ocean temperature is positively correlated with hurricane incidence as one goes up, the other goes up. If hurricanes occur less often when sun spots are numerous, then you say that sun spots are inversely correlated with hurricane incidence. Meaning the two variables go in opposite direction as one goes up, the other goes down.

A correlation coefficient is a symmetric scale-invariant measure of the correlation between two variables. It is symmetric because the correlation between variables x and y is the same as the correlation between variables y and x. It is scale-invariant because the value does not depend on the units of either variable.

Correlation coefficients range from -1 to $+1$, where the extremes indicate perfect correlation and 0 means no correlation. The sign is negative when large values of one variable are associated with small values of the other and positive if both tend to be large or small together. Different metrics of correlation lead to different correlation coefficients. Most common is Pearson's product-moment correlation coefficient followed by Spearman's rank and Kendall's τ.

3.9.1 Pearson's Product-Moment Correlation

Pearson's product-moment correlation coefficient is derived from the bivariate normal distribution of two variables, where the theoretical correlation describes contour ellipses about the two-dimensional densities. It is the workhorse of climatological studies. If both variables are scaled to have unit variance, then a correlation of zero corresponds to circular contours and a correlation of one corresponds to a line segment.

Figure 3.7 shows two examples: one where the variables x and y have a small positive correlation and the other where they have a fairly large negative correlation. The points are generated from a sample of bivariate normal values with a Pearson product-moment correlation of 0.2 and -0.7. The contours enclose the 75 and 95 percent probability region for a bivariate normal distribution with mean of zero, unit variances, and corresponding correlations.

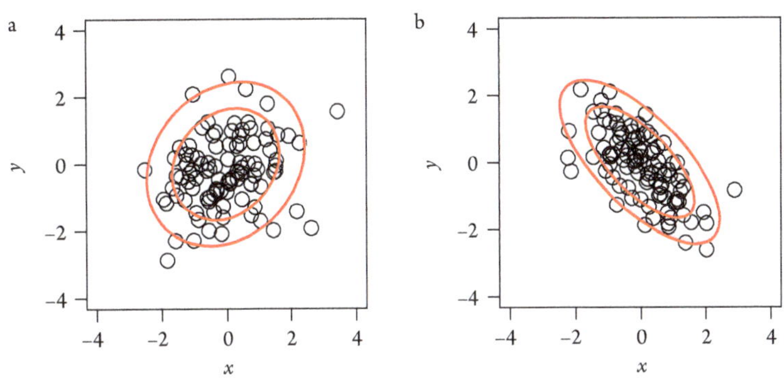

Figure 3.7 Scatter plots of correlated variables with (a) $r = 0.2$ and (b) $r = -0.7$.

The Pearson correlation coefficient between x and y is

$$r(x,y) = \frac{\sum (x_i - \bar{x})(y_i - \bar{y})}{\sqrt{\sum (x_i - \bar{x})^2 \sum (y_i - \bar{y})^2}} \tag{3.12}$$

The Pearson correlation is often called the "linear" correlation since the absolute value of r will be one when there is a perfect linear relationship between x_i and y_i.

The function `cor` is used to compute the correlation between two or more vectors. For example, to get the linear correlation between the May and June values of the NAO, type

```
> cor(NAO$May, NAO$Jun)
[1] 0.0368
```

The value indicates weak positive correlation. Note that the order of the vectors in the function is irrelevant as $r(x,y) = r(y,x)$. You can verify in this case by typing

```
> cor(NAO$May, NAO$Jun) == cor(NAO$Jun, NAO$May)
[1] TRUE
```

If there are missing values, the function will return `NA`. The argument `na.rm=TRUE` works for one-vector functions like `mean`, `sd`, `max`, and others to indicate that missing values should be removed before computation. However, with `cor` function there are additional ways to handle the missing values, so you need the `use` argument. As an example, to handle the missing values in the SST data frame by case-wise deletion, type

```
> cor(SST$Aug, SST$Sep, use="complete.obs")
[1] 0.944
```

Here the value indicates strong positive correlation between August and September SST.

This value of r estimated from the data is a random variable and is thus subject to sampling variation. For instance, adding another year's worth of data will result in an r value that is somewhat different. Typically, your hypothesis is that the population correlation is zero. As might be guessed from the differences in r, your conclusions about this hypothesis will likely be different for the SST and NAO data.

You can ask the question differently. For example, in 1,000 samples of x and y each of size 30 from a population with zero correlation, what is the largest value of r? You answer this question using simulations by typing

```
> set.seed(3042)
> n = 30
> cc = numeric()
> for(i in 1:1000){
+   x = rnorm(n); y = rnorm(n)
+   cc[i] = cor(x, y)
```

```
+ }
> mean(cc); max(cc)
[1] -0.0148
[1] 0.569
```

The variable `n` sets the sample size and you simulate 1,000 different correlation coefficients from different independent samples of x and y. The average correlation is close to zero as expected, but the maximum correlation is quite large.

High correlation can arise by chance. Thus when you report a correlation coefficient, a CI on your estimate or a test of significance should be included. This is done with the `cor.test` function. The test is based on transforming r to a statistic that has a t-distribution using

$$t = \sqrt{\nu}\frac{r}{\sqrt{1-r^2}} \tag{3.13}$$

where $\nu = n - 2$ is the degrees of freedom and n is the sample size.

Returning to the NAO example, to obtain a confidence interval on the correlation between the May and June values of the NAO and a test of significance, type

```
> cor.test(NAO$May, NAO$Jun)
    Pearson's product-moment correlation

data:  NAO$May and NAO$Jun
t = 0.463, df = 158, p-value = 0.6439
alternative hypothesis: true corr is not equal to 0
95 percent confidence interval:
 -0.119  0.191
sample estimates:
   cor
0.0368
```

The type of correlation and the data used in the test are output in the preamble. The correlation of 0.037 (given as the last bit of output) is transformed to a t value of 0.463 with 158 degrees of freedom providing a 95 percent CI of $[-0.119, 0.191]$. This means that if the procedure used to estimate the CI was repeated 100 times, 95 of the intervals would contain the true correlation coefficient. The output also gives a p-value of 0.644 as evidence in support of the null hypothesis of no correlation.

Repeat this example using the January and September values of SST. What is the CI on the correlation estimate? How would you describe the evidence against the null hypothesis of zero correlation in this case?

3.9.2 Spearman's Rank and Kendall's τ Correlation

Inferences based on the Pearson correlation assume the variables are adequately described by normal distributions. An alternative is Spearman's rank (ρ) correlation, which overcomes the effect of outliers and skewness by considering the rank of the data rather than the magnitude. The Spearman correlation coefficient is defined as the Pearson correlation coefficient between the ranked variables.

The Pearson correlation is the default in the `cor.test` function. You change this with the `method` argument. To obtain Spearman's rank correlation and the associated test of significance, type

```
> cor.test(H$G, H$FL, method="spearman")
     Spearman's rank correlation rho

data:  H$G and H$FL
S = 551867, p-value = 0.01524
alternative hypothesis: true rho is not equal to 0
sample estimates:
  rho
0.192
```

The correlation is 0.192 with a p-value of 0.015, providing suggestive evidence against the null hypothesis of no correlation.

Kendall's τ is another alternative to Pearson correlation. It is based on counting the number of concordant and discordant point pairs from your data. For two data vectors x and y each of length n, a point at location i is given in two-dimensional space as (x_i, y_i). A point pair is defined as $[(x_i, y_i); (x_j, y_j)]$ for $i \neq j$.

A point pair is concordant if the difference in the x values is of the same sign as the difference in the y values, otherwise it is discordant. The value of Kendall's τ is the number of concordant pairs minus the number of discordant pairs divided by the total number of unique point pairs, which is $n(n-1)/2$ where n is the sample size. For a perfect correlation, either all point pairs are concordant or all pairs are discordant. Under zero correlation, there are as many concordant pairs as discordant pairs.

Repeat the call to function `cor.test` on coastal hurricane activity, but now use the `kendall` method and save the resulting estimate.

```
> x = cor.test(H$G, H$FL, method="kendall")
> tau.all = x$estimate
```

The correlation is 0.174 with a p-value of 0.015, again providing suggestive evidence against the null hypothesis of no correlation.e

3.9.3 Bootstrap Confidence Intervals

Kendall's τ and Spearman's rank correlations are computed without confidence intervals. You should always report a CI. In this case, you use a procedure called *bootstrapping*, which is a resampling technique.

The idea is to sample the values from your data with replacement using the `sample` function. The sample size is the size of your data. The bootstrap sample is called a replicate. You compute the statistic of interest using the values in your replicate. The bootstrap value will be different than the value computed from your data because the replicate contains repeats and not all data are included. You repeat the procedure many times collecting all the bootstrap statistics. You then use the `quantile` function to determine the lower and upper quantiles corresponding to the 0.025 and 0.975 probabilities.

Bootstrapping is widely used for assigning measures of accuracy to sample estimates (Efron and Tibshirani, 1986). The function `boot` from the package **boot** generates bootstrap replicates of any statistic applied to your data. It has options for parametric and nonparametric resampling.

For example, to implement a bootstrap procedure for Kendall's τ using data on Florida and Gulf coast hurricane frequencies, you first create a function as follows:

```
> mybootfun = function(x, i){
+     Gbs = x$G[i]
+     Fbs = x$FL[i]
+ return(cor.test(Gbs, Fbs,method="k")$est)
+ }
```

Your function has two variables: the data (`x`) and an index variable (`i`). Next you generate 1,000 bootstrap samples and calculate the CIs by typing

```
> require(boot)
> tau.bs = boot(data=H, statistic=mybootfun, R=1000)
> ci = boot.ci(tau.bs, conf=.95)
```

The `boot` function must be run prior to running the `boot.ci` function. The result is a 95 percent CI of (0.032, 0.314) about the estimated τ of 0.174.

3.9.4 Causation

If you compute the correlation between June SST and U.S. hurricane counts using Kendall's method, you find a positive value for τ of 0.151 with a p-value of 0.011 indicating suggestive, but inconclusive, evidence against the null hypothesis of no correlation.

A positive correlation between ocean warmth and hurricane activity does not prove causality. Moreover, since the association is symmetric, it does not say that x causes y any more than it says y causes x. This is why you frequently hear "correlation does not equal causation." The problem with this adage is that it ignores the fact that correlation is needed for causation. It is necessary, but insufficient. When correlation is properly interpreted, it is indispensable in the study of hurricanes and climate.

Your correlation results are more meaningful if you explain how the variables are physically related. Particularly if your explanation comes before you look at your data. Studies showing a consistent correlation between two variables using different time and space scales, and over different time periods and different regions, provide greater evidence of an association than a single study. However, if you want proof that a single factor causes hurricanes, then correlation is not enough.

3.10 LINEAR REGRESSION

Correlation is the most widely used statistic in climatology, but linear regression is the most important statistical model. When you say that variable x has a linear relationship to variable y, you mean $y = a + bx$, where a is the y-intercept and b is the slope of the line. You call x the independent variable and y the dependent variable because the value of y depends on the value of x.

But in statistics, you do not assume these variables have a perfect linear relationship. Instead, in describing the relationship between two random vectors x_i and y_i, you add an error term (ε) to the equation such that

$$y_i = \alpha + \beta x_i + \varepsilon_i \tag{3.14}$$

You assume the values ε_i are iid $N(0,\sigma^2)$.

The slope of the line is the regression coefficient β, which is the change in the *average* value of y per unit change in x. The line intersects the y-axis at the *intercept* α. The vector x is called the explanatory variable and the vector y is called the response variable.

Equation 3.14 describes a regression of the variable y *onto* the variable x. This is always the case. You regress your response variable onto your explanatory variable(s), where the word "regression" refers to a model for the mean of the response variable.

The model consists of three parameters α, β, and σ^2. For a set of explanatory and response values, the parameters are estimated using the *method of least squares*. The method finds a set of α and β values that minimize the sum of squared residuals given as

$$\text{SS}_{res} = \sum_i [y_i - (\alpha + \beta x_i)]^2 \tag{3.15}$$

The solution to this minimization is a set of equations given by

$$\hat{\beta} = \frac{\sum(x_i - \bar{x})(y_i - \bar{y})}{\sum(x_i - \bar{x})^2} \tag{3.16}$$

$$\hat{\alpha} = \bar{y} - \hat{\beta}\bar{x} \tag{3.17}$$

that define estimates for α and β. The residual variance is $\text{SS}_{res}/(n-2)$, where $\hat{\alpha}$ and $\hat{\beta}$ are used in Eq. 3.15.

The regression line is written as

$$\hat{y}_i = \hat{\alpha} + \hat{\beta} x_i \tag{3.18}$$

The method of least squares to determine the $\hat{\alpha}$ and $\hat{\beta}$ values is performed by the function `lm` (linear model). If you are interested in regressing the August values of Atlantic SST onto the preceding January SST values, you type

```
> lm(SST$Aug ~ SST$Jan)
Call:
lm(formula = SST$Aug ~ SST$Jan)
```

```
Coefficients:
(Intercept)        SST$Jan
       7.11           0.84
```

The argument to `lm` is a model formula where the tilde symbol ($\sim$) is read as "described by." Or to be more complete in this particular case, we state that the August SST values are *described by* the January SST values *using a linear regression model.* In this case, you have a single explanatory variable so the formula is simply `y~x` and you call the model *simple* linear regression. In the next section, you will see that with two covariates the model is `y~x+z`.

The response variable is the variable you are interested in modeling. This is crucial. You must decide which variable is your response variable before hard. Unlike correlation, a regression of y onto x is not the same as a regression of x onto y. Your choice depends on the question you want answered. You get no guidance by examining your data nor will R tell you. Here you choose August SST as the response since it is natural to consider using information from an earlier month to predict what might happen in a later month.

The output from `lm` includes a preamble of the call that includes the model structure and the data used. Parameter estimates are given in the table of coefficients. The estimated intercept value ($\hat{\alpha}$) is given under `(Intercept)` and the estimated slope value ($\hat{\beta}$) under `SST$Jan`. The output allows you to state that the best-fitting straight line (regression line) is defined as August SST = 7.11 + 0.84 $\times$ January SST.

The units on the intercept parameter are the same as the units of the response variable, here $^\circ$C. The units on the slope parameter are the units of the response divided by the units of the explanatory variable, here $^\circ$C per $^\circ$C. Thus you interpret the slope value in this example as follows: for every 1°C increase in January SST, the August SST increases on average by 0.84°C.

The slope and intercept values will deviate somewhat from the true values due to sampling variation. One way to examine how much deviation is to take samples from the data and, with each sample, use the `lm` function to determine the parameter. The code below does this for the slope parameter using January SST values as the explanatory variable and August SST values as the response.

```
> sl = numeric()
> for (i in 1:1000) {
+   id = sample(1:length(SST$Aug), replace=TRUE)
+   sl[i] = lm(SST$Aug[id] ~ SST$Jan[id])$coef[2]
+ }
> round(quantile(sl), digits=2)
  0%  25%  50%  75% 100%
0.52 0.78 0.83 0.89 1.06
```

Note, you sample from the set of row indices and use the same index for the January and the August values. Results indicate that 50 percent of the slopes fall between the values 0.78 and 0.89°C per $^\circ$C.

Although illustrative, sampling is not really needed. Recall you calculated the s.e.($\bar{x}$) from a single sample to describe the variability of the sample mean. You do the same to calculate the standard error of the slope (and intercept) from a sample of x and y values. These standard errors, denoted s.e.($\hat{\beta}$) and s.e.($\hat{\alpha}$), are used for inference and to compute CI.

Typically the key inference is a test of the null hypothesis that the population value for the slope is zero. A zero slope implies that the line is horizontal and there is no relationship between the response and explanatory variables. The test statistic in this case is

$$t = \frac{\hat{\beta}}{\text{s.e.}(\hat{\beta})} \tag{3.19}$$

which follows a t distribution with $n-2$ degrees of freedom if the true slope is zero. Similarly you can test the null hypothesis that the intercept is zero, but this often has little physical meaning because it typically involves an extrapolation outside the range of your x values.

The value for the test statistic (t value) is not provided as part of the raw output from the `lm` function. The result of `lm` is a *model* object. This is a key concept. In Chapter 2, you encountered data objects. You created structured data vectors and input data frames from a spreadsheet. The saved objects are listed in your working session by typing `objects()`. Functions, like `table`, are applied to these objects to extract information.

In the same way, a model object contains a lot of information. This information is extracted using functions. The `summary` function is one example. You saw previously that applied to a data frame object, the `summary` function extracts statistics about the values in each column. When applied to a model object, it extracts information about the model.

For instance, to obtain the information about the regression model of August SST onto January SST, type

```
> summary(lm(SST$Aug ~ SST$Jan))
Call:
lm(formula = SST$Aug ~ SST$Jan)

Residuals:
    Min      1Q  Median      3Q     Max
-0.4773 -0.1390 -0.0089  0.1401  0.4928

Coefficients:
            Estimate Std. Error t value Pr(>|t|)
(Intercept)   7.1117     1.4597    4.87  2.7e-06
SST$Jan       0.8400     0.0765   10.98  < 2e-16

Residual standard error: 0.197 on 153 df
  (5 observations deleted due to missingness)
```

```
Multiple R-squared: 0.441,    Adjusted R-squared: 0.437
F-statistic:   121 on 1 and 153 DF,   p-value: <2e-16
```

The output includes the call function, summary of residuals, table of coefficients, residual standard error, R^2 values, and the *F*-statistic and the associated *p*-value. The output starts with a repeat of the function call. This is useful if the object is saved and examined later.

Summary statistics on the set of model residuals follow. A residual is the difference between the response value at a particular explanatory value and the modeled value. Residuals are important in helping you diagnose how well your model fits the data. The average of the residuals is zero by definition of least squares, so the median should not be far from zero and the minimum and maximum should roughly be equal in absolute magnitude. This is true here for the first (`1Q`) and third (`3Q`) quartile values.

Next, the table of coefficients shows the intercept and slope values as in the raw output, but here the accompanying standard errors, *t* values, and *p*-values are also provided. The output is in tabular form with the each row corresponding to a separate parameter. The first row is the intercept and the second row is the slope associated with the explanatory variable (`SST$Jan`). The vector `SST$Jan` is the explanatory variable and the slope of 0.84 is the amount of change in the mean response associated with a unit change in the explanatory variable.

The first labeled column is the sample estimate (`Estimate`), the second is the standard error (`Std. Error`), the third is the *t* value (`t value`), and the fourth is the *p*-value (`Pr(>|t|)`). Note that according to Eq. 3.19, the *t* value is the ratio of the estimated value to its standard error. The *p*-value is the probability of finding a *t* value as large or larger (in absolute value) by chance assuming the slope is zero. Here the *p*-value on the January SST slope is less than 2×10^{-16}, which is output in exponential notation as `2e-16`. This indicates a near-zero chance of no relationship between January SST and August SST given your sample of data.

By default, symbols are placed to the right of the *p*-values as indicators of the level of significance, and a line below the table provides the definition. Here we turned them off using an argument in the `options` function. The *p*-value should always be reported rather than simply providing a categorical signifiance level.

Note that the interpretation of a *p*-value as evidence in support of the null hypothesis is the same as for the *t*-test you encountered earlier. Your job is to determine the null hypothesis. In the context of regression, the assumption of no relationship between the explanatory and response variables is typically your null hypothesis. Therefore, a low *p*-value indicates evidence of a relationship between your explanatory and response variables.

Continuing with the output, the residual standard error quantifies the variation of the observed values about the regression line. It is computed as the square root of the sum of the squared residuals divided by the square root of the degrees of freedom. The degrees of freedom is the sample size minus the number of coefficients. It provides an estimate of the model parameter σ.

Next are the R-squared values. The "multiple R squared," is the proportion of variation in the response variable that can be explained by the explanatory variable. So here you state that the model explains 44.1 percent of the variation in August SST values. With only a single explanatory variable (simple linear regression), the multiple R squared is equal to the square of the Pearson correlation coefficient, which you verify by typing

```
> cor(SST$Jan, SST$Aug, use="complete")^2
[1] 0.441
```

The adjusted R squared ($\bar{R}^2$) is a modification to the multiple R squared for the number of explanatory variables. The adjusted R squared increases only if the new variable improves the model. It can be negative and will always be less than or equal to the multiple R squared. It is defined as

$$\bar{R}^2 = 1 - (1 - R^2)\frac{n-1}{n-p-1} \tag{3.20}$$

where n is the sample size and p is the number of explanatory variables. In small samples with many explanatory variables, the difference between R^2 and $\bar{R}^2$ will be large.

The final bit of output is related to an F test, which is a test concerning the entire model. The output includes the F statistic, the degrees of freedom (in this case, two of them), and the corresponding p-value as evidence in support of the null hypothesis that the model has no explanatory power. In the case of simple regression, it is equivalent to the test on the slope so it is only interesting when there is more than one explanatory variable. Note that the F statistic is equal to the square of the t statistic, which is true of any linear regression model with one explanatory variable.

Other extractor functions provide useful information about your model. The function `resid` takes a model object and extracts the vector of residual values. For example, type

```
> lrm = lm(Aug ~ Jan, data=SST)
> resid(lrm)[1:10]
      6       7       8       9      10      11
-0.0629 -0.2690  0.0278  0.0892 -0.1426  0.2165
     12      13      14      15
-0.3297 -0.2614  0.4356 -0.2032
```

First the model object is saved with name `lrm`. Here only the column names are referenced in the model formula because you specify the data frame with the `data` argument. Then the extractor function `resid` lists the residuals. Here using the subset function, you list only the first 10 residuals.

Similarly, the function `fitted` computes the mean response value for each value of the explanatory variable. For example, type

```
> fitted(lrm)[1:10]
   6    7    8    9   10   11   12   13   14   15
23.2 23.2 22.8 22.9 23.1 23.0 23.0 22.9 22.8 23.2
```

These fitted values lie along the regression line and are obtained by solving for $\hat{y}$ in Eq. 3.18.

Note that the residuals and fitted values are labeled with the row numbers of the SST data frame. In particular, note that they do not contain rows 1 through 5, which are missing in the response and explanatory variable columns.

Your model can be used to make predictions. The `predict` function is similar to the `fitted` function but allows you to predict values of the response for arbitrary values of the explanatory variable. The caveat is that you need to specify the explanatory values as a data frame using the `newdata` argument. For example, to make an SST prediction for August given a January value of 19.4°C, type

```
> predict(lrm,newdata=data.frame(Jan=19.4))
   1
23.4
```

A note on terminology. The word "predictor" is the generic term for an explanatory variable in a statistical model. A further distinction is sometimes made between covariates, which are continuous-valued predictors and factors, which can take on only a few values that may or may not be ordered.

A prediction is not worth much without an estimate of uncertainty. Assuming the model is correct, a predicted value from a statistical model has two sources of uncertainty. One is the uncertainty about the mean of the response *conditional* on the value of the explanatory variable. It is the precision with which the conditional mean is known. It is known as a *confidence interval.* To obtain the CI on the predicted value, type

```
> predict(lrm, data.frame(Jan=19.4), int="c")
   fit  lwr  upr
1 23.4 23.3 23.5
```

The argument `int="c"` tells the function `predict` to provide a CI on the predicted value. The output includes the predicted value in the column labeled `fit` and the lower and upper confidence limits in the columns `lwr` and `upr`, respectively. By default, the limits define the 95 percent CI. This can be changed with the `level` argument.

The interpretation is the same as before. Given the data and the model, there is a 95 percent chance that the interval defined by the limits will cover the true (population) mean when the January SST value is 19.4°C. The other source of uncertainty arises from the distribution of a particular value given the conditional mean. That is, even if you know the conditional mean exactly, the distribution of particular values about the mean will have a spread.

The *prediction interval* provides a bound on a set of new values from the model that contains both sources of uncertainty. As a consequence, for a given confidence level, the prediction interval will always be wider than the CI. The prediction interval relies on the assumption of normally distributed errors with a constant variance across the values of the explanatory variable.

To obtain the prediction interval on the predicted value, type

```
> predict(lrm, data.frame(Jan=19.4), int="p")
   fit lwr  upr
1 23.4  23 23.8
```

Given the data and the model, there is a 95 percent chance that the interval defined by these limits will cover any future values of August SST given that the January SST value is 19.4°C.

3.11 MULTIPLE LINEAR REGRESSION

Multiple regression extends simple linear regression by allowing more than one explanatory variable. Everything from simple regression carries over. Each additional explanatory variable contributes a new term to the model. However, an issue now arises because of possible relationships between the explanatory variables.

As an illustration, we continue with a model for predicting August SST values over the North Atlantic using SST values from earlier months. Specifically, for this example, you are interested in making predictions with the model at the end of March. You have January, February, and March SST values plus Year as the set of explanatory variables.

The first step is to plot your response and explanatory variables. This is done with the `pairs` function. By including the `panel.smooth` function as the argument to `panel`, a local smoother is used on the set of points that allows you to more easily see possible relationships. Smoothing is discussed in more detail in Chapter 5. Here you specify the August values (column 9 in `SST`) to be plotted in the first row (and column) followed by year and then the January through March values.

```
> pairs(SST[, c(9,1:4)], panel=panel.smooth)
```

The scatter plots are arranged in a two-dimensional matrix (Fig. 3.8). The response variable is August SST and the four explanatory variables include Year, and the SST values during January, February, and March. A locally weighted polynomial smoother with a span of 67 percent of the points is used to draw the red lines.

The diagonal elements of the matrix are the variable labels. The plot in the first row and second column is the August SST values on the vertical axis and the year

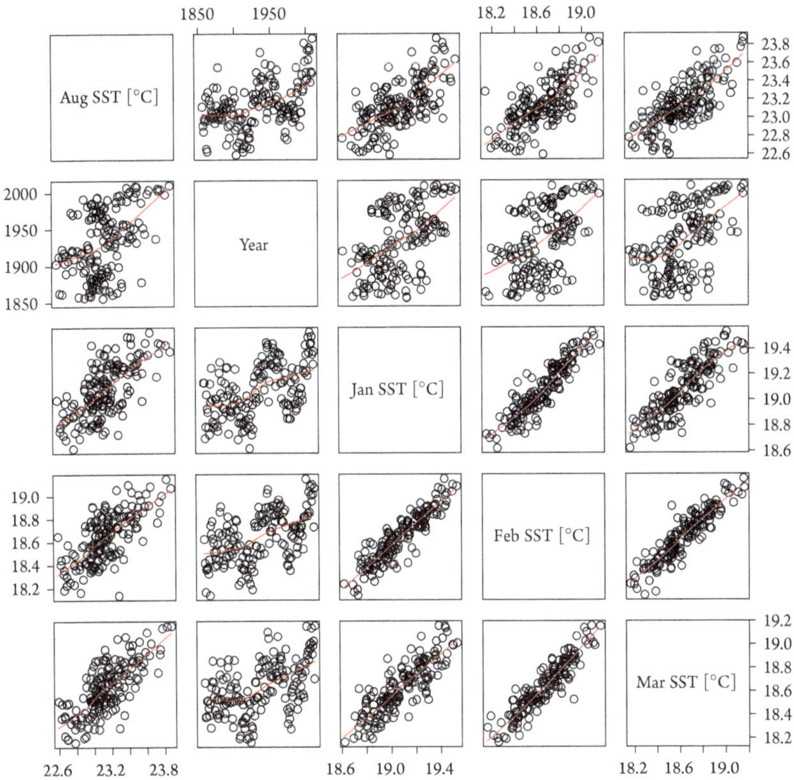

Figure 3.8 Scatter plot matrix of monthly SST values.

on the horizontal axis. The plot in row 1 of column 3 is the August SST values on the vertical axis and January SST values on the horizontal axis, and so on. The lower left set of plots are the same as the upper right set of plots except that the axes are reversed.

The matrix of plots is useful in drawing attention to what explanatory variables might be important in your model of the response variable. Here you see all relationships are positive. Specifically, August SST values increase with increasing year and increasing January through March SST values. Based on these bivariate relationships, you might expect that all four explanatory variables, with the exception of perhaps Year, will be important in the model of August SST.

Importantly, the plots also reveal the relationships between the covariates. Here you see a tight linear relationship between each month's SST values. This warrants attention as a model that includes all three SST variables contains a large amount of redundant information. Information contained in the February SST values is about the same as the information contained in the January SST values, and the information contained in the March SST values is about the same as the information contained in the February SST values.

Table 3.2 Coefficients of the multiple regression model.

	Estimate	*Std. Error*	*t value*	*Pr(> |t|)*
(Intercept)	6.1749	1.3154	4.69	0.0000
Year	0.0007	0.0004	1.90	0.0597
Jan	0.1856	0.1667	1.11	0.2673
Feb	−0.1174	0.2187	−0.54	0.5921
Mar	0.7649	0.1611	4.75	0.0000

To fit a multiple regression model to these data with August SST as the response variable, type

```
> m1 = lm(Aug ~ Year + Jan + Feb + Mar, data=SST)
```

Then to examine the model coefficients, type

```
> summary(m1)
```

The model coefficients and associated statistics are shown in Table 3.2. As expected, March SST is positively related to August SST, and significantly so. Year is also positively related to August SST. The Year term has a p value on its coefficient that is marginally significant (suggestive, but inconclusive) against the null hypothesis of a zero trend. However, you can see that the coefficient on January SST is positive, but not statistically significant, and the coefficient on February SST is negative.

From Figure 3.8 you can see that there is a *positive* relationship between February SST and August SST, so the fact that the relationship is negative in the context of multiple regression indicates a problem. The problem stems from the correlation between explanatory variables (multicollinearity). High correlation results in an unstable model because the standard errors on the coefficients are not estimated with enough precision.

As long as two variables are not perfectly correlated estimation of the regression coefficients is possible, but the estimates and standard errors become sensitive to even the slightest change in the data, when variables are correlated at levels above 0.6. Prior understanding of the partial correlation (here the correlation between February SST and August SST controlling for March SST) may help argue in favor of retaining two highly correlated explanatory variables, but in the usual case it is better to eliminate the variable whose relationship with the response variable is harder to explain physically or to eliminate the variable that has the smaller correlation with the response variable.

Here February SST has a smaller correlation with August SST, so you remove it and reestimate the coefficients. You create a new linear model object and summarize it by typing

```
> m2 = lm(Aug ~ Year + Jan + Mar, data=SST)
> summary(m2)
```

You see all the remaining explanatory variables have a positive relationship with the response variable, consistent with their bivariate plots, but the coefficient on January SST is not statistically significant.

Thus you need to try a third model with the insignificant term removed.

```
> m3 = lm(Aug ~ Year + Mar, data=SST)
> summary(m3)
```

The model makes sense. August SST values are higher when March SST values are higher. This relationship holds after accounting for the upward trend in August SST values.

Note that the order of the explanatory variables on the right side of the ~ does not matter. That is, you get the same output by typing

```
> summary(lm(Aug ~ Mar + Year, data=SST))
```

Note also that the multiple R-squared value is slightly lower in the final model with fewer variables. This is always the case and is the reason why the R squared should not be used for comparing models. Instead, use the adjusted R squred, which increases only when a statistically significant variable is added to the model.

The final model is checked for adequacy by examining the distribution of model residuals. The five-number summary of the residuals given as part of the summary output gives you no reason to suspect the assumption of normally distributed residuals. However, the residuals are likely to have some autocorrelation violating the assumption of independence. We will revisit this topic in Chapter 5.

3.11.1 Predictor Choice

Suppose H is your variable of interest and $X_1, \ldots, X_p$, a set of potential explanatory variables, are vectors of n observations. The problem of predictor selection arises when you want to model the relationship between H and a subset of the explanatory variables, but there is uncertainty about which subset to use. The situation is particularly interesting when p is large and the variables contain redundant information.

You want a model that fits the data well and has small variance. The problem is these two goals are in conflict. An additional predictor in a model will improve the fit (reduce the bias), but will increase the variance due to a loss in the number of degrees of freedom. This is known as the bias variance trade-off.

A commonly used statistic that helps with this trade-off is called the Akaike Information Criterion (AIC), given by

$$\mathrm{AIC} = 2(p+1) + n[\log(\mathrm{SSE}/n)], \tag{3.21}$$

where p is the number of predictors and SSE is the residual sum of squares. You can compare the AIC values when each predictor is added or removed from a given model. For example, if after adding a predictor, the AIC value for the model increases, then the trade-off is in favor of the extra degree of freedom and against retaining the predictor.

Returning to your original model for August SST, the model is saved in the object `m1`. The `drop1` function takes your regression model object and returns a table showing the effects of dropping, in turn, each variable from the model. To see this, type

```
> drop1(m1)
Single term deletions

Model:
Aug ~ Year + Jan + Feb + Mar
       Df Sum of Sq  RSS  AIC
<none>              4.61 -535
Year    1     0.111 4.72 -533
Jan     1     0.038 4.65 -536
Feb     1     0.009 4.62 -537
Mar     1     0.693 5.30 -515
```

Here the full model (all four covariates) has a residual sum of squares (RSS) of 4.61 (in the <none> row; none dropped). If you drop the Year variable, the RSS increases to 4.72 (you add 0.11 to 4.61) and you gain one degree of freedom. That is too much increase in RSS for the gain of only a single degree of freedom, thus the AIC increases to 4.72 from 4.61. You conclude that Year is too important to drop from the model. This is true of March SST, but not of January or February SST.

Therefore, to help you choose variables, you compare the AIC values for each variable against the AIC value for the full model. If the AIC value is less than the AIC for the full model, then the trade-off favors removing the variable from the model. If you repeat the procedure after removing the January and February SST variables, you will conclude that there is no statistical reason to make the model simpler.

Stepwise regression is a procedure for automating the `drop1` functionality. It is efficient in finding the best set of predictors. It can be done in three ways: forward selection, backward deletion, or both. Each uses the AIC as a criterion for choosing or deleting a variable and for stopping. To see how this works with your model, type

```
> step(m1)
```

The output is a series of tables showing the RSS and AIC values with successive variables removed. The default method is backward deletion, which amounts to a successive application of the `drop1` function. It is a good strategy to also try forward selection methods to see if the results are the same. If the results are the same using forward selection and backward deletion then you have greater confidence in your model.

3.11.2 Cross-validation

A cross-validation is needed if your statistical model will be used to make actual forecasts. Cross-validation is a procedure to assess how well your model will do in forecasting the unknown future. In the case of independent hurricane seasons, the procedure involves withholding a season's worth of data, developing the model on the remaining data, then using the model to predict data from the season that was withheld. Note that if your model-building algorithm involves stepwise regression or machine learning (Chapter 7), then the predictor selection component must be part of the cross-validation. That is, after removing a season's worth of data, you must run your selection procedure and then make a single-season prediction using the final model(s). And this needs to be done for each season removed.

The result of cross-validation is an estimate of out-of-sample error that accurately estimates the average forecast error when the model is used in predicting the future. The out-of-sample prediction error is then compared with the prediction error computed out of sample using a simpler model. If the error is larger with the simpler model, then your model is considered skillful. Note that the prediction error from the simpler model, even if it is long-run climatology, also needs to be obtained using cross-validation. More details on this important topic including some examples are given in Chapter 7.

In this chapter we reviewed classical statistics with examples from hurricane climatology. Topics included descriptive statistics, probability and distributions, one- and two-sample tests, statistical formula in R, correlation, and regression. Next, we give an introduction to Bayesian statistics.

4

BAYESIAN STATISTICS

"Probability does not exist."
—Bruno de Finetti

Classical statistics involves ways to test hypotheses and estimate confidence intervals. Bayesian statistics involves methods to calculate probabilities associated with your hypotheses. The result is a posterior distribution that combines information from your data with prior beliefs. The term "Bayesian" comes from Bayes's theorem—a single formula for obtaining the posterior distribution. It is the cornerstone of modern statistical methods.

In this chapter we introduce Bayesian statistics. We begin by considering the problem of learning about the population proportion (percentage) of all hurricanes that make landfall. We then consider the problem of estimating how many days it takes your manuscript to get accepted for publication. Again, we encourage you to follow along by typing the code into your R console.

4.1 LEARNING ABOUT THE PROPORTION OF LANDFALLS

Models that have skill at forecasting hurricane counts are more relevant to society if they include an estimate of the proportion that make landfall. Before examining the data, you hold a belief about the value of this proportion. You model your belief in terms of a prior distribution. Then after examining some data, you update your belief about the proportion by computing the posterior distribution (Albert, 2009).

This setting allows you to predict the likely outcomes of a new sample taken from the population, for example, the proportion of landfalls for next year. The use of the pronoun "you" focuses attention on the Bayesian viewpoint that probabilities do not exist but are how much *you* personally believe that something is true. Said another way, probabilities are subjective and based on all the relevant information available to you.

Here you think of a population consisting of past and future hurricanes in the North Atlantic. Then let π represent the proportion of this population that hit the United States at hurricane intensity. The value of π is unknown. You are interested in learning about what the value of π could be.

Bayesian statistics requires that you represent your belief about the uncertainty in this percentage with a probability distribution. The distribution reflects your subjective prior opinion about plausible values of π. Before you examine a sample of hurricanes, you think about what the value of π might be. If all hurricanes make landfall π is one, and if none make landfall π is zero. This is the nature of proportions, bounded below by zero and above by one. Also while the count of hurricanes is an integer the percentage that make landfall is a real number.

As a climatologist you are also aware that not all hurricanes make it to land. The nature of percentages together with your background provides you with information about π that is called "your prior." From this information, suppose you believe that the percentage of hurricanes making landfall in the United States is equally likely to be smaller or larger than 0.3. Moreover, suppose you are 90 percent confident that π is less than 0.5.

A convenient family of densities for a proportion is the beta. The beta density, here representing your prior belief about the population percentage of all hurricanes making landfall π, is proportional to

$$g(\pi) \propto \pi^{a-1}(1-\pi)^{b-1} \tag{4.1}$$

where the parameters a and b are chosen to reflect your prior beliefs about π.

The mean of a beta density is $m = a/(a+b)$ and the variance is $v = m(1-m)/(a+b+1)$. Unfortunately, it is difficult to assess values of m and v for distributions like the beta that are not symmetric. It is easier to obtain a and b indirectly through statements about the percentiles of the distribution. Recall from before you have a belief about the median (.3) and the 90th percentile (.5).

The `beta.select` function in the **LearnBayes** package (Albert, 2011) is used to find the two parameters (shape and scale) of the beta density that match this prior knowledge. The inputs are two lists, `q1` and `q2`, that define these two percentiles, and the function returns values for the beta parameters, a and b as respective elements of a vector.

```
> require(LearnBayes)
> q1 = list(p=.5, x=.3)
> q2 = list(p=.9, x=.5)
> beta.select(q1, q2)
[1] 3.26 7.19
```

Note the argument `p` is the distribution percentile and the argument `x` is a value for π.

Now you have your prior specified as a continuous distribution. Next you plot your prior using the `curve` function to see if it looks reasonable relative to your beliefs.

```
> a = beta.select(q1, q2)[1]
> b = beta.select(q1, q2)[2]
> curve(dbeta(x, a, b), from=0, to=1,
+   xlab="Proportion of Landfalls",
+   ylab="Density", lwd=4, las=1, col="green")
> abline(v=.5, lty=2)
> abline(v=.3, lty=2)
```

As seen in Figure 4.1, the distribution appears to adequately reflect your prior knowledge of landfall percentages. For reference, the vertical dashed lines are the values you specified for the median and the 90th percentile. The `abline` function is used after the curve is plotted. The `v` argument specifies where along the horizontal axis to draw the vertical line on the graph. Recall, to learn more about the function use `?abline`.

This is a start, but you need to take your prior distribution and combine it with a likelihood function. The likelihood function comes from your data. It is a function that describes how likely it is, given your model, that you observed the data you actually have. It might sound a bit confusing, but consider that if you have a good model for your data, then the probability that your model will replicate your data should be relatively high.

For example, consider the number of hurricanes over the most recent set of years. Read the data files containing the annual basin-wide (`A`) and landfall (`US`) counts and compute the sum.[1]

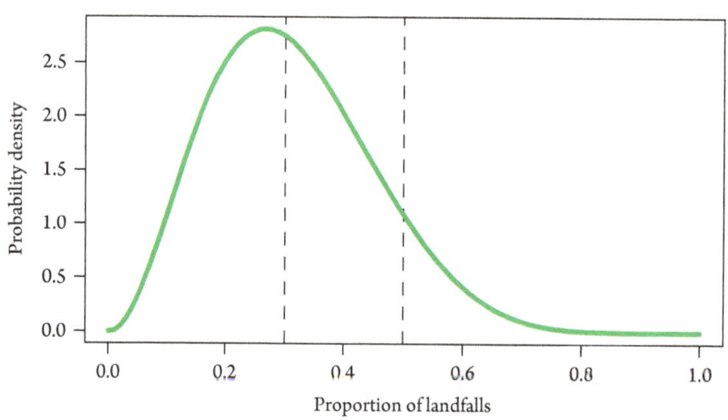

Figure 4.1 Beta density describing hurricane landfall proportion.

[1] The basin counts were obtained from `http://www.aoml.noaa.gov/hrd/tcfaq/E11.html` on January 17, 2012.

```
> A = read.table("ATL.txt", header=TRUE)
> US = read.table("H.txt", header=TRUE)
> Yr = 2006
> sum(A$H[A$Year >= Yr])
[1] 34
> sum(US$All[US$Year >= Yr])
[1] 4
```

You find 4 of the 34 hurricanes made landfall in the United States since 2006. You could simply report that $\pi = 0.12$ or 12 percent of hurricanes make landfall. However, this is the single sample estimate and it does not consider your prior belief about p. It also does not have an estimate of uncertainity. In addition, you might be interested in predicting the number of landfalls in a new sample of next year's hurricanes.

If you regard a "success" as a hurricane making landfall and you take a random sample of n hurricanes with s successes and $f = n - s$ failures, then the *likelihood function* (or simply, the likelihood) is given by

$$L(s,f|\pi) \propto \pi^s(1-\pi)^f, \qquad 0 < \pi < 1 \tag{4.2}$$

The likelihood depends on the model parameters, here only π, and is defined by your observations written L(data$|\pi$). Specifically, the likelihood of π given that you observed four landfalls in a set of 34 hurricanes is equal to the probability of that particular observed set of outcomes.

Here the model is the beta density with parameters $a = s + 1$ and $b = f + 1$. You can convince yourself that the likelihood is maximized when $\pi = 0.12$ for $s = 4$ and $f = 30$ by plotting $L(4,30|\pi)$ for π from 0 to 1.

Then with a likelihood function describing your data along with your prior beliefs, your posterior density for π is obtained up to a proportionality constant, by multiplying the prior density ($g(\pi)$) by the likelihood ($L(\text{data}|\pi)$), which is Bayes's rule given as

$$g(\pi|\text{data}) \propto g(\pi)L(\text{data}|\pi) \tag{4.3}$$

Your prior represents your thinking about the parameter (in terms of probability) of interest before examining data. Afterward, you have a likelihood function representing the probability of your data, given values for the parameter. Finally, the posterior probability distribution is computed using Bayes's rule.

It can be shown that if the prior is a beta density with parameters a and b and the likelihood is a beta density with parameters s and f, then the posterior density is also a beta density with parameters $a + s$ and $b + f$. This is an example of a conjugate model, where the prior and posterior densities belong to the same family of densities. To compute and plot the prior, likelihood, and posterior together, type

```
> s = sum(US$All[US$Year >= Yr])
> f = sum(A$H[A$Year >= Yr]) -
+   sum(US$All[US$Year >= Yr])
```

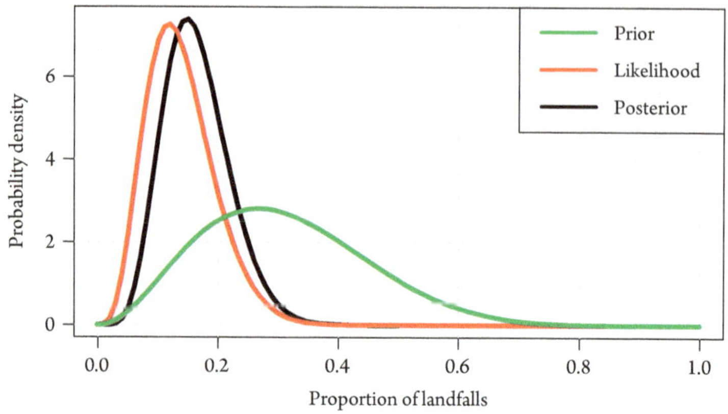

Figure 4.2 Densities describing hurricane landfall proportion.

```
> curve(dbeta(x, a + s, b + f), from=0, to=1,
+   xlab="Proportion of Landfalls", ylab="Density",
+   col=1, lwd=4, las=1)
> curve(dbeta(x, s + 1, f + 1), add=TRUE, col=2,
+   lwd=4)
> curve(dbeta(x, a, b), add=TRUE, col=3, lwd=4)
> legend("topright", c("Prior", "Likelihood",
+   "Posterior"), col=c(3, 2, 1), lwd=c(3, 3, 3))
```

The densities are shown in Figure 4.2. Note that the posterior density resembles the likelihood, but it is shifted in the direction of the prior. This is always the case. The posterior is a weighted average of the prior and the likelihood where the weights are proportional to the precision. The greater the precision the more weight it carries in determining the posterior.

For your prior, the precision is related to how committed you are to a particular value. Believing that there is a 90 percent chance that the proportion is less than 0.5 provides a quantitative level of commitment. The less committed you are, the flatter the prior distribution and the less weight it plays in the posterior. For your likelihood, the precision is inversely related to the standard error, so directly related to the sample size. The more data you have, the more weight the likelihood plays in determining the posterior.

4.2 INFERENCE

A benefit of Bayesian statistics is that the posterior distribution contains the information you need to make any inference. In this example, since the posterior is a beta distribution, the `pbeta` (cumulative distribution) and `qbeta` (quantile) functions can be used. For example, given your prior beliefs and your sample of data, how likely is it that the population percentage of landfalls is less than or equal to 25 percent? This is answered by computing the posterior probability $P(\pi \leq 0.25|\text{data, prior})$, which is given by

```
> pbeta(q=.25, a + s, b + f)
[1] 0.93
```

You interpret this probability in a natural way. You can state that given the evidence in hand (data and your prior belief), there is a 93 percent chance that less than or equal to a quarter of all hurricanes hit the United States (at hurricane intensity). Or you can state that it is quite unlikely (1.5%) that more than 3 in 10 hurricanes make it to the United States. (`1-pbeta(.3,a+s,b+f)`).

You cannot do this with classical statistics. Instead, the p-value resulting from a significance test is the evidence in support of a null hypothesis. Suppose the null hypothesis is that the population proportion exceeds 3 in 10. You state $H_o : \pi > 0.3$ against the alternative $H_a : \pi \leq 0.3$ and decide between the two on the basis of a p-value obtained by typing

```
> prop.test(s, s + f, p=.3, alt="less")$p.value
[1] 0.0165
```

where the argument `p` specifies the value of the null hypothesis and the argument `alt` specifies the direction of the alternative hypothesis.

The p-value of 0.016 indicates that if the null is true (the proportion is greater than 0.3), the evidence seems unusual. You conclude that there is moderate evidence against the null. However, it is incorrect to conclude from the p-value that there is only a 1.6 percent chance that the proportion of landfalls is greater than 0.3.

4.3 CREDIBLE INTERVAL

Bayesian interpretation extends to a posterior summaries. For instance, the 95 percent interval estimate for the percentage of landfalls is obtained from the 2.5th and 97.5th percentiles of the posterior density. This is done with the `qbeta` function by typing

```
> qbeta(c(.025, .975), a + s, b + f)
[1] 0.0713 0.2839
```

You state the 95 percent *credible interval* for the proportion is [0.071, 0.284] and conclude you are 95 percent confident that the true proportion lies inside this interval.

Note the distinction. With a credible interval you say that given the data and your prior beliefs, there is a 95 percent chance that population proportion lies within this particular interval. With a confidence interval you say you are 95 percent confident that the method produces an interval that contains the true proportion. In the former the population parameter is the random variable, and in the latter the interval is the random variable.

The 95 percent confidence interval estimate for the proportion π using a large enough sample of data, where $\hat{p}$ is the sample proportion and n is the sample size, is given by

$$\hat{p} \pm \sqrt{\frac{\hat{p}(1-\hat{p})}{n}} \times \Phi^{-1}(0.975) \tag{4.4}$$

where $\Phi^{-1}(0.975)$ (inverse of the cumulative distribution function) is the 97.5th percentile value from a standard normal distribution. The confidence interval is obtained by typing

```
> prop.test(s, s + f)$conf.int
[1] 0.0384 0.2839
attr(,"conf.level")
[1] 0.95
```

The result allows you to state that the 95 percent confidence interval for the proportion is [0.038, 0.284] and conclude that if you had access to additional samples and computed the interval in this way for each sample, 95 percent of the intervals will cover the true proportion.

In some cases, the intervals produced (confidence and credible) from the same data set are identical. They are almost certainly different if an informative prior is included and they may be different even if the prior is relatively uninformative. However, the interpretations are always different.

With Bayesian statistics inferential statements are easier to communicate. It is natural to talk about the probability that π falls within an interval or the probability that a hypothesis is true. Also Bayes's rule (Eq. 4.3) is the only thing you need to remember. It is used for small and large samples. It is the only consistent way to update your probabilities. The cost is that you need to specify your prior beliefs.

4.4 PREDICTIVE DENSITY

So far we focused on learning about the population proportion of hurricanes that make landfall. Suppose your interest is predicting the number of U.S. landfalls ($\tilde{l}$) in a future sample of seven hurricanes. If your current understanding of π is contained in the density $g(\pi)$ (e.g., the posterior density), then the predictive density for $\tilde{l}$ is given by

$$f(\tilde{l}) = \int f(\tilde{l}|\pi)g(\pi)d\pi \tag{4.5}$$

With a beta distribution, you can integrate to get an expression for the predictive density. The beta predictive probabilities are computed using the function `pbetap` from the **LearnBayes** package. The inputs are a vector `ab` containing the beta parameters, the size of the future sample m, and a vector of the number of future landfalls `lf`.

```
> ab = c(a + s, b + f)
> m = 7; lf = 0:m
```

```
> plf = pbetap(ab, m, lf)
> round(cbind(lf, plf), digits=3)
     lf   plf
[1,]  0 0.312
[2,]  1 0.367
[3,]  2 0.216
[4,]  3 0.081
[5,]  4 0.021
[6,]  5 0.004
[7,]  6 0.000
[8,]  7 0.000
```

Simulation provides a convenient way to compute a predictive density. For example, to obtain $\tilde{l}$, first simulate p from $g(p)$ and then simulate $\tilde{l}$ from the binomial distribution, $f_B(l|n,p)$. You can use this approach on your beta posterior. First simulate 1,000 draws from the posterior and store them in `p`.

```
> p = rbeta(n=1000, a + s, b + f)
```

Then simulate values of $\tilde{l}$ for these random values using the `rbinom` function and tabulate them.

```
> lc = rbinom(n=1000, size=7, prob=p)
> table(lc)
lc
  0   1   2   3   4   5
319 351 220  83  20   7
```

The table indicates that, of the 1,000 simulations, 319 of them resulted in no landfalls, 351 of them resulted in one landfall, and so on.

You save the frequencies in a vector and then convert them to probabilities by dividing by the total number of simulations. Finally, you plot the probabilities using the histogram argument (`type="h"`).

```
> freq = table(lc)
> pp = freq/sum(freq)
> plot(pp, type="h", xlab="Number of U.S. Hurricanes",
+    las=1, lwd=3, ylab="Predictive Probability")
```

The plot is shown in Figure 4.3. It is most likely that one of the seven hurricanes will hit the United States. The probability of four or more doing so is 2.7 percent.

The cumulative sum of the probabilities is found by typing

```
> cumsum(pp)
    0     1     2     3     4     5
0.319 0.670 0.890 0.973 0.993 1.000
```

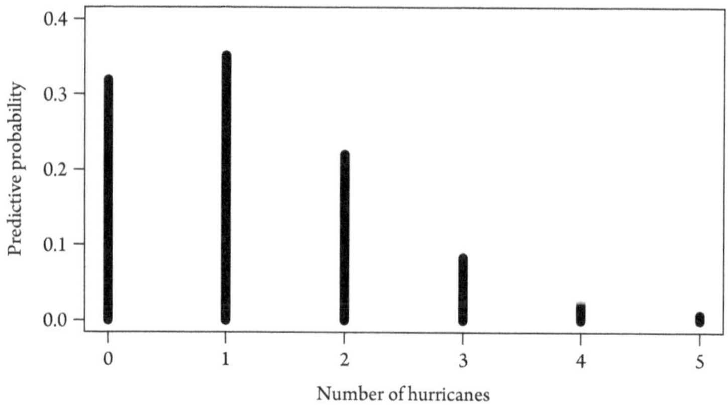

Figure 4.3 Predictive probabilities for the number of landfalling hurricanes.

The probability of three or fewer landfalls is 0.973. Suppose you wish to summarize this discrete predictive distribution by an interval that covers a certain amount of the probability.

This can be done using the function `discint` from the **LearnBayes** package. You combine the vector of probabilities with a vector of counts and specify a coverage probability as

```
> mc = length(pp) - 1
> int = discint(cbind(0:mc, pp), .95)
> int
$prob
    3
0.973

$set
0 1 2 3
0 1 2 3
```

The output has two lists, the minimum probability (`$prob`) greater than the specified coverage probability (here 95%) and the list of counts (`$set`) over which the individual probabilities are summed. You can see the probability is 97.3 percent that the number of U.S. hurricanes is one of these counts.

The interval covers a range of four counts, which is necessarily wider than the range of counts computed by using the bounds on the 97.3 percent credible interval around the population proportion. You check this by typing

```
> lp = (1 - as.numeric(int$prob))/2
> up = 1 - lp
> (qbeta(up, a + s, b + f) -
+   qbeta(lp, a + s, b + f)) * 7
[1] 1.67
```

This is because in predicting a specific value, as you saw in Chapter 3 with the prediction intervals from a linear regression model, there are two sources of uncertainty. Here the uncertainty about the population proportion π and the uncertainty about the count given an estimate of π.

4.5 IS BAYES'S RULE NEEDED?

Since probabilities are subjective, why is Bayes's rule needed? Why not simply examine your data and subjectively determine your posterior probability distribution, updating what you already know? Would not this save you trouble? Strange as it sounds, it would actually cause you more trouble.

The catch is that it is hard to assess probabilities rationally. For example, probabilities must be nonnegative and sum to one. If these conditions are not met, then your probabilities are said to be inconsistent. In fact they should obey certain axioms of coherent assessment (Winkler, 2003). For example, if you consider event A to be more likely than event B, and event B to be more likely than event C, then you should consider event A to be more likely than event C. If your probabilities fail to obey transitivity, then your assessment will be inconsistent in a decision-making sense.

This is not a serious impediment since you can easily remove your inconsistency once you are made aware of it. If someone notes an arithmetic error, you would certainly correct it. Similarly, if someone notes that your prior probabilities are not transitive, you would change them accordingly. In this regard, it is sometimes useful to attempt to assess the same set of probabilities in more than one way in order to "check" your assessments. You might assess probabilities using the mean and variance and then compare these to an assessment based on quantiles.

Importantly, if you obey the axioms of coherence, then to revise your probabilities you must use Bayes's rule. To do otherwise would be inconsistent. Numerous psychological studies have indicated that people do not always intuitively revise probabilities on the basis of this rule. People tend to give too little "weight" to the data. In general, this means that they do not change their probabilities as much as Bayes's theorem tells them that they should. By using the rule, you ensure consistency in updating your beliefs based on the evolving evidence.

4.6 BAYESIAN COMPUTATION

The model presented earlier is simple: inference for an unknown proportion. For this problem and others like it, the likelihood functions are standard, and, if you use a conjugate model where the prior and posterior densities come from the same family, deriving the posterior density poses no computational burden. Indeed this is why conjugate models like the beta are widely used in Bayesian analysis (Jackman, 2009).

As you have seen the probability of a random hurricane hitting the United States can be modeled using a beta prior. In this case, computation is nothing but addition. But Bayesian computation quickly becomes challenging when working with more complicated models or when you use nonconjugate models. Characterizing the posterior density in these cases becomes nontrivial. Many statistical models in geography and the social sciences have this feature (see the models in Chapter 12).

Options still exist. One is brute force. When the posterior distribution is complicated, you can compute values over a grid and then approximate the continuous posterior by a discrete one (Albert, 2009). This is possible for models with one or two parameters. In situations where you can directly sample from the posterior, a Monte Carlo (MC) algorithm gives an estimate for the posterior mean for any function of your parameters. In the situation where you cannot directly sample, you can use rejection sampling, provided you have a suitable proposal density or, in the most general case, you can adopt a Markov chain Monte Carlo (MCMC) approach.

4.6.1 Time-to-Acceptance

Here you assume that information about the time it takes a research paper to get reviewed might be useful to authors and editors, especially if it can say something about future submissions. This motivates you to collect data on publication times from recent journals and to model them. Here you cannot exploit conjugacy so you approximate the posterior with a discrete distribution. You sample directly from the distribution and use an MC algorithm for summarizing your samples.

We use the American Meteorological Society (AMS) journals and the keyword "hurricane" appearing in published titles over the years 2008–2010 (see Hodges et al. (2012)). The search is done from the web site `http://journals.ametsoc.org/action/doSearch`. Selecting "Full Text" on a particular article brings up the abstract, keywords, and two dates: received and accepted—in month, day, year format.

The data were originally entered manually but are available to you by typing.

```
> art = read.csv("hurart.txt", header=TRUE)
```

For each article both dates are provided along with the lead author's last name and of the journal. There are 100 articles with the word "hurricane" in the title over the 3-year period. Of these 34, 41, and 25 had "accepted" years of 2008, 2009, and 2010, respectively. Articles appearing in 2008, but with accepted years before 2008 are not included.

Journals publishing these articles are shown in Figure 4.4. Ten different journals are represented. Thirty-seven of these articles are published in *Monthly Weather Review*, 22 in *Journal of the Atmospheric Sciences*, 13 in the *Journal of Climate*, and 12 in *Weather and Forecasting*.

First you compute the time period between received and accepted dates. This is done by converting the numeric values of year, month, and day to an actual date using the `ISOdate` function and then using the `difftime` to compute the time difference in days.

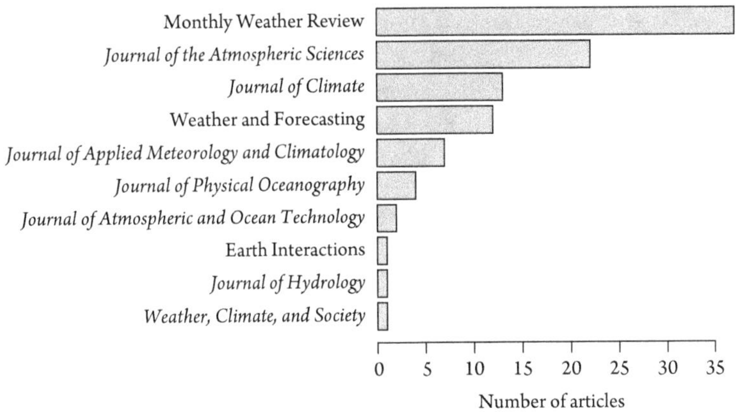

Figure 4.4 AMS journals publishing articles with 'hurricane' in the title.

```
> rec = ISOdate(art$RecYr, art$RecMo, art$RecDay)
> acc = ISOdate(art$FinYr, art$FinMo, art$FinDay)
> tau = difftime(acc, rec, units="days")
```

You call the temporal difference the time-to-acceptance (τ). This is your statistic of interest.

The mean τ is 195 days in 2008 (type `mean(tau[art$FinYr==2008])`), 168 days in 2009, and 193 days in 2010. The mean τ for each of the four journals with the most articles is 166 (*Monthly Weather Review*), 169 (*Journal of the Atmospheric Sciences*), 189 (*Journal of Climate*), and 181 (*Weather and Forecasting*). On average, the *Journal of Climate* is slowest and *Monthly Weather Review* is fastest, but the difference is less than 3.5 weeks.

Undoubtedly, there are many factors that influence the value of τ. On the editor's side, there are prereview, review requests, and dispensation decisions among others. On the reviewer's side, there is workload as well as breaks for travel and vacation, and on the author's side, there is the effort needed to revise and resubmit.

The goal for you is a predictive distribution for τ. This will allow you to make inferences about the time-to-acceptance for your future manuscript submissions. You assume that is τ a random variable having a gamma density given by

$$f(\tau|\alpha,\beta) = \frac{\tau^{\alpha-1}\exp(\frac{-\tau}{\beta})}{\beta^{\alpha}\Gamma(\alpha)} \tag{4.6}$$

where α and β are the shape and scale parameters, respectively, and $\Gamma(\alpha) = (\alpha-1)!$. The gamma density is commonly used to model time periods (wait times, etc.). If you place a uniform prior distribution on the parameter vector (α,β), the posterior density is given by

$$g(\alpha,\beta|\tau) \propto g(\alpha,\beta)f(\tau|\alpha,\beta) \tag{4.7}$$

The uniform prior is consistent with a judgment that the parameters are the same regardless of author or journal. Random draws from this joint posterior density are summarized and used as parameters to a gamma density to draw predictive samples for τ.

To make the computation easier, the posterior density is reformulated in terms of $\log\alpha$ and $\log\mu$, where $\mu = \alpha \times \beta$ is the posterior mean of τ (Albert, 2009). The posterior density is available in the `gsp.R` file (from Jim Albert). Here you source the function by typing

```
> source("gsp.R")
```

Remember to include the quotes around the file name.

Given a pair of parameter values $(\log\alpha, \log\mu)$, the function computes the posterior probability given the data and the posterior density using

$$\log g(\alpha, \frac{\mu}{\alpha}|\tau) = \log\mu + \sum_{i=1}^{n} \log f(\tau_i|\alpha, \frac{\mu}{\alpha}) \quad (4.8)$$

It performs this computation using pairs of parameters defined as a two-dimensional grid spanning the domain of the posterior.

The `gsp` function is used to determine the posterior density as defined in Eq. 4.8 over a two-dimensional grid of parameter values (on log scales) using the `mycontour` function from the **LearnBayes** package. It is also used to sample from the posterior using the `simcontour` function from the same package.

Using these functions, you contour the joint posterior of the two parameters and then add 1,000 random draws from the distribution by typing

```
> lim = c(.8, 2.1, 5, 5.45)
> mycontour(gsp, limits=lim, data=tau,
+    xlab="log alpha", ylab="log mu")
> s = simcontour(gsp, lim, data=tau, m=1000)
> points(s$x, s$y, pch=19, col="gray", cex=.5)
```

These computations may take several seconds to complete. The result is shown in Figure 4.5. The contours from inside-out are the 10 percent, 1 percent, and 0.1 percent probability intervals on a log scale.

The graph shows that the mean time-to-acceptance is about 181 days (on the ordinate exp(5.2)). This might be useful to an editor. But the posterior gives additional information. For example, the editor can use your model to estimate the probability that the average time-to-acceptance will exceed 200 days for the set of manuscripts arriving next month. Since the model provides random values, the question is answered by finding the percentage of values exceeding the logarithm of 200. Assuming that the set of manuscripts is a random sample, the model predicts 4.3 percent.

Averages are not particularly useful to you as an author. You would like to know if a recently submitted manuscript will be accepted in less than 120 days. To answer this question, you take random draws of τ's from a gamma density using

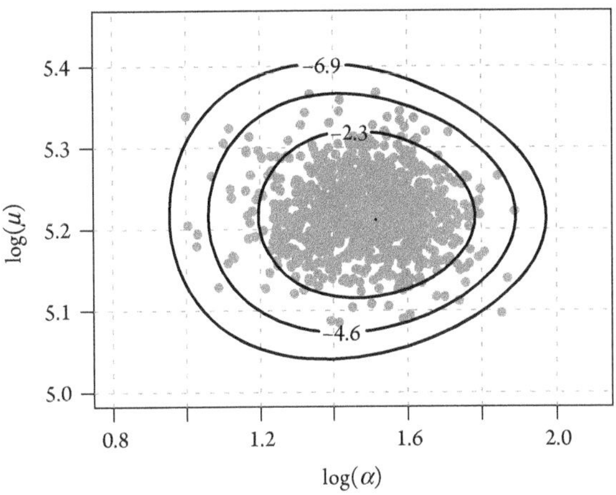

Figure 4.5 Posterior density of the gamma parameters for a model describing time-to-acceptance.

the random draws from the posterior parameter distribution. This is done with the `rgamma` function and then finding the percentage of these draws less than this many days.

```
> alpha = exp(s$x)
> mu = exp(s$y)
> beta = mu/alpha
> taum = rgamma(n=1000, shape=alpha, scale=beta)
```

Here the model predicts a probability of 26 percent. Note that this probability is lower than the average percentage less than 120 days as it includes additional uncertainty associated with modeling an individual estimate.

Changes to review rules and manuscript timetables and tracking will influence time-to-acceptance. To the extent that these changes occur during the period of data collection or subsequently, they will influence your model's ability to accurately anticipate time-to-acceptance.

Model fit is checked by examining quantile statistics from the data against the same statistics from the posterior draws. For instance, the percentage of articles in the data with τ less than 90 days is 11, which compares with a percentage of 13.3 from the posterior draws. Continuing, the percentage of articles with τ longer than 360 days from the data is 6, which compares with a percentage of 4.4 from the posterior draws.

The model has practical applications if you wish to meet a deadline. As an example, in order for research to be considered by the 5th Assessment Report of the Intergovernmental Panel on Climate Change (IPCC), you must have your manuscript accepted for publication by March 15, 2013.

The graph in Figure 4.6 shows your model's predictive probability of meeting this deadline versus date. Results are based on 1,000 random draws from a gamma distribution where the scale and shape parameters are derived from 1,000 draws from the joint posterior given the data. The probability is the percentage of posterior τ draws less than τ_o days from March 15, 2013. The 95 percent credible interval shown by

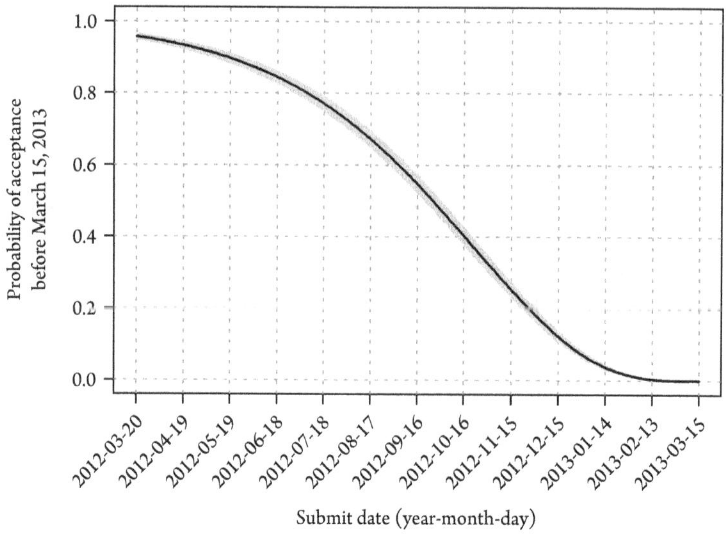

Figure 4.6 Probability of paper acceptance as a function of submit date.

the gray band is obtained by repeating the 1,000 draws from the gamma distribution 1,000 times and taking the 0.025 and 0.975 quantile values of the probabilities. The probability is high in the early months of 2012 when the deadline is still a year in the future. However, by mid-September of 2012, the probability drops below 50 percent and by mid-January 2013, the probability is less than 10 percent.

The predictive probabilities from your model reflect what can be expected if you submit to an arbitrary AMS journal with "hurricane" in the title under the assumption that your paper will be accepted. Certain journals and authors could have a faster or slower turnaround time. In this case, you need to use a hierarchical model (see Chapter 12) to accommodate author and journal differences. With a hierarchical model, you use an MCMC approach to obtain the posterior probabilities.

4.6.2 Markov Chain Monte Carlo approach

MCMC is a class of algorithms for sampling from a probability distribution using a Markov chain. It is a way to obtain samples from your posterior distribution. It is flexible, easy to implement, and requires little input from you. It is one reason behind the recent surge in popularity of Bayesian statistics.

Gibbs sampling is an example of an MCMC algorithm. Suppose your parameter vector of interest is $\theta = (\theta_1, \theta_2, \ldots, \theta_p)$. The joint posterior distribution of theta, which is denoted $[\theta|\text{ data}]$ may be of high dimension and difficult to summarize. Instead, suppose you define the set of conditional distributions as

$$\left[\theta_1|\theta_2, \ldots, \theta_p, \text{data}\right], \left[\theta_2|\theta_1, \theta_3 \ldots, \theta_p, \text{data}\right], \cdots \left[\theta_p|\theta_1, \ldots, \theta_{p-1}, \text{data}\right] \quad (4.9)$$

where $[X|Y,Z]$ represents the distribution of X conditional on values of random variables Y and Z. The idea is that you can set up a Markov chain from the joint posterior distribution by simulating parameters from the set of p conditional distributions.

Drawing one value for each parameter from these distributions in turn is called one update (iteration) of the Gibbs sampling. Under general conditions, draws will converge to the target distribution (joint posterior of θ). Unfortunately, this theoretical result provides no practical guidance on how to decide if the simulated sample provides a reasonable approximation to the posterior density (Jackman, 2009).

4.6.3 JAGS

A popular general purpose MCMC software that implements Gibbs sampling is WinBUGS (Windows version of Bayesian inference Using Gibbs Sampling) (Lunn et al, 2000). It is stand-alone software with a GUI. JAGS (Just Another Gibbs Sampler) is an open-source project written in C++. It runs on any computing platform and can be called from R with functions from the **rjags** package (Plummer, 2011).

Here you use JAGS on the earlier problem of making inferences about the proportion of U.S. landfalls. The example is simple enough and an MCMC algorithm is not really necessary, but it is instructive for you to see the work flow as you will see it again in Chapter 12. To begin, download and install the latest version of JAGS. This is C++ code that gets installed outside of R. Next open a text file and write the model code in a language that JAGS understands. You can copy and paste from here. Call the file **JAGSmodel.txt**.

```
___JAGS code___
model {
  h ~ dbin(pi, n)   #data
  pi ~ dbeta(a, b) #prior
  a <- 3.26
  b <- 7.19
  }
_______________
```

Note the similarity to R syntax.

Every JAGS (and BUGS) program begins with a model statement followed by an open curly brace and ends with an closed curly brace. The second line specifies the landfall count `h` using the "twiddle" or "tilde" character, which indicates a stochastic relation. The stochastic relation is of the form `node~ddens(arguments)`, where `node` is your parameter and `ddens` is the name of a function calling one of the statistical distributions supported in JAGS. These functions start with the letter `d`, as here in `dbin` for the binomial mass function and `dbeta` for the beta density used in the third line for your prior. Here the values of the parameters `a` and `b` are the same as you used earlier in the chapter and are set using the deterministic operator ($<-$).

JAGS and BUGS are declarative languages. R is a procedural language. The JAGS syntax specifies the model, but it does not define the computational steps. When the

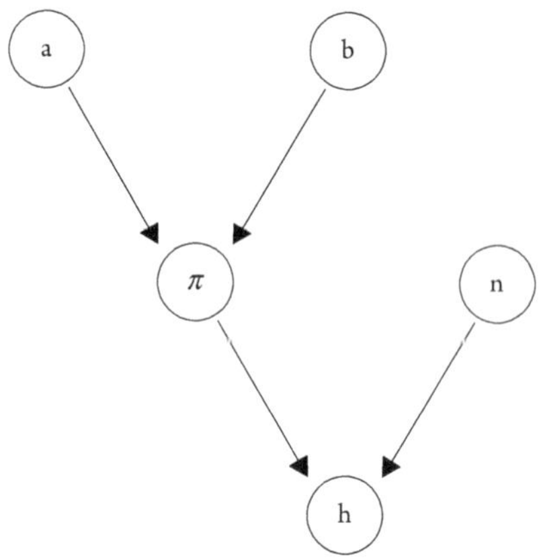

Figure 4.7 Directed acyclic graph of the landfall proportion model.

JAGS model is compiled, the model syntax is turned into a set of sequential instructions, but this is hidden. Because of this, the statement order is not important. What matters is that the compiler can resolve the names and links in the model.

Think of the JAGS model as a directed acyclic graph (DAG) as shown in Figure 4.7. Nodes are the parameters and data and arrows indicate the conditional dependency, either stochastic or deterministic. Hurricane landfall h depends on the number of hurricanes n and the proportion π, where the proportion π depends on your prior assumptions encoded in the beta parameters a and b. Landfall proportion π is conditionally dependent on the scale (a) and shape (b) values through the beta density. The number of landfalls h is dependent on the basin-wide count n and the proportion of landfalls π through the binomial distribution. The conditional independence structure of your model is clear in the DAG. Note that BUGS allows you to generate model code from a DAG.

Next open R and type

```
> require(rjags)
```

You call JAGS from R with the `jags.model` function by typing,

```
> model = jags.model('JAGSmodel.txt',
+    data = list('h'=4, 'n'=34),
+    inits = list('pi'=.5, .RNG.seed=3042,
+    .RNG.name="base::Super-Duper"))
Compiling model graph
   Resolving undeclared variables
   Allocating nodes
   Graph Size: 7

Initializing model
```

The function sends the model in file **JAGSmodel.txt** to JAGS for parsing and compiling. Make sure the file is in your working directory. The `data` argument specifies the number of landfalling hurricanes h and the number of hurricanes n as a list. The `inits` argument directs JAGS to initialize the MCMC although that is not strictly necessary with this simple model. The last two elements in the `inits` list specify the random number generator (RNG) and an initial seed for the generator. This allows the results to be replicated as the sequence of random numbers will be identical if the same seed value is used. The object `model` is of class `jags`.

By default, the function generates 1,000 samples (updates) of the MCMC algorithm. This is enough for successive values of π to depart from the initial value and reach the posterior density. In fact, this "burn-in" is instantaneous with this model. If more iterations are needed, you specify the number in the function with the argument `n.adapt` or you can use the `update` method on the model object by typing,

```
> update(model, 1000)
```

An additional 1,000 iterations from the MCMC are generated but not saved.

Finally you use the `coda.samples` function to generate posterior samples of π and save them. It continues updating the chain for the number of iterations specified by `n.iter`, but this time it saves them if the parameter is listed in the `variable.names` argument.

```
> out = coda.samples(model, variable.names='pi',
+     n.iter=1000)
```

CODA stands for COnvergence Diagnostic and Analysis. It describes a suite of functions for analyzing outputs generated from BUGS software.

The object returned is of class `mcmc.list`. You use the `summary` method on this object to obtain a summary of the posterior distribution for π.

```
> summary(out)
Iterations = 1001:2000
Thinning interval = 1
Number of chains = 1
Sample size per chain = 1000

1. Empirical mean and standard deviation for each variable,
   plus standard error of the mean:

       Mean            SD        Naive SE
    0.16294       0.05342         0.00169
Time-series SE
    0.00134
```

```
2. Quantiles for each variable:

  2.5%    25%    50%    75%  97.5%
0.0739 0.1228 0.1595 0.1957 0.2878
```

The output shows first the characteristics of your MCMC sampler and then provides statistics on the samples from the posterior distribution. The posterior mean indicating the proportion of all North Atlantic hurricanes that make landfall in the United States is 0.16. Quantiles of π from the posterior samples indicate that the 95 percent credible interval for the proportion is $(0.07, 0.29)$.

You obtain other statistics from the posterior samples by extracting the array corresponding to π from the MCMC list and performing the appropriate computations. For instance, given your prior and your data, how likely is it that the population percentage of landfalls is less than or equal to 0.25? This is answered by typing,

```
> pi = as.array(out)
> sum(pi <= .25)/length(pi)
[1] 0.941
```

Here, because there is only one chain and one variable, you extract the vector from the array using `as.array`. The answer (94 percent) compares well with what you obtained in §4.2 where you used the conjugate model.

Additional analysis can be done with the posterior samples, and the plot method will produce a trace and density plot of your π values by typing

```
> plot(out)
```

The trace plot shows your samples versus the simulation index. The trace plot is useful in assessing whether your chain has converged. Although the samples vary, the mean and variance are relatively constant indicating convergence. This is expected with a simple model. The density plot indicates the distribution of all the samples.

The MCMC approach is flexible and it allows you to easily answer other related inferential questions. For instance, suppose you want to know the probability that any two consecutive hurricanes that form in the North Atlantic will hit the United States. You can add a node to your model of the form `pi2 <- pow(pi,2)`. This is changed in your file **JAGSmodel.txt**. You then monitor this node and draw inferences from the posterior samples. The median posterior probability indicates only a 16 percent chance of consecutive hurricanes hitting the United States. The result assumes that consecutive hurricanes are independent.

4.6.4 WinBUGS

WinBUGS and OpenBUGS have functions for modeling spatial data that are not yet available in JAGS. In Chapter 12, you will build a Bayesian space-time model for your hurricane data. Here we show you the work flow for implementing a Bayesian model in WinBUGS through R. If you are using Windows, there is no problem. If you are using a Mac, your options are limited. One is to run WinBUGS with Wine. Another is to use OpenBUGS without using R.

Here we assume that you are using a Windows machine or running Windows on a partition of your Mac (e.g., bootcamp). First download and install WinBUGS?[2] Make sure you have permission to write files in the directory where the executable file is stored. Next, write a WinBUGS model describing the time-to-acceptance for an article on hurricanes submitted to the AMS (see §4.6.1). The code is given as follows.

```
___WinBUGS code___
model {
  for (i in 1:N) {
    TTA[i] ~ dgamma(shape, rate)
    }
  ttastar ~ dgamma(shape, rate)
  ht <- step(120-ttastar)
    shape ~ dgamma(.1, .1)
    rate ~ dgamma(.1, .1)
}
_______________
```

The syntax is nearly identical to JAGS, and in this case you can run this as a JAGS model using the work flow described previously. Save the code in **WinBUGSmodel.txt** outside of R. The `step` function is 1 if its argument is positive and 0 otherwise. The parameter `ht` thus indicates whether the sampled time-to-acceptance is less than 120 days.

Next prepare the data inputs required by the `bugs` function from the **R2WinBUGS** package. This is a list containing the name of each data vector.

```
> require(R2WinBUGS)
> N = length(tau)
> TTA = as.numeric(tau)
> data = list("TTA","N")
```

Using these data together with your model, you run an MCMC sampler to get estimates for `ttastar`. Beforehand you decide how many chains to run for how many iterations. If the length of the burn-in is not specified, then the burn-in is taken as half the number of iterations. You also specify the starting values for the chains. Here you do this by writing the following function

```
> inits=function(){
+   list(shape=runif(1, .3, .5), rate=runif(1, .3, .5),
+     ttastar = runif(1, 100, 120))
+ }
```

[2] http://www.mrc-bsu.cam.ac.uk/bugs/winbugs/contents.shtml

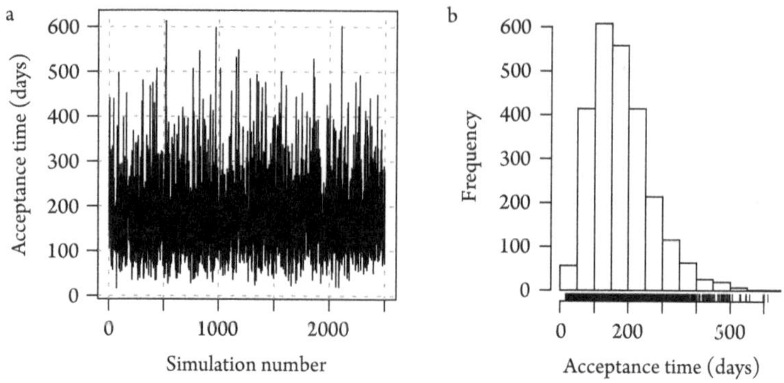

Figure 4.8 Posterior samples of time-to-acceptance. (a) Trace plot and (b) histogram.

You then start the MCMC by typing

```
> model = bugs(data, inits,
+     model.file="WinBUGSmodel.txt",
+     parameters=c("ttastar", "ht", "shape", "rate"),
+     n.chains=3, n.iter=5000, n.thin=1,
+     bugs.directory="C:/Program Files/WinBUGS14")
```

The argument `bugs.directory` must point to the directory containing the WinBUGS executable, and you must have write permission in this directory.

The results are saved in the object `model`. The `bugs` function uses the parameter names you gave it in the WinBUGS text file to structure the output into scalar, vector, and arrays containing the samples. The object `model$sims.array`, as example, contains an array of the posterior samples. In addition, the samples are stored in a vector. You use the `print` method to get a summary of your results. Additionally, you use the `plot` method to create graphs summarizing the posterior statistics of your parameters (see Figure 4.8) and to display diagnostics relating to chain convergence. The chains are convergent when your samples are taken from the posterior.

By default, `bugs` removes the first half of the iterations as "burn-in", where the samples are moving away from the initial set of values toward the posterior distribution. Samples can be thinned if successive values are highly correlated (see Chapter 12). The burn-in and thinning determine the final number of samples (saved in `model$n.keep`) available for inference.

Here you are interested in the posterior samples of the time-to-acceptance `ttastar`. To plot the values as a sequence (trace plot) and as a distribution, type

```
> plot(model$sims.array[, 1, 1], type="l")
> plot(density(model$sims.array[, 1, 1]))
```

The trace plot shows the sequence of sample values by the simulation number. The mean and variance of the values do not appear to change across the sequence indicating the samples are coming from the posterior distribution. A view of the marginal distribution of time-to-acceptance is shown with a histogram.

Because your model contains a node indicating whether the sample time-to-acceptance is greater than 120 days, you determine the posterior probability of this occurrence by typing

```
> sum(model$sims.array[, 1, 2])/model$n.keep
[1] 0.279
```

The answer of 27.9 percent is close to what you obtained in §4.6.1.

Greater flexibility in summarizing and plotting your results are available with functions in the **coda** package (Plummer et al., 2010). You turn on the `codaPkg` switch and rerun the simulations.

```
> modelc = bugs(data, inits,
+     model.file="WinBUGSmodel.txt",
+     parameters=c("ttastar", "ht", "shape", "rate"),
+     n.chains=3, n.iter=5000, n.thin=1, codaPkg=TRUE,
+     bugs.directory="C:/Program Files/WinBUGS14")
```

The saved object `modelc` is a character vector of file names with each file containing coda output for one of the three chains. You create an MCMC list object (see §4.6.3) with the `read.bugs` function.

```
> out = read.bugs(modelc, quiet=TRUE)
```

Additional plotting options are available using the **lattice** package. To plot the density of all the model parameters, type

```
> require(lattice)
> densityplot(out)
```

The plots contain distributions of the MCMC samples from all the model nodes and for the three chains. Density overlap across the chains is another indication that the samples are from the posterior.

This chapter introduced Bayesian statistics for hurricane climate. The focus was on the proportion of landfalling hurricanes. We examined a conjugate model for this proportion. We also looked at the time-to-acceptance for a manuscript. The conjugate model revealed the mechanics of the Bayesian approach. We also introduced MCMC samplers. This innovation allows flexibility in creating models for your data. The next chapter shows you how to graph and plot your data.

5

GRAPHS AND MAPS

"The ideal situation occurs when the things that we regard as beautiful are also regarded by other people as useful."
—Donald Knuth

Graphs and maps help you reason with data. They also help you communicate results. A good graph gives you the most information in the shortest time, with the least ink in the smallest space (Tufte, 1997). In this chapter, we show you how to make graphs and maps using R.

A good strategy is to follow along with an open session, typing (or copying) the code as you read. Before you begin make sure you have the following data sets available in your workspace. Do this by typing

```
> SOI = read.table("SOI.txt", header=TRUE)
> NAO = read.table("NAO.txt", header=TRUE)
> SST = read.table("SST.txt", header=TRUE)
> A = read.table("ATL.txt", header=TRUE)
> US = read.table("H.txt", header=TRUE)
```

Not all the code is shown but all is available on our Web site.

5.1 GRAPHS

It is easy to make a graph. Here we provide guidance to help you make informative graphs. It is a tutorial on how to create publishable figures from your data. In R you have several choices. With the standard (base) graphics environment, you can produce a variety of plots with fine details. Most of the figures in this book use the standard graphics environment. The grid graphics environment is even more flexible.

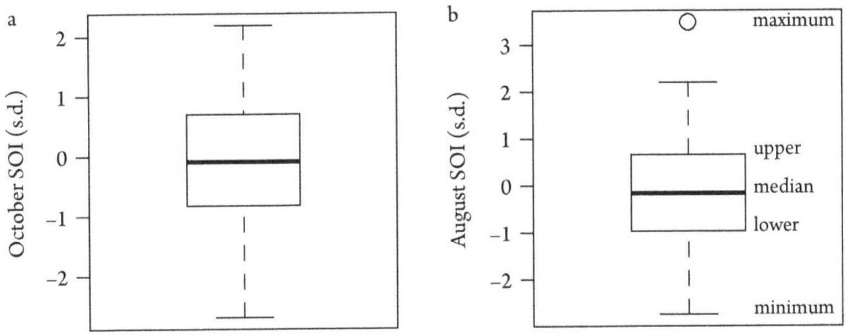

Figure 5.1 Box plot of the October SOI.

It allows you to design complex layouts with nested graphs where scaling is maintained upon resizing. The **lattice** and **ggplot2** packages use grid graphics to create more specialized graphing functions and methods. The `spplot` function for example is plot method built with grid graphics that you will use to create maps. The **ggplot2** package is an implementation of the grammar of graphics combining advantages from the standard and lattice graphic environments. It is worth the effort to learn. We begin with the standard graphics environment.

5.1.1 Box Plot

A box plot is a graph of the five-number summary. The `summary` function applied to data produces the sample mean along with five other statistics including the minimum, the first quartile value, the median, the third quartile value, and the maximum. The box plot graphs these numbers. This is done using the `boxplot` function. For example, to create a box plot of your October SOI data, type

```
> boxplot(SOI$Oct, ylab="October SOI (s.d.)")
```

Figure 5.1 shows the results. The line inside the box is the median value. The bottom of the box (lower hinge) is the first quartile value and the top of the box (upper hinge) is the third quartile. The vertical line (whisker) from the top of the box extends to the maximum value and the vertical line from the bottom of the box extends to the minimum value.

Hinge values equal the quartiles exactly when there is an odd number of observations. Otherwise, hinges are the middle value of the lower (or upper) half of the observations if there is an odd number of observations below the median and are the middle of two values if there is an even number of observations below the median. The `fivenum` function gives the five numbers used by `boxplot`. The height of the box is the interquartile range (`IQR`) and the range is the distance from the bottom of the lower whisker to the top of the upper whisker.

By default, the whiskers are drawn as a dashed line extending from the box to the minimum and maximum data values. Convention is to make the length of the whiskers no longer than 1.5 times the height of the box. The outliers, data values larger or smaller than this range, are marked separately with points. Figure 5.1 also shows the box plot for the August SOI values. The text identifies the values. Here there is a single outlier. In this case, the upper whisker extends to the last data value less than 1.5 × IQR.

For example, if you type

```
> Q1 = fivenum(SOI$Aug)[2]
> Q2 = fivenum(SOI$Aug)[3]
> Q3 = fivenum(SOI$Aug)[4]
> Q2 + (Q3 - Q1) * 1.5
[1] 2.28
```

you see one observation greater than 2.3. In this case, the upper whisker ends at the next highest observation value less than 2.3. Observations above and below the whiskers are considered outliers. You can find the value of the single outlier of the August SOI by typing

```
> sort(SOI$Aug)
```

The largest observation in the data less than 2.3 is 2.2.

Your observations are said to be symmetric if the median is near the middle of the box with the two whiskers of equal lengths. A symmetric set of observations will also have the same number of high and low outliers.

Twenty-five percent of all your observations are below the lower quartile (below the box), 50% are below (and above) the median, and 25% are above the upper quartile. The box contains 50% of all your data. The upper whisker extends from the upper quartile to the maximum and the lower whisker extends from the lower quartile value to the minimum except if they exceed 1.5 times the interquartile range above the upper or below the lower quartiles. In this case, outliers are plotted as points. This outlier option can be turned off by setting the `range` argument to zero.

The box plot is an efficient graphical summary of your data. By removing the box lines altogether, the same information is available with less ink. Figure 5.2 is series of box plots representing the SOI for each month. The dot represents the median; the ends of the lines toward the dot are the lower and upper quartiles, respectively; the ends of the lines toward the bottom and top of the graph are the minimum and maximum values, respectively.

5.1.2 Histogram

A histogram is a graph of the distribution of your observations. It shows where the values tend to cluster and where they tend to be sparse. The histogram is similar but not identical to a bar plot (see Chapter 2). The histogram uses bars to indicate frequency (or proportion) in data intervals, whereas a bar plot uses bars to indicate the frequency of data by categories. The `hist` function creates a histogram.

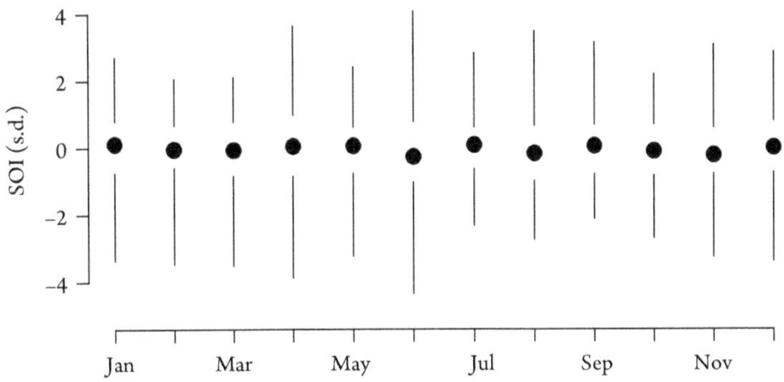

Figure 5.2 Five-number summary of the monthly SOI.

Consider NOAA's annual values of accumulated cyclone energy (ACE) for the North Atlantic and July SOI values. Annual ACE is calculated by squaring the maximum wind speed for each six-hour tropical cyclone observation and summing over all cyclones in the season. The values obtained from NOAA (`http://www.aoml.noaa.gov/hrd/tcfaq/E11.html`) are expressed in units of knots squared $\times 10^4$. You create the two histograms and plot them side by side. First set the plotting parameters with the `par` function. Details on plotting options are given in Murrell (2006). After your histogram is plotted, the function `rug` adds tick marks along the horizontal axis at the location of each observation (like a floor carpet).

```
> par(mfrow=c(1, 2), pty="s")
> hist(A$ACE)
> rug(A$ACE)
> hist(SOI$Jul)
> rug(SOI$Jul)
```

Figure 5.3 shows the result. Here we added an axis label, turned off the default title, and placed text ("a" and "b") in the figure margins. Plot titles are useful in presentations, but are redundant in publication. The default horizontal axis label is the name of the data vector. The default vertical axis is frequency and is labeled accordingly.

Default values for the function provide a good starting point, but you might want to make adjustments. It helps to know how the histogram is assembled. First a contiguous collection of disjoint intervals, called bins (or classes), is chosen that cover the range of data values. The default for the number bins is the value $\lceil \log_2(n) + 1 \rceil$, where n is the sample size and $\lceil \rceil$ indicates the ceiling value (next largest integer). If you type

```
> n = length(SOI$Jul)
> ceiling(log(n, base=2) + 1)
[1] 9
```

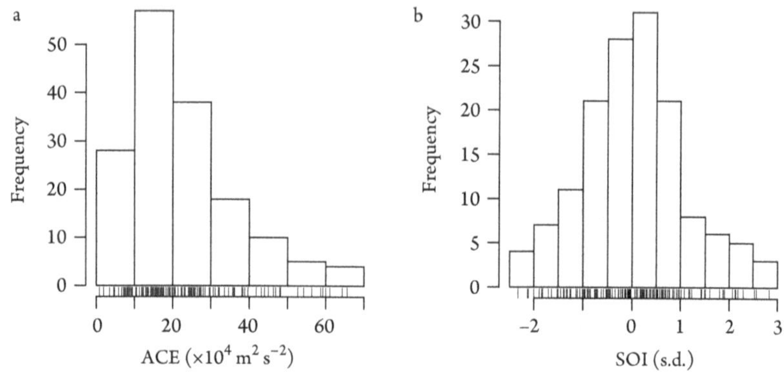

Figure 5.3 Histograms of (a) ACE and (b) SOI.

you can see that adjustments are made to this number so that the cut points correspond to whole number data values. In the case of ACE, the adjustment results in 7 bins and in the case of the SOI it results in 11 bins. Thus the computed number of bins is a suggestion that gets modified to make for nice breaks.

The bins are contiguous and disjoint so the intervals look like $(a, b]$ or $[a, b)$ where the interval $(a, b]$ means from a to b including b but not a. Next, the number of data values in each of the intervals is counted. Finally, a bar is drawn above the interval so that the bar height is the number of data values (frequency). A useful argument to make histograms comparable is `prob=TRUE`, which allows you to set the bar height to the density, where the sum of the densities times the bar interval width equals one.

You conclude that ACE is positively skewed with a few years having very large values. By contrast, the SOI appears symmetric with short tails as you would expect from a normal distribution.

5.1.3 Density Plot

A histogram outlines the general shape of your data. Usually that is sufficient. You can adjust the number of bins (or bin width) to get more or less detail on the shape. An alternative is a density plot. A density plot captures the distribution shape by smoothing the histogram. Instead of specifying the bin width, you specify the amount (and type) of smoothing. There are two steps. First you use the `density` function to obtain a set of kernel density estimates from your observations. Second you plot these estimates using the `plot` method.

A kernel density is a function that provides an estimate of the average number of values at any location in the space defined by your data. This is illustrated in Figure 5.4, where the October SOI values in the period 2005–2010 are indicated as a rug, and a kernel density function is shown as the black curve. The height of the function, representing the local density, is a sum of the heights of the individual kernels shown in red.

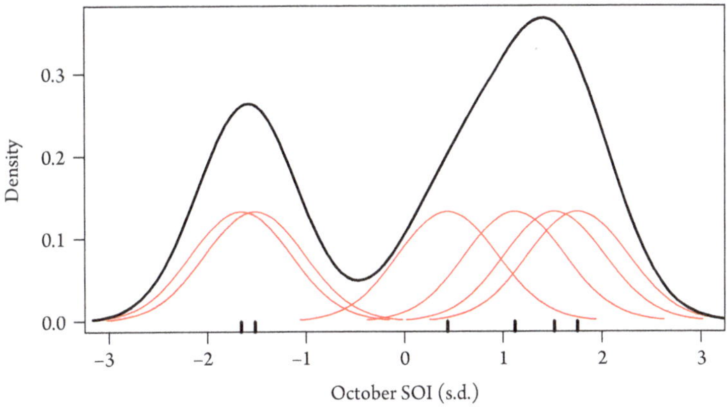

Figure 5.4 Density of October SOI (2005–2010).

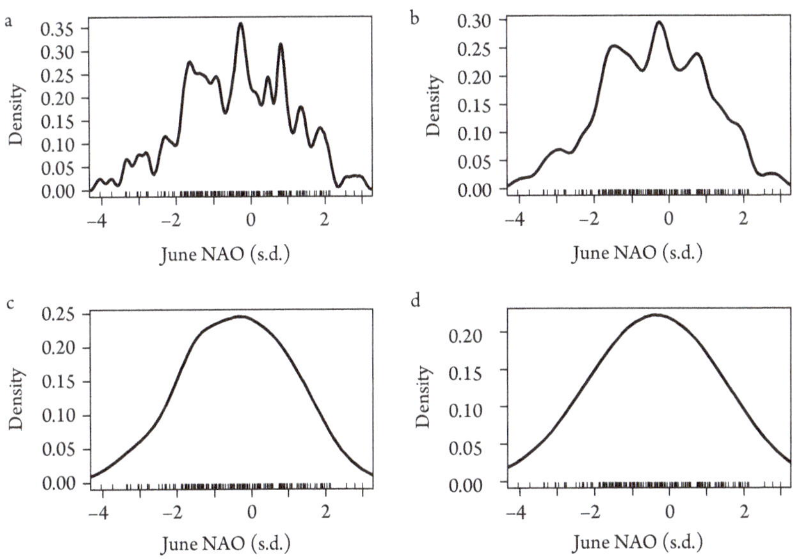

Figure 5.5 Density of June NAO. (a) .1, (b) .2, (c) .5, and (d) 1 s.d. bandwidth.

The kernel is a Gaussian (normal) distribution centered at each data value. The width of the kernel, called the bandwidth, controls the amount of smoothing. The bandwidth is the standard deviation of the kernel in the `density` function. This means the inflection points on the kernel occur one bandwidth away from the data location in units of the data values. Here with the SOI in units of standard deviation, the bandwidth equals .5 s.d.

A larger bandwidth produces a smoother density plot for a fixed number of observations because the kernels have greater overlap. Figure 5.5 shows the density plot of June NAO values from the period 1851 to 2010 using bandwidths of .1, .2, .5, and 1.

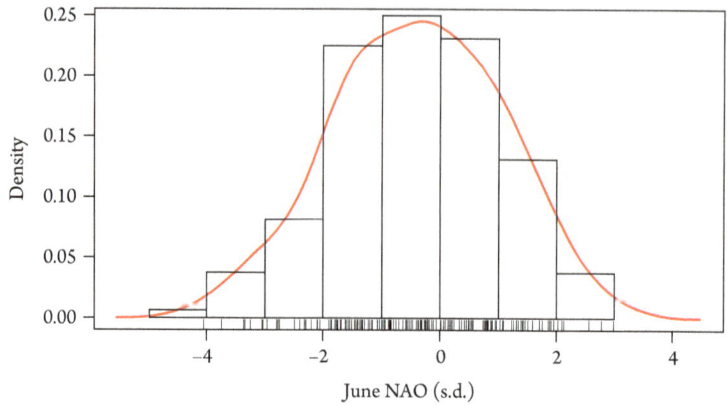

Figure 5.6 Density and histogram of June NAO.

The smallest bandwidth produces a density plot that has spikes as it captures the fine-scale variability in the distribution of values. As the bandwidth increases, the spikes disappear and the density gets smoother. The largest bandwidth produces a smooth symmetric density centered on the value of zero.

To create a density plot for the NAO values with a histogram overlay, type

```
> d = density(NAO$Jun, bw=.5)
> plot(d, main="", xlab="June NAO [s.d.]",
+   lwd=2, col="red")
> hist(NAO$Jun, prob=TRUE, add=TRUE)
> rug(NAO$Jun)
```

The density function takes your vector of data values as input and allows you to specify a bandwidth using the `bw` argument. Here you are using the vector of June NAO values and a bandwidth of .5 s.d. The bandwidth units are the same as the units of your data, here s.d. for the NAO. The output is saved as a density object, here called `d`. The object is then plotted using the `plot` method. You turn off the default plot title with the `main=""` and you specify a label for the values to be plotted below the horizontal axis. You specify the line width as 2 and the line color as red.

You then overlay the histogram using the `hist` function (see Figure 5.6). You use the `prob=TRUE` argument to make the bar height proportional to the density. The `add=TRUE` argument is needed so that the histogram plots on the same graph. One reason for plotting the histogram or density is to see whether your data can be described by a normal distribution. The Q–Q plot provides another way to make this assessment.

5.1.4 Q–Q Plot

A Q–Q plot is a way to compare distributions. It does this by plotting quantile (Q) values of one distribution against the corresponding quantile (Q) values of the other

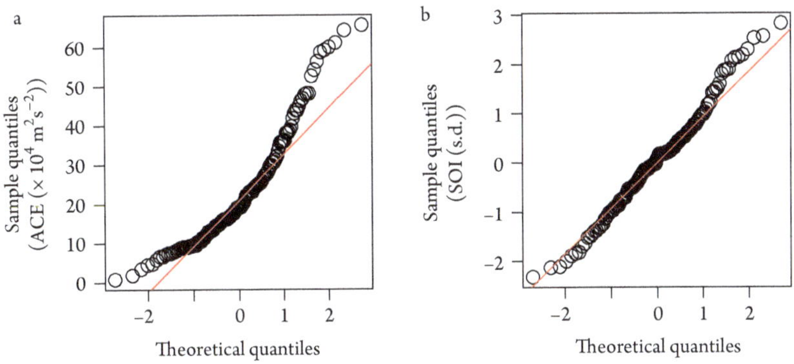

Figure 5.7 Q-Q normal plot of (a) ACE and (b) July SOI.

distribution. In the case of assessing whether or not your data are normally distributed, the sample quantiles are plotted on the vertical axis and quantiles from a standard normal distribution are plotted along the horizontal axis. In this case, it is called a Q–Q normal plot.

That is, the kth smallest observation is plotted against the expected value of the kth smallest random value from an $N(0, 1)$ sample size equal to the length of your data vector. The pattern of points in the plot is then used to compare your data against a normal distribution. If your data are normally distributed then the points align along the $y = x$ line shown on the plot.

This is done using the `qqnorm` function. To make side-by-side Q–Q normal plots for the ACE values and the July SOI values, you type

```
> par(mfrow=c(1, 2), pty="s")
> qqnorm(A$ACE)
> qqline(A$ACE, col="red")
> qqnorm(SOI$Jul)
> qqline(SOI$Jul, col="red")
```

The plots are shown in Figure 5.7. The quantiles are nondecreasing. The $y = x$ line is added to the plot using the `qqline` function. Additionally, we adjusted the vertical axis label and turned the default title off.

The plots show that July SOI values appear to have a normal distribution while the seasonal ACE does not. For observations that have a positive skew, like the ACE, the pattern of points on a Q–Q normal plot is concave upward. For observations that have a negative skew, the pattern of points is concave downward. For values that have a symmetric distribution but with fatter tails than the normal (e.g., the t-distribution), the pattern of points resembles an inverse sine function.

The Q–Q normal plot is useful in checking the residuals from a regression model. The assumption is that the residuals are independent and identically distributed characterized by a normal distribution centered on zero. In Chapter 3, you created a multiple linear regression model for August SST using March SST and year as explanatory variables. To examine the assumption of normally distributed residuals with a Q–Q normal plot, type

```
> model = lm(Aug ~ Year + Mar, data=SST)
> qqnorm(model$residuals)
> qqline(model$residuals, col="red")
```

Points align along the $y = x$ axis indicating a normal distribution.

5.1.5 Scatter Plot

The `plot` function (method) is used to create a scatter plot. The values of one variable are plotted against the values of the other variable as points in a Cartesian plane (see Chapter 2). The values named in the first argument are plotted along the horizontal axis.

This pairing is useful in generating and testing hypotheses about a relationship between the two variables. In the context of correlation, which variable gets plotted on which axis is not of concern. Either way, the scatter of points illustrates the amount of correlation. However, in the context of a statistical model, by convention, the dependent variable (the variable you are interested in explaining) is plotted on the vertical axis and the explanatory variable is plotted on the horizontal axis. For example, if your interest is whether ACE is related to pre-hurricane season ocean warmth (e.g., June SST), your model is

```
> ace = A$ACE*.5144^2
> sst = SST$Jun
> model = lm(ace ~ sst)
```

and you plot ACE on the vertical axis. Since your slope and intercept coefficients from the linear regression model are saved as part of the object `model`, you can first create a scatter plot and then use the `abline` function to add the linear regression line. Here the function extracts the intercept and slope coefficient values from the model object and draws the straight line using the point-intercept formula.

Here you use the model formula syntax (`ace ~ sst`) as the first argument in the `plot` function.

```
> plot(ace ~ sst, ylab=expression(
+    paste("ACE [x", 10^4," ", m^2, s^-2,"]")),
+    xlab=expression(paste("SST [",degree,"C]")))
> abline(model, col="red", lwd=2)
```

Figure 5.8 is the result. The relationship between ACE and SST is summarized by the linear regression model shown by the straight line. The slope of the line indicates that for every 1°C increase in SST, the average value of ACE increases by 27×10^4 m^2/s^2 (type `coef(model[2])`).

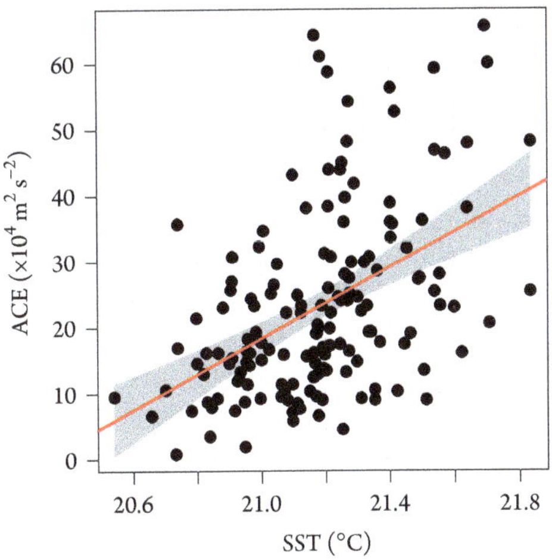

Figure 5.8 Scatter plot and linear regression line of ACE and June SST.

Since the regression line is based on a sample of data, you should display it inside a band of uncertainty. As we saw in Chapter 3, there are two types of uncertainty: a confidence band (narrow) and a prediction band (wide). The confidence band reflects the uncertainty about the line itself, which like the standard error of the mean indicates the precision by which you know the mean. Here the mean is not constant but rather a function of the explanatory variable.

The 95 percent confidence band is shown in Figure 5.8. The width of the band is inversely related to the sample size. In a large sample of data, the confidence band will be narrow reflecting a well-determined line. Note that it is impossible to draw a horizontal line that fits completely within the band. This indicates that there is a significant relationship between ACE and SST.

The band is narrowest in the middle, which is understood by the fact that the predicted value at the mean SST will be the mean of ACE, whatever the slope, and thus the standard error of the predicted value at this point is the standard error of the mean of ACE. At other values of SST, the variability associated with the estimated slope is included. This variability is larger for values of SST farther from the mean, which is why the band looks like a bow tie.

The prediction band adds another layer of uncertainty, the uncertainty about *future* values of ACE. The prediction band captures the majority of the observed points in the scatter plot. Unlike the confidence band, the width of the prediction band depends on the assumption of normally distributed errors with a constant variance across the values of the explanatory variable.

5.1.6 Conditional Scatter Plot

Scatter plots *conditional* on the values of a third variable can be quite informative. This is done with the `coplot` function. The syntax is the same as above except you add the name of the conditioning variable after a vertical bar.

For example, as SST increases so does ACE. The conditioning plot answers the question: is there a change in the relationship depending on values of a third variable? Here you use August SOI values as the conditioning variable and type

```
> soi = SOI$Aug
> coplot(ace ~ sst | soi, panel=panel.smooth)
```

The syntax is read "conditioning plot of ACE versus SST given values of SOI." The function divides the range of the conditioning variable (SOI) into six intervals (by default) with each interval having approximately the same number of years. The range of SOI values in each interval overlaps by 50 percent. The conditioning intervals are plotted in the top panel as horizontal bars (shingles). The plot is shown in Figure 5.9.

The scatter plots of ACE and SST are arranged in a matrix of panels below the shingles. The panels are arranged from lower left to upper right. The lower left panel corresponds to the lowest range of SOI values (less than about −1 s.d.) and the upper right panel corresponds to the highest range of SOI values (greater than about +.5 s.d.). Half of the data points in a panel are shared with the panel to the left and half of the data points are shared with the panel to the right. This is indicated by the amount of shingle overlap.

Results show a positive, nearly linear, relationship between ACE and SST for all ranges of SOI values. Over the SOI range between −1.5 and 0, the relationship is somewhat curved. ACE is least sensitive to SST when SOI is the most negative (El Niño years) as indicated by the nearly flat line in the lower left panel. The argument `panel` adds a local linear curve (red line) through the set of points in each plot.

5.2 TIME SERIES

Hurricane data often take the form of a time series. A time series is a sequence of data values measured at successive times and spaced at uniform intervals. You can treat a time series as a vector and use structured data functions (see Chapter 2) to generate time series.

However, additional functions are available for data that are converted to time-series objects. Time series objects are created using the `ts` function. You do this with the monthly NAO data frame as follows. First create a matrix of the monthly values, skipping the year column in the data frame. Second take the transpose of this matrix (switch the rows with the columns) using the `t` function and save the matrix as a vector. Finally, create a time series object, specifying the frequency of values and the start month. Here the first value is from January 1851.

```
> nao.m = as.matrix(NAO[, 2:13])
> nao.v = as.vector(t(nao.m))
> nao.ts = ts(nao.v, frequency=12, start=c(1851, 1))
```

Also create a time series object for the cumulative sum of the monthly SOI values. The is done with the `cumsum` function applied to your data vector.

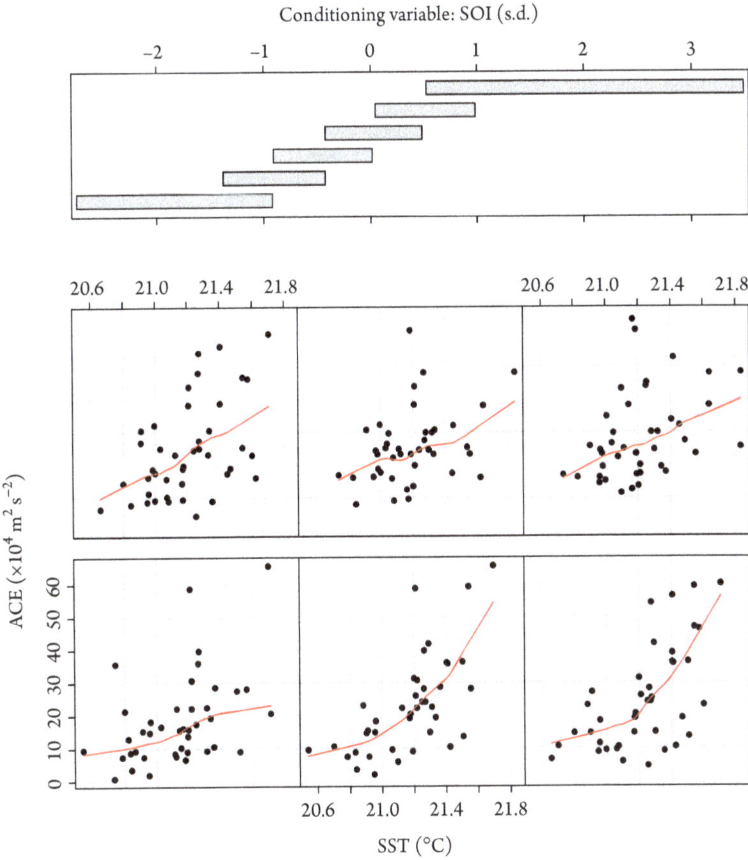

Figure 5.9 Scatter plots of ACE and SST conditional on the SOI.

```
> nao.cts = ts(cumsum(nao.v),
+     frequency=12, start=c(1851, 1))
```

This results in objects of class `ts`, which is used for time series having numeric time information. Additional classes for working with time series data that can handle dates and other types of time information are available. For example, the **fts** package implements regular and irregular time series based on `POSIXct` time stamps (see §5.2.3), and the **zoo** package provides functions for most time series classes.

5.2.1 Time Series Graph

The objects of class `ts` make it easy to plot your data as a time series. For instance, you plot the cumulative sum of the NAO values using the `plot` method. The method recognizes the object as a time series and plots it accordingly eliminating the need to specify a separate time variable.

```
> plot(nao.cts)
```

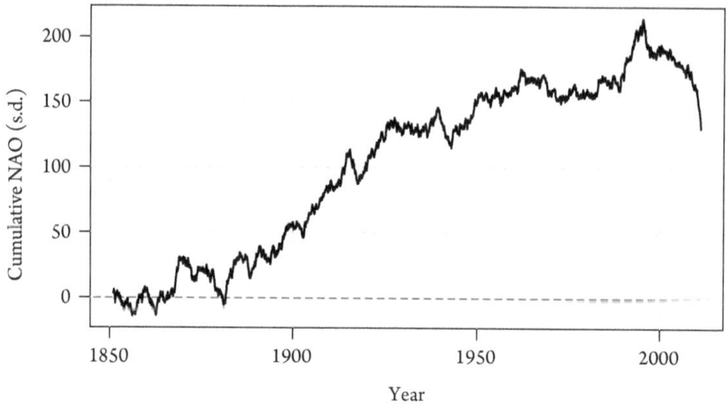

Figure 5.10 Time series of the cumulative sum of NAO values.

Figure 5.10 shows the result. The cumulative sum indicates a pattern typical of a random walk. That is, over the long term there is a tendency for more positive-value months leading to a "wandering" of the cumulative sum away from the zero line. This tendency begins to reverse in the late twentieth century.

5.2.2 Autocorrelation

Autocorrelation is correlation between values of a single variable. For time data it, refers to single series correlated with itself as a function of temporal lag. For spatial data, it refers to single variable correlated with itself as a function of spatial lag, which can be a vector of distance and orientation (see Chapter 9). In both cases, the term "autocorrelation function" is used, but with spatial data, the term is often qualified with the word "spatial."

As an example, save 30 random values from a standard normal distribution in a vector where the elements are considered ordered in time. First, create a time series object. Then use the `lag.plot` function to create a scatter plot of the time series against a lagged copy where the lagged copy starts one time interval earlier.

```
> t0 = ts(rnorm(30))
> lag.plot(t0, lag=1)
```

With n values, the plot for lag one contains $n - 1$ points. The points are plotted using the text number indicating the temporal order so that the first point labeled "1" is given by the coordinates (`t0[1]`,`t0[2]`). The correlation at lag one can be inferred by the scatter of points. The plot can be repeated for any number of lags, but with higher lags, the number of points decreases.

You use the autocorrelation function (`acf`) to quantify the correlation at various temporal lags. The function accepts univariate and multivariate numeric time series objects and produces a plot of the autocorrelation values as a function of lag. For example, to create a plot of the autocorrelation function for the NAO time series object from the previous section, type

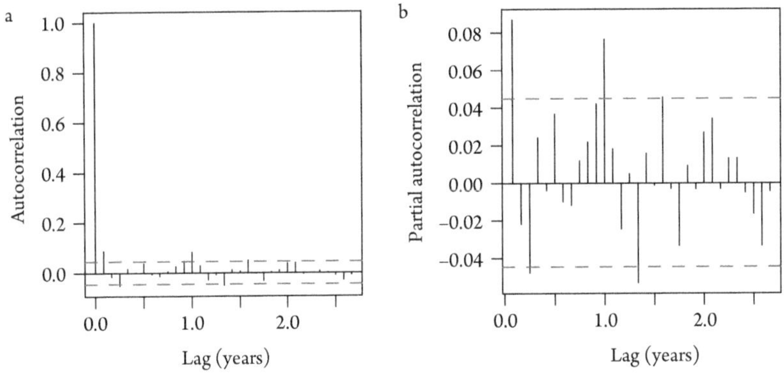

Figure 5.11 Autocorrelation and partial autocorrelation functions of monthly NAO.

```
> acf(nao.ts, xlab="Lag [Years]",
+    ylab="Autocorrelation")
```

The lag values on the horizontal axis are plotted in units of time rather than numbers of observations (see Fig. 5.11). Dashed lines are the 95 percent confidence limits. Here the time series object uses monthly frequency, so the lags are given in fractions of 12 with 1.0 corresponding to a year. The maximum lag is calculated as $10 \times \log_{10} n$, where n is the number of observations. This can be changed using the argument `lag.max`.

The lag-zero autocorrelation is fixed at 1 by convention. The nonzero autocorrelations are all less than 0.1 in absolute value indicative of an uncorrelated process. By default, the plot includes 95 percent confidence limits computed as $\pm 1.96/\sqrt{n}$.

The partial autocorrelation function `pacf` computes the autocorrelation at lag k after the linear dependencies between lags 1 to $k-1$ are removed. The partial autocorrelation is used to identify the temporal extent of the autocorrelation. Here the partial autocorrelation vacillates between positive and negative values indicative of a moving-average process.[1]

If your regression model uses time series data, it is important to examine the autocorrelation in the model residuals. If residuals from your regression model have significant positive autocorrelation, then the assumption of independence is violated. This violation does not bias the coefficient estimates, but the standard errors on the coefficients tend to be too small making you over confident in your inferences.

5.2.3 Dates and Times

Various options exist for working with date and time data in R. The `as.Date` function gives you flexibility in handling dates through the `format` argument. The default

[1] A moving-average process is one in which the mean of the current value of the series is linearly related to previous white noise (uncorrelated) errors.

Table 5.1 Format codes for dates.

Code	*Value*
%d	Day of the month (decimal number)
%m	Month (decimal number)
%b	Month (abbreviated, e.g., Jan)
%B	Month (full name)
%y	Year (2 digit)
%Y	Year (4 digit)

format is a four-digit year, a month, then a day, separated by dashes or slashes. For example, the character string "1992-8-24" is accepted as a date by typing

```
> Andrew = as.Date("1992-8-24")
```

Although the print method displays it as a character string, the object is a Date class stored as the number of days since January 1, 1970, with negative numbers for earlier dates.

If your input dates are not in the standard year-month-day order, a format string can be composed using the elements shown in Table 5.1. For instance, if your date is specified as August 29, 2005, then you type

```
> Katrina = as.Date("August 29, 2005",
+     format="%B %d, %Y")
```

You can find the number of days between hurricanes Andrew and Katrina by typing

```
> difftime(Katrina, Andrew, units="days")
Time difference of 4753 days
```

Or you can obtain the number of days from today since Andrew by typing

```
> difftime(Sys.Date(), Andrew, units="days")
```

The function Sys.Date with no arguments gives the current day in year-month-day format as a Date object.

The portable operating system interface (POSIX) has formats for dates and times, with functionality for converting between time zones (Spector, 2008). The POSIX date/time classes store times to the nearest second. There are two such classes differing only in the way the values are kept internally. The POSIXct class stores date/time values as the number of seconds since January 1, 1970, while the POSIXlt class stores them as a list. The list contains elements for second, minute, hour, day, month, and year among others.

The default input format for POSIX dates consists of the year, month, and day, separated by slashes or dashes with time information followed after a space. The time information is in the format hour:minutes:seconds or simply hour:minutes. For example, according to the U.S. National Hurricane Center, Hurricane Andrew hit Homestead Air Force Base at 0905 UTC on August 24, 1992. You add time information to your Andrew date object and convert it to a `POSIXct` object by typing.

```
> Andrew = as.POSIXct(paste(Andrew, "09:05"),
+   tz="GMT")
```

You then retrieve your local time from your operating system as a character string and use the date–time conversion `strptime` function to convert the string to a POSIXlt class.

```
> mytime = strptime(Sys.time(), format=
+   "%Y-%m-%d %H:%M:%S", tz="EST5EDT")
```

Our time zone is U.S. Eastern standard time, so we use `tz="EST5EDT"`. You then find the number of hours since Andrew's landfall by typing,

```
> difftime(mytime, Andrew, units="hours")
Time difference of 171482 hours
```

Note that time zones are not portable, but `EST5EDT` comes pretty close.

Additional functionality for working with times is available in the **chron** and **lubridate** packages. In particular, **lubridate** (great package name) makes it easy to work with dates and times by providing functions to identify and parse date–time data, extract and modify components (years, months, days, hours, minutes, and seconds), perform math on date–times, and handle time zones and Daylight Savings Time (Grolemund and Wickham, 2011).

For example, to return the day of the week from your object `Andrew`, you use the `wday` function in the package by typing

```
> require(lubridate)
> wday(Andrew, label=TRUE, abbr=FALSE)
[1] Monday
7 Levels: Sunday < Monday < ... < Saturday
```

If you lived in south Florida, what a Monday it was. Other examples of useful functions in the package related to the Andrew time object include, the year, was it a leap year, what week of the year was it, and what local time was it. Finally, what is the current time in Chicago?

```
> year(Andrew)
> leap_year(Andrew)
> week(Andrew)
> with_tz(Andrew,tz="America/New_york")
> now(tz="America/Chicago")
```

5.3 MAPS

A great pleasure in working with graphs is the chance to visualize patterns. Maps are among the most compelling graphs as the space they map is the space in which hurricanes occur. We can use them to find interesting, sometimes hidden, structure. Various packages are available for creating maps. Here we look at a few examples.

5.3.1 Boundaries

Sometimes all you need is a reference map to show your study location. This can be created using state and country boundaries. For example, the **maps** package is used to draw country and state borders. To draw a map of the United States with state boundaries, type

```
> require(maps)
> map("state")
```

The call to `map` creates the country outline and adds the state boundaries. The map is shown in Figure 5.12. The package contains outlines for countries around the world (e.g., type `map()`).

The coordinate system is latitude and longitude, so you can overlay other spatial data. As an example, first input the track of Hurricane Ivan (2004) as it approached the U.S. Gulf coast. Then list the first six rows of data.

```
> Ivan = read.table("Ivan.txt", header=TRUE)
> head(Ivan)
  Year Mo Da Hr   Lon  Lat Wind WindS Pmin Rmw   Hb
1 2004  9 15  8 -87.6 25.9  118   118  917  37 1.27
2 2004  9 15  9 -87.7 26.1  118   117  917  37 1.27
3 2004  9 15 10 -87.7 26.3  117   116  917  37 1.26
4 2004  9 15 11 -87.8 26.5  117   116  918  37 1.26
```

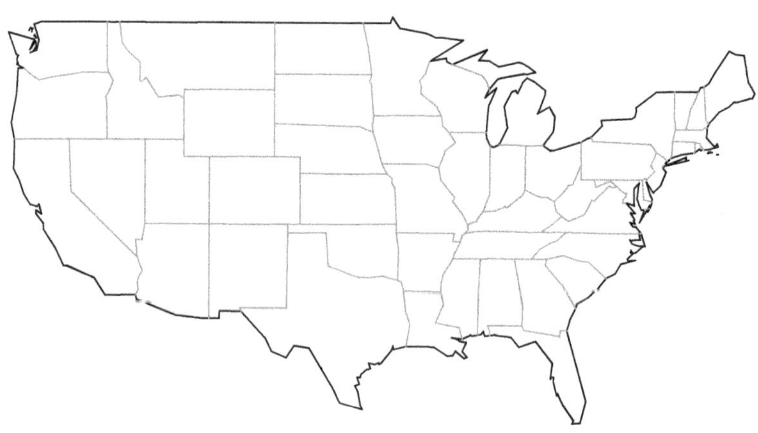

Figure 5.12 Map with state boundaries.

```
5 2004   9 15 12 -87.9 26.7  117    115   918   37 1.26
6 2004   9 15 13 -88.0 26.9  116    115   919   37 1.26
  Speed L Lhr
1  11.6 0 -24
2  12.1 0 -23
3  12.4 0 -22
4  12.6 0 -21
5  12.7 0 -20
6  12.8 0 -19
```

Among other attributes, the data frame `Ivan` contains the latitude and longitude position of the hurricane center every hour from 24 hours before landfall until 12 hours after landfall.

Here your geographic domain is the southeast, so first create a character vector of state names.

```
> cs = c('texas', 'louisiana', 'mississippi',
+   'alabama', 'florida', 'georgia', 'south carolina')
```

Next use the `map` function with this list to plot the state boundaries and fill the state polygons with a gray shade. Finally, connect the hourly location points with the `lines` function and add an arrowhead to the last two locations.

```
> map("state", region=cs, boundary=FALSE, col="gray",
+   fill=TRUE)
> Lo = Ivan$Lon
> La = Ivan$Lat
> n = length(Lo)
> lines(Lo, La, lwd=2.5, col="red")
> arrows(Lo[n - 1], La[n - 1], Lo[n], La[n], lwd=2.5,
+    length=.1, col="red")
```

The result is shown in Figure 5.13. Hurricane Ivan moved northward from the central Gulf of Mexico and made landfall in the western panhandle region of Florida before moving into southeastern Alabama.

The scale of the map is defined as the ratio of the map distance in a particular unit (e.g., centimeters) to the actual distance in the same unit. Small scale describes maps of large regions where this ratio is small and large scale describes maps of small regions where the ratio is larger. The boundary data in the **maps** package is sufficient for use with small-scale maps but the number of boundary points is not sufficient for large-scale maps (close-up or high resolution). Higher-resolution boundary data are available in the **mapdata** package.

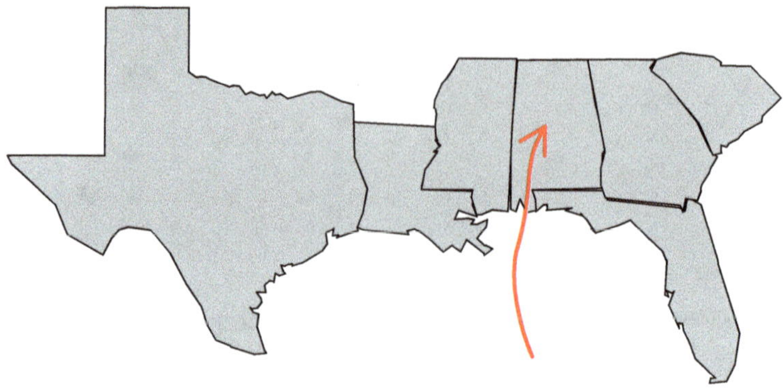

Figure 5.13 Track of Hurricane Ivan (2004) before and after landfall.

5.3.2 Data Types

The type of map you make will depend on the type of spatial date you have. Broadly speaking, there are three types: point, areal, and field data. Point data are event locations. Any location in a continuous spatial domain may have an event. The events may carry additional information, called "marks." Interest centers on the distribution of events and on whether there are event clusters. The set of all locations where hurricanes first reached maximum intensity is an example of point data. The events are the location of the hurricane at maximum intensity, and a mark is the corresponding wind speed.

Areal data are aggregated within fixed polygons. The set of polygons form a lattice so areal data are called "lattice data." Interest centers on how the values change across the domain and on how much correlation exists within neighborhoods defined by contiguity or distance. County-wide population is an example of areal data. The values may be the number of people living in the county or a population density indicating the average number of people per area.

Field data are observations of a spatially continuous variable, like pressure or temperature. The values are given at certain locations and the interest centers on using these values to create a continuous surface from which inferences can be made at any location. Sea-level pressure is an example of field data.

Point Data

Consider the set of events defined by the location at which a hurricane first reaches lifetime maximum intensity. The data are available in the file *LMI.txt* and are input by typing

```
> LMI.df = read.table("LMI.txt", header=TRUE)
> LMI.df$WmaxS = LMI.df$WmaxS * .5144
> head(LMI.df[, c(4:10, 11)])
                 name    Yr Mo Da hr   lon  lat  Wmax
30861.5 DENNIS       1981  8 20 23 -70.8 37.0  70.4
```

```
30891.4 EMILY          1981  9  6 10 -58.1 40.6  80.6
30930.2 FLOYD          1981  9  7  2 -69.1 26.8 100.4
30972.2 GERT           1981  9 11 14 -71.7 29.4  90.5
31003.5 HARVEY         1981  9 14 23 -62.6 28.3 115.1
31054.4 IRENE          1981  9 28 16 -56.4 27.9 105.5
```

The `Wmax` column is a spline-interpolated maximum wind speed and `WmaxS` (not shown) is first smoothed then spline interpolated to allow time derivatives to be computed. Chapter 6 provides more details and explains how this data set is constructed.

The raw wind speed values are given in 5-kt increments. Although knots (kt) are the operational unit used for reporting tropical cyclone intensity to the public in the United States, here you use the SI units of m s^{-1}. We use the term "intensity" as shorthand for "maximum wind speed," where maximum wind speed refers to the estimated fastest wind velocity somewhere in the core of the hurricane. Lifetime maximum refers to the fastest wind during the life of the hurricane.

You draw a map of the event locations with the `plot` method using the longitude coordinate as the *x* variable and the latitude coordinate as the *y* variable by typing

```
> with(LMI.df, plot(lon, lat, pch=19))
> map("world", col="gray", add=TRUE)
> grid()
```

Adding country borders and latitude/longitude grid lines (`grid` function) enhances the geographic information. The argument `pch` specifies a point character using an integer code. Here 19 refers to a solid circle (type `?points` for more information). The `with` function allows you use the column names from the data frame in the `plot` method.

Note the order of function calls. By plotting the events first, then adding the country borders, the borders are clipped to the plot window. The dimensions of the plot window are slightly larger than the range of the longitude and latitude coordinates. The function chooses a number of axis tics that are placed along the range of coordinate values at reasonable intervals.

Since the events are marked by storm intensity, it is informative to add this information to the map. Hurricane intensity, as indexed by an estimate of the wind speed maximum, is a continuous variable. You can choose a set of discrete intensity intervals and group the events by these class intervals. For example, you might want to choose the Saffir–Simpson hurricane intensity scale.

To efficiently communicate differences in intensities with colors, you should limit the number classes to six or less. The package **classInt** is a collection of functions for choosing class intervals. Here you require the package and create a vector of lifetime maxima. You then obtain class boundaries using the `classIntervals` function. Here the number of class intervals is set to five and the method of determining the interval breaks is based on Jenks optimization (`style="jenks"`). Given the number of classes, the optimization minimizes the variance of the values within the intervals while maximizing the variance between the intervals.

```
> require(classInt)
> lmi = LMI.df$WmaxS
> q5 = classIntervals(lmi, n=5, style="jenks",
+     dataPrecision=1)
```

The `dataPrecision` argument sets the number of digits to the right of the decimal place.

Next you choose a palette of colors. This is best left to someone with an understanding of hues and color separation schemes. The palettes described and printed in Brewer et al. (2003) for continuous, diverging, and categorical variables can be examined on maps at `http://colorbrewer2.org/`. Select the HEX radio button for a color palette of your choice and then copy and paste the hex code into a character vector preceded by the pound symbol.

For example, here you create a character vector (`cls`) of length 5 containing the hex codes from the color brewer Web site from a sequential color ramp ranging among yellow, orange, and red.

```
> cls = c("#FFFFB2", "#FECC5C", "#FD8D3C", "#F03B20",
+    "#BD0026")
```

To use your own set of colors, simply modify this list. A character vector of color hex codes is generated automatically with functions in the **colorRamps** package (see Chapter 9).

The empirical cumulative distribution function of cyclone intensities with the corresponding class intervals and colors is plotted by typing

```
> plot(q5, pal=cls, main="", xlab=
+    expression(paste("Wind Speed [m ", s^-1,"]")),
+    ylab="Cumulative Frequency")
```

The graph is shown in Figure 5.14. The points (with horizontal dashes) are the lifetime maximum intensities in rank order from lowest to highest. You can see that half of all hurricanes have lifetime intensities greater than 46 m s^{-1}.

Once you are satisfied with the class intervals and color palette, you can plot the events on a map. First you need to assign a color for each event depending on its wind speed value. This is done with the `findColours` function as

```
> q5c = findColours(q5, cls)
```

Now, instead of black dots with a color bar, each value is assigned a color corresponding to the class interval. For convenience, you create the axis labels and save them as expression objects. You do this with the `expression` and `paste` functions to get the degree symbol.

```
> xl = expression(paste("Longitude [",{}^o,"E]"))
> yl = expression(paste("Latitude [",{}^o,"N]"))
```

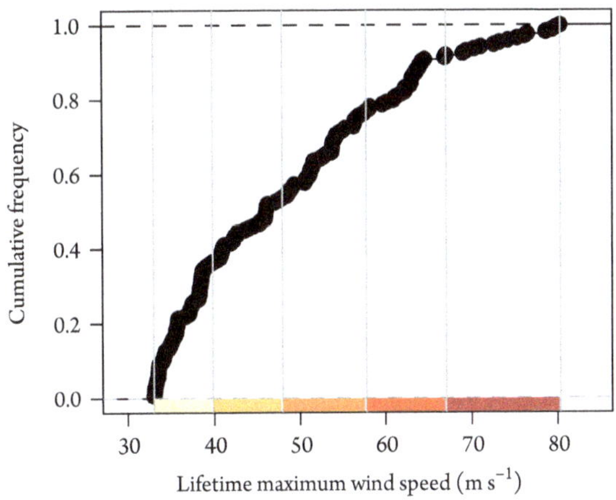

Figure 5.14 Cumulative distribution of lifetime maximum intensity. Vertical lines and corresponding color bar mark the class intervals with the number of classes set at five.

Since the degree symbol is not attached to a character, you use `{}` in front of the superscript symbol. You again use the `plot` method on the location coordinates, but this time set the color argument to the corresponding vector of colors saved in `q5c`.

```
> plot(LMI.df$lon, LMI.df$lat, xlab=xl, ylab=yl,
+     col=q5c, pch=19)
> points(LMI.df$lon, LMI.df$lat)
```

To improve the map, you add country boundaries, place axis labels in the top and right margins, and add a coordinate grid.

```
> map("world", add=TRUE)
> axis(3)
> axis(4)
> grid()
```

To complete the map, you add a legend by typing

```
> legend("bottomright", bg="white",
+     fill=attr(q5c, "palette"),
+     legend=names(attr(q5c, "table")),
+     title=expression(paste("Wind Speed [m "
+     , s^-1, "]")))
```

Note that fill colors and names for the legend are obtained using the `attr` function on the `q5c` object. The function retrieves the table attribute of the object. The result is shown in Figure 5.15. Colors indicate the wind speed in five classes as described in Figure 5.14.

The spatial distribution of lifetime maxima is fairly uniform over the ocean for locations west of the −40°E longitude. Fewer events are noted over the eastern Caribbean Sea and southwestern Gulf of Mexico. Events over the western Caribbean tend to

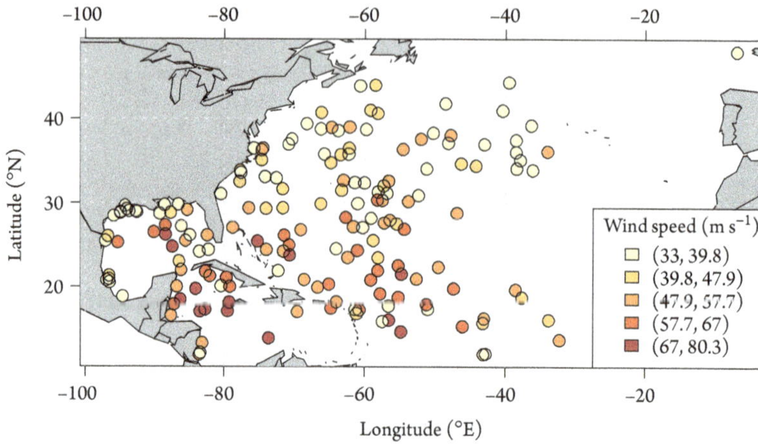

Figure 5.15 Location of lifetime maximum wind speed.

have the highest intensities. Also there is a tendency for hurricanes reaching lifetime maxima at lower latitudes to have higher intensities.

Areal Data

A shapefile stores geometry and attribute information for spatial data. The geometry is a set of vector coordinates. Shapefiles support point, line, and area data. Area data are represented as closed-loop polygons. Each attribute record has a one-to-one relationship with the associated shape record. For example, a shapefile might consist of the set of polygons for the counties in Florida and an attribute might be population. Associated with each county population record (attribute) is a shape record.

The shapefile is actually a set of several files in a directory. The three files with extensions `*.shp` (file of geometries), `*.shx` (index file to the geometries), and `*.dbf` (file for storing attribute data) form the core of the directory. Note that there is no standard for specifying missing attribute values. The `*.prj` file, if present, contains the coordinate reference system (CRS; see §5.4).

Information in a shapefile format makes it easy to map. As an example, consider the U.S. Census Bureau boundary file for the state of Florida `http://www.census.gov/cgi-bin/geo/shapefiles/national-files`. Browse to Current State and Equivalent, Select State, then Florida. Download the zipped file. Unzip it to your R working directory. To make things a bit easier for typing, rename the directory and the shapefiles to `FL`.

The `readShapeSpatial` function from the **maptools** package reads in the polygon shapefile consisting of the boundaries of the 67 Florida counties.

```
> require(maptools)
> FLpoly = readShapeSpatial("FL/FL")
> class(FLpoly)
```

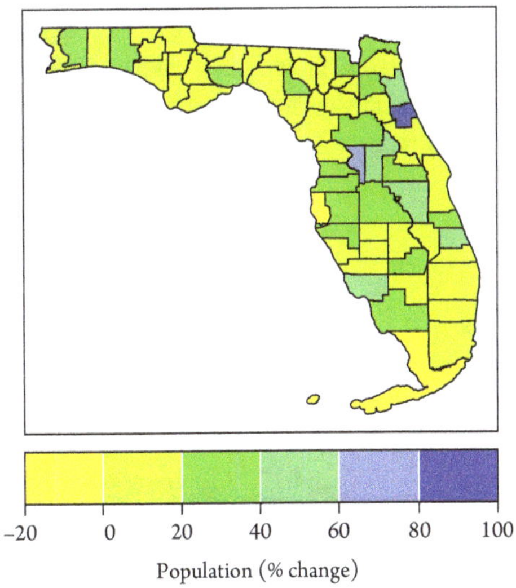

Figure 5.16
Population change in Florida counties (2000 to 2010).

```
[1] "SpatialPolygonsDataFrame"
attr(,"package")
[1] "sp"
```

Note the shapefiles are in directory `FL` with file names the same as the directory name. The object `FLpoly` you created is a `SpatialPolygonsDataFrame`. It extends the class `data.frame` by adding geographic information (for more details see Bivand et al. (2008)).

You can use the `plot` method to produce a map of the polygon borders. More interestingly is a map displaying an attribute of the polygons. For instance, demographic data at the county level are important for emergency managers. First read in a table of the percentage change in population over the 10-year period 2000–2010.

```
> FLPop = read.table("FLPop.txt", header=TRUE)
> names(FLPop)
[1] "Order"   "County"  "Pop2010" "Pop2000" "Diff"
[6] "Change"
```

Here the table rows are arranged in the same order as the polygons. You assign the column `Change` to the data slot of the spatial data frame by typing

```
> FLpoly$Change = FLPop$Change
```

Then use the function `spplot` to create a choropleth map of the attribute `Change`.

```
> spplot(FLpoly, "Change")
```

Results are shown in Figure 5.16. With the exception of Monroe and Pinellas counties, population throughout the state increased over this period. Largest population increases are noted over portions of central Florida.

The `spplot` method is available in the **sp** package. It is an example of a lattice plot method (Sarkar, 2008) for spatial data with attributes. The function returns a plot of class `trellis`. If the function does not automatically bring up your graphics device, you need to wrap it in the `print` function. Missing values in the attributes are not allowed.

Field Data

Climate data are often presented as valves on a grid. For example, NOAA-CIRES 20th Century Reanalysis version 2 provides monthly sea-surface temperatures at latitude–longitude intersections. A portion of these data are available in the file *JulySST2005.txt*. The data are the SST values on a 2° latitude–longitude grid for the month of July 2005. The grid is bounded by −100° and 10°E longitudes and the equator and 70°N latitude.

First input the data and convert the column of SST values to a matrix using the `matrix` function specifying the number of columns as the number of longitudes. The number of rows is inferred based on the length of the vector. Next create two structured vectors, one of the meridians and the other of the parallels using the `seq` function. Specify the geographic limits and an interval of 2° in both directions.

```
> sst.df = read.table("JulySST2005.txt", header=TRUE)
> sst = matrix(sst.df$SST, ncol=36)
> lo = seq(-100, 10, 2)
> la = seq(0, 70, 2)
```

To create a map of the SST field, first choose a set of colors. Since the values represent temperature, you want the colors to go from blue (cool) to red (warm). R provides a number of color palettes including `rainbow`, `heat.colors`, `cm.colors`, `topo.colors`, `grey.colors`, and `terrain.colors`. The palettes are functions that generate a sequence of color codes interpolated between two or more colors. The `cm.colors` is the default palette in `sp.plot` and the colors diverge from white to cyan and magenta.

More color options from the Web site are given in §5.3.2. The package **RColorBrewer** provides the palettes described in Brewer et al. (2003). Palettes are available for continuous, diverging, and categorical variables and for choices of print and screen projection. The **sp** package has the `bpy.colors` function that produces a range of colors from blue to yellow that work for color and black-and-white print. You can create your own palette using the `colorRampPalette` function. Here you save the function as `bwr` and use a set of three colors. The number of colors to interpolate is the argument to the `bwr` function.

```
> bwr = colorRampPalette(c("blue", "white", "red"))
```

The `image` function creates a grid of rectangles with colors corresponding to the values in the third argument as specified by the palette and the number of colors set here at 20. The first two arguments correspond to the two-dimensional location of the rectangles. The x and y labels use the `expression` and `paste` functions to get the degree symbol. You add country boundaries and place axis labels in the top and right margins (margins 3 and 4) to complete the graph.

```
> image(lo, la, sst, col=bwr(20), xlab=xl, ylab=yl)
> map("world", add=TRUE)
> axis(3)
> axis(4)
```

Note that `image` interprets the matrix of SST values as a table with the x-axis corresponding to the row number and the y-axis to the column number, with column one at the bottom. This is an orthogonal counterclockwise rotation of the conventional matrix layout.

Overlay a contour plot of the SST data using the `contour` function. First determine the range of the SST values and round to the nearest whole integer. There are missing values (over land) so you need to use the `na.rm` argument in the `range` function.

```
> r = round(range(sst, na.rm=TRUE))
```

Next create a string of temperature values at equal intervals within this range. Contours will be drawn at these values.

```
> levs = seq(r[1], r[2], 2)
> levs
 [1] -2  0  2  4  6  8 10 12 14 16 18 20 22 24 26 28
[17] 30
```

Then paste the character string "C" onto the interval labels. The corresponding list is used as contour labels.

```
> cl = paste(levs, "C")
> contour(lo, la, sst, levels=levs, labels=cl,
+   add=TRUE)
```

The result is shown in Figure 5.17. Ocean temperatures above about 28°C are warm enough to support the development of hurricanes. This covers a large area from the west coast of Africa westward through the Caribbean and Gulf of Mexico and northward toward Bermuda.

5.4 COORDINATE REFERENCE SYSTEMS

For data covering a large geographic area, you need a map with a projected coordinate reference system (CRS). A geographic CRS includes a model for the shape of the earth (oblate spheroid) plus latitudes and longitudes. Longitudes and latitudes can be used to create a two-dimensional coordinate system for plotting hurricane data, but this framework is for a sphere and not a flat map.

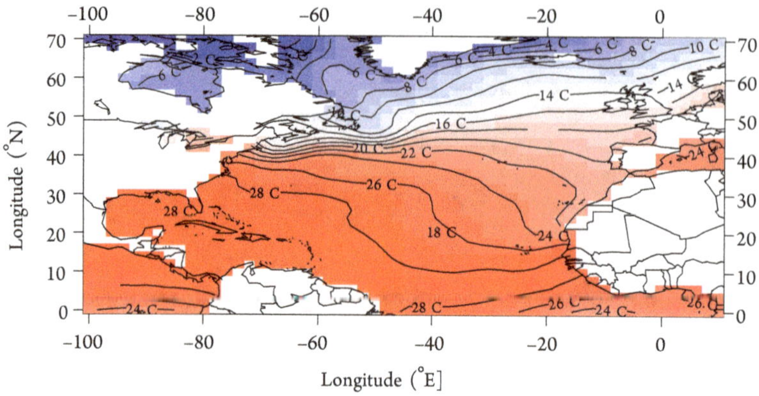

Figure 5.17 Sea-surface temperature field from July 2005.

A projected CRS is a two-dimensional approximation of the earth as a flat surface. It includes a model for the earth's shape plus a specific geometric model for projecting coordinates onto the plane. The PROJ.4 Cartographic Projections library uses a `tab = value` representation of a CRS, with a tag and value pair within a single character string. The Geospatial Data Abstraction Library (GDAL) contains code for translating between different CRSs. Both the PROJ.4 and GDAL libraries are available in the **rgdal** package (Keitt et al., 2012). Here you specify a geographic CRS and save it in a CRS object called `ll_crs` (lat-lon coordinate reference system).

```
> require(rgdal)
> require(mapdata)
> ll_crs = CRS("+proj=longlat +ellps=WGS84")
```

The only values used autonomously in CRS objects are whether the string is a character `NA` (missing) for an unknown CRS, and whether it contains the string `longlat`, in which case the CRS is geographic (Bivand et al., 2008).

There are a number of different tags, beginning with "+", and separated from the value with "=", using white space to separate the tag/value pairs. Here you specify the earth's shape using the World Geodetic System (WGS) 1984, which is the reference coordinate system used by the Global Positioning System to reference the earth's center of mass.

As an example, you create a `SpatialPoints` object called `LMI_ll` by combining the matrix of event coordinates (location of lifetime maximum intensity) in native longitude and latitude degrees with the CRS object defined above.

```
> LMI_mat = cbind(LMI.df$lon, LMI.df$lat)
> LMI_ll = SpatialPoints(LMI_mat,
+    proj4string=ll_crs)
> summary(LMI_ll)
Object of class SpatialPoints
Coordinates:
```

```
              min    max
coords.x1 -97.1 -6.87
coords.x2  11.9 48.05
Is projected: FALSE
proj4string : [+proj=longlat +ellps=WGS84]
Number of points: 173
```

Here you are interested in transforming the geographic CRS into a Lambert conformal conic (LCC) planar projection. The projection superimposes a cone over the earth, with two reference parallels secant to the globe. The LCC projection is used for aeronautical charts. It is used by the U.S. NHC in their seasonal summary maps. Other projections, ellipsoids, and datum are available, and a list of the various tag options can be generated by typing

```
> projInfo(type = "proj")
```

Besides the projection tag (lcc), you need to specify the two secant parallels and a meridian. The NHC summary maps use the parallels 30 and 60°N and a meridian of 60°W. First save the CRS as a character string, then use the `spTransform` function to transform the latitude–longitude coordinates to coordinates of the LCC planar projection.

```
> lcc_crs = CRS("+proj=lcc +lat_1=60 +lat_2=30
+    +lon_0=-60")
> LMI_lcc = spTransform(LMI_ll, lcc_crs)
```

This transforms the original set of longitude–latitude coordinates to a set of projected coordinates. You need to repeat this transformation for each of the map components. For instance, to transform the country borders, first save them from a call to the `map` function. The function includes arguments to specify a longitude–latitude bounding box. Second, convert the returned map object to a spatial lines object with the `map2SpatialLines` function using a geographic CRS. Finally, transform the coordinates of the spatial lines object to the LCC coordinates.

```
> brd = map('world', xlim=c(-100, 0), ylim=c(5, 50),
+    interior=FALSE, plot=FALSE)
> brd_ll = map2SpatialLines(brd, proj4string=ll_crs)
> brd_lcc = spTransform(brd_ll, lcc_crs)
```

To include longitude–latitude grid lines, you use the `gridlines` function on the longitude–latitude borders and then transform them to LCC coordinates. Similarly, to include grid labels, you convert the locations in longitude–latitude space to LCC space.

```
> grd_ll = gridlines(brd_ll)
> grd_lcc = spTransform(grd_ll, lcc_crs)
> at_ll = gridat(brd_ll)
> at_lcc = spTransform(at_ll, lcc_crs)
```

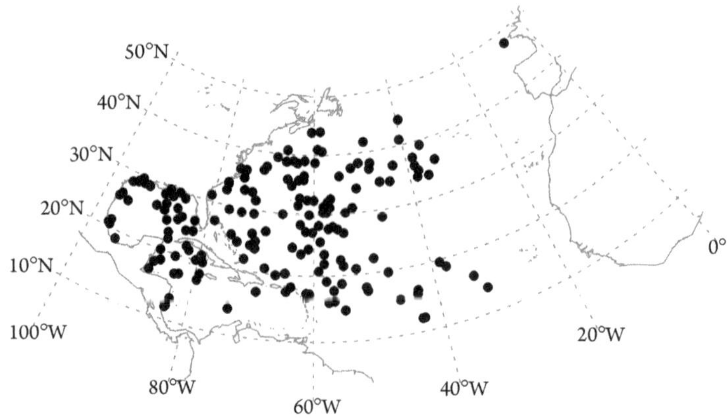

Figure 5.18 Lifetime maximum intensity events on a Lambert conic conformal map.

Finally, to plot the events on a projected map, first plot the grid and then add the country borders and event locations. Use the `text` function to add grid labels and include a box around the plot.

```
> plot(grd_lcc, col="grey60", lty="dotted")
> plot(brd_lcc, col="grey60", add=TRUE)
> plot(LMI_lcc, pch=19, add=TRUE, cex=.7)
> text(coordinates(at_lcc), pos=at_lcc$pos,
+     offset=at_lcc$offset-.3, labels=
+     parse(text=as.character(at_lcc$labels)),
+     cex=.6)
```

The result is shown in Figure 5.18. Conformal maps preserve angles and shapes of small figures, but not size. The size distortion is zero at the two reference latitudes. These features are useful for hurricane tracking maps.

The `spplot` method for points, lines, and polygons has advantages over successive calls to `plot`. Chapter 9 contains examples.

5.5 EXPORT

The **rgdal** package has drivers for reading and writing spatial vector data using the OGR[2] Simple Features Library modeled on the OpenGIS simple features data model supported by the Open Geospatial Consortium, Inc.®. If the data have a CRS, it will be read and written. The availability of OGR drivers depends on your computer. To get a list of the drivers available type `ogrDrivers()`.

[2] Historically, OGR was an abbreviation for "OpenGIS Simple Features Reference Implementation." However, since OGR is not fully compliant with the OpenGIS Simple Feature specification and is not approved as a reference implementation, the name was changed to "OGR Simple Features Library." OGR is the prefix used everywhere in the library source for class names, filenames, and so on.

Here you consider two examples. First export the lifetime maximum intensity events as a Keyhole Markup Language (KML) for overlay using Google Earth™ and then export the events as an ESRI™ shapefile suitable for input into ArcMap®, the main component of ESRI™'s Geographic Information System (GIS).

First create a spatial points data frame from the spatial points object. This is done using the `SpatialPointsDataFrame` function. The first argument is the coordinates of the spatial points object. The underlying CRS for Google Earth™ is geographical (latitude/longitude spatial coordinates) in the WGS84 datum, so you use the `LMI_ll` object defined above and specify the argument `proj4string` as the character string `ll_crs`, also defined earlier.

```
> LMI_sdf = SpatialPointsDataFrame(coordinates(LMI_ll),
+     proj4string=ll_crs, data=as(LMI.df, "data.frame")
+     [c("WmaxS")])
> class(LMI_sdf)
[1] "SpatialPointsDataFrame"
attr(,"package")
[1] "sp"
```

The resulting spatial points data frame (`LMI_sdf`) contains a data slot with a single variable `WmaxS` from the `LMI.df` data frame, which was specified by the `data` argument.

To display the structure of the object, type

```
> str(LMI_sdf, max.level=3)
```

The argument `max.level` specifies the level of nesting (e.g., lists containing sub lists). By default, all nesting levels are shown, and this can produce too much output for spatial objects. Note that there are five slots with names `data`, `coords.nrs`, `coords`, `bbox`, and `proj4string`. The data slot is a regular data frame, here containing a single variable.

The `writeOGR` function takes as input the spatial data frame object and the name of the data layer and outputs a file in your working directory with a name given by the `dsn` argument and in a format given by the `driver` argument.

```
> writeOGR(LMI_sdf, layer="WmaxS", dsn="LMI.kml",
+   driver="KML", overwrite_layer=TRUE)
```

The resulting file can be viewed in Google Earth™ with pushpins for event locations. The pins can be selected revealing the layer values. You will see how to create an overlay image in Chapter 9.

You can also export to a shapefile. First transform your spatial data frame into the Lambert conformal conic used by the NHC.

```
> LMI_sdf2 = spTransform(LMI_sdf, lcc_crs)
> str(LMI_sdf2, max.level=2)
```

Note that the coordinate values are not longitude and latitude and neither are the dimensions of the bounding box (bbox slot).

You export using the driver ESRI Shapefile. The argument dsn is a directory name.

```
> drv = "ESRI Shapefile"
> writeOGR(LMI_sdf2, layer="WmaxS", dsn="WmaxS",
+   driver=drv, overwrite_layer=TRUE)
```

The output contains a set of four files in the WmaxS directory including a .prj file with the fully specified CRS. The data can be imported as a layer to ArcMap®.

5.6 OTHER GRAPHIC PACKAGES

R's traditional (standard) graphics offer a nice set of tools for making statistical plots including box plots, histograms, and scatter plots. The plot are produced using a single function. Yet some plots require a lot of work and even simple changes can be tedious. This is particularly true when you want to make a series of related plots for different partitions of your data. Two alternatives to the standard graphics are worth mentioning.

5.6.1 lattice

The **lattice** package (Sarkar, 2008) contains functions for creating trellis graphs for a variety of plot types. A trellis graph displays a variable or the relationship between variables, conditioned on another variable(s).

In simple usage, lattice functions work like traditional graphics functions. As an example of a lattice graphic function that produces a density plot of the June NAO values, type

```
> require(lattice)
> densityplot(~ Jun, data=NAO)
```

The function's syntax includes the name of the variable and the name of the data frame. The variable is preceded by the tilde symbol. By default, the density plot includes the values as points jittered above the horizontal axis.

The power of trellis graphs comes from being able to easily create a series of plots with the same axes (trellis) as you did with the coplot function in §5.1.6. For instance, in an exploratory analysis, you might want to see if the annual U.S. hurricane count is related to the NAO. You first create a variable that splits the NAO into four groups.

```
> steer = equal.count(NAO$Jun, number=4, overlap=.1)
```

The grouping variable has class shingle and the number of years in each group is the same. The overlap argument indicates the fraction of overlap in the data used to group the years. If you want to leave gaps, you specify a negative fraction. You can type `plot(steer)` to see the range of values for each group.

Next you use the `histogram` function to plot the percentage of hurricanes by count conditional on your grouping variable.

```
> histogram(~ US$All | steer, breaks=seq(0, 8))
```

The vertical line indicates that the conditioning variable follows. The `breaks` argument is used on the hurricane counts. The resulting four-panel graph is arranged from lower left to upper right with increasing values of the grouping variable. Each panel contains a histogram of U.S. hurricane counts drawn using an identical scale for the corresponding range of NAO values. The relative range is shown above each panel in a strip (shingle).

Lattice functions produce an object of class `trellis` that contains a description of the plot. First assign it to the object *dplot* then print it. The print method for objects of this class does the actual drawing of the plot. For example, the following code does the same as shown previously.

```
> dplot = densityplot(~Jun, data=NAO)
> print(dplot)
```

Now you can use the `update` function to modify the plot design. For example, to add an axis label, type

```
> update(dplot, xlab="June NAO (s.d.)")
```

To save the modified plot for additional changes, you need to reassign it.

5.6.2 ggplot2

The **ggplot2** package (Wickham, 2009) contains plotting functions that are more flexible than the traditional R graphics. The `gg` stands for the "Grammar of Graphics," a theory of how to create a graphics system (Wilkinson, 2005). The grammar specifies how a graphic maps data to attributes of geometric objects. The attributes are things like color, shape, and size, and the geometric objects are things like points, lines, bars, and polygons.

The plot is drawn on a specific coordinate system (which can be geographic) and it may contain statistical manipulations of the data. Faceting can be used to replicate the plot using subsets of your data. Here we give a few examples to help you get started.

Returning to your October SOI values. To create a histogram with a bin width of one standard deviation (units of SOI), type

```
> require(ggplot2)
> qplot(Oct, data=SOI, geom="histogram", binwidth=1)
```

The `geom` argument (short for geometric object) represents what you see on the plot, here a histogram. The default geometric objects are points and lines.

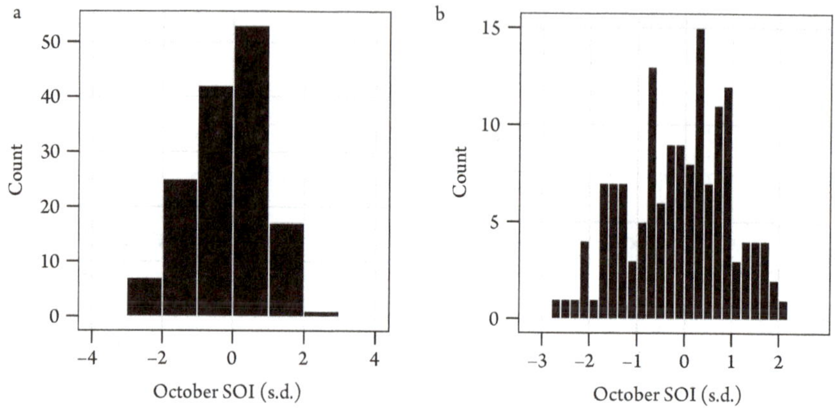

Figure 5.19 Histograms of October SOI.

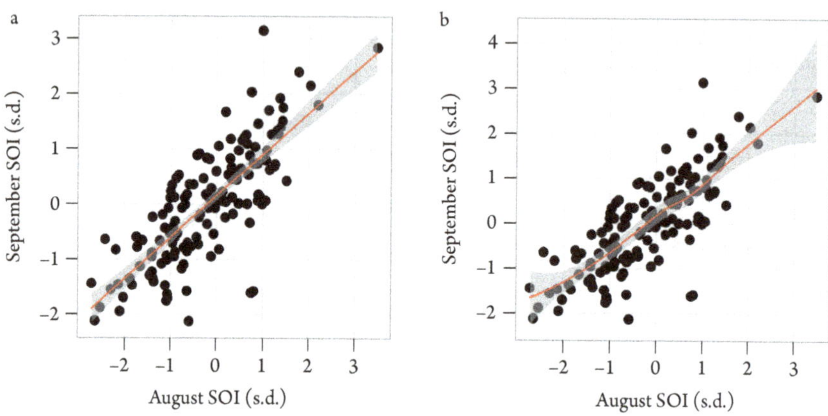

Figure 5.20 Scatter plots of August and September SOI.

Figure 5.19 shows histograms of the October SOI for two different bin widths. Note the use of grids and a background gray shade. This can be changed with the `theme_set` function.

You create a scatter plot using the same `qplot` function and in the same way as `plot`. Here you specify the data with an argument. The default geometric object in this case is the point.

```
> qplot(Aug, Sep, data=SOI)
```

You add a smoothing function (an example of a statistical manipulation of your data) with `smooth` as a character string in the `geom` argument.

```
> qplot(Aug, Sep, data=SOI, geom=c("point", "smooth"))
```

The default method for smoothing is local regression. You can change this to a linear regression by specifying `method="lm"`. Scatter plots with both types of smoothers are shown in Figure 5.20. The graph on the right uses the default local smoothing and the graph on the left uses a linear regression. The geom plots the points and adds a best-fit line through them. The line is drawn by connecting predictions of the best-fit model at a set of equally spaced values of the explanatory variable (here August SOI) over the range of data values. A 95 percent confidence band about the prediction line is included.

Plots are built layer by layer. Layers are regular R objects and so can be stored as variables. This makes it easy for you to write clean code with a minimal amount of duplication. For instance, a set of plots can be enhanced by adding new data as a separate layer. As an example, here is code to produce the left plot in Figure 5.20.

```
> bestfit = geom_smooth(method="lm", color='red')
> pts = qplot(Aug, Sep, data=SOI)
> pts + bestfit
```

The bestfit layer is created and saved as a **geom** object and the pts layer is created from the qplot function. The two layers are added and then rendered to your graphics device in the third line of code.

Finally, consider again the NAO time series object you created in §5.2. You create a vector of times at which the series was sampled using the times function. Here you use the line **geom** instead of the default point.

```
> tm = time(nao.ts)
> qplot(tm, nao.ts, geom="line")
```

Results are shown in Figure 5.21. The values fluctuate widely from one month to the next, but there is no long-term trend. A local regression smoother (geom_smooth) using a span of 10 percent of the data indicates a tendency for a greater number of negative NAO values since the start of the twenty-first century.

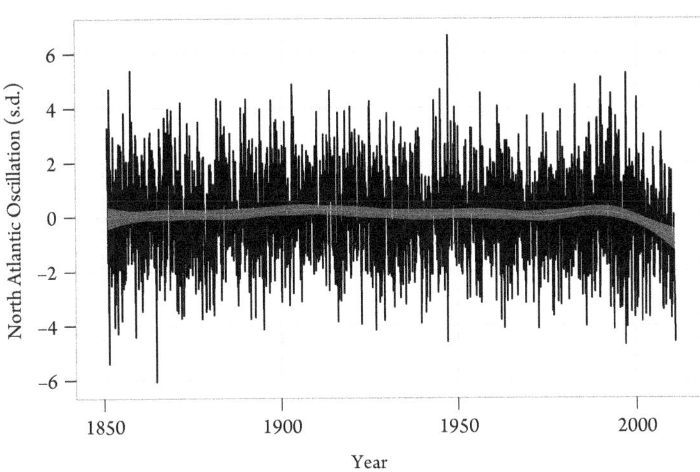

Figure 5.21 Time series of the monthly NAO. The red line is a local smoother.

As with the `plot` function, the first two arguments to `qplot` are the ordinate and abscissa data vectors, but you can use the optional argument `data` to specify column names in a data frame. The `ggplot` function, which allows even greater flexibility, accepts only data frames. Functions in **plyr** and **reshape** packages help you create data frames from other data objects (Teetor, 2011).

5.6.3 ggmap

The **ggmap** package (Kahle and Wickham, 2012) extends the grammar of graphics to maps. The function `get_map` queries the Google Maps OpenStreetMap, or Stamen Map server for a map at a specified location and spatial zoom. For example, to grab a map of Tallahassee, Florida, type

```
> require(ggmap)
> Tally = get_map(location = "Tallahassee", zoom=13)
> str(Tally)
```

The result is an object of class get_map and raster with a matrix (640×640) of character strings specifying the fill color for each raster.

The level of zoom ranges from 0 for the entire world to 19 for the individual city blocks highest. The default zoom is 10. The default map type (`maptype`) is terrain with options for "roadmap", "mobile", "hybrid", among others. To plot the map on your graphics device, type

```
> ggmap(Tally)
```

To determine a center for your map, you use the `geocode` function to get a location. For example, to determine the location of Florida State University, type.

```
> geocode("Florida State University")
```

This chapter showed you how to produce graphs and maps with R. A good graph helps you understand your data and communicate your results. We began by looking at how to make bar charts, histograms, density plots, scatter plots, and graphs involving time. We then looked at utilities for drawing maps and described the various types of spatial data. We showed you how to create coordinate reference systems and transform between them. We also showed you how to export your graphs and maps. We ended by taking a look at two additional graphics systems within R. You will get more practice with these tools as you work through the book.

6

DATA SETS

"Data, data, data, I cannot make bricks without clay."
—Sherlock Holmes

Hurricane data originate from careful analysis of past storms by operational meteorologists. The data include estimates of the hurricane position and intensity at 6-hourly intervals. Information related to landfall time, local wind speeds, damages, and deaths, as well as cyclone size, are included. The data are archived by season.

Some effort is needed to make the data useful for hurricane climate studies. In this chapter, we describe the data sets used throughout this book. We show you a work flow that includes importing, interpolating, smoothing, and adding attributes. We also show you how to create subsets of the data. Code in this chapter is more complicated and it can take longer to run. You can skip this material on first reading and continue with model building in Chapter 7. You can return here when you have an updated version of the data that includes the most recent years.

6.1 BEST-TRACKS DATA

Most statistical models in this book use the best-track data. Here we describe these data and provide original source material. We also explain how to smooth and interpolate them. Interpolations are needed for regional hurricane analyses.

6.1.1 Description

The *best-track* data set contains the 6-hourly center locations and intensities of all known tropical cyclones across the North Atlantic basin, including the Gulf of Mexico and Caribbean Sea. The data set is called HURDAT for HURricane DATa. It is maintained by the U.S. National Oceanic and Atmospheric Administration (NOAA) at the National Hurricane Center (NHC).

Center locations are given in geographic coordinates (in tenths of degrees) and the intensities, representing the one-minute near-surface ($\sim$ 10 m) wind speeds, are given in knots (1 kt $=$ .5144 m s^{-1}) and the minimum central pressures are given in millibars (1 mb $=$ 1 hPa). The data are provided in 6-hourly intervals starting at 00 UTC (Universal Time Coordinate). The version of HURDAT file used here contains cyclones over the period 1851 through 2010 inclusive.[1] Information on the history and origin of these data is found in Jarvinen et al (1984).

The file has a logical structure that makes it easy to read with a FORTRAN program. Each cyclone contains a header record, a series of data records, and a trailer record. Original best-track data for the first cyclone in the file is shown here.

```
00005 06/25/1851 M= 4  1 SNBR=   1 NOT NAMED   XING=1 SSS=1
00010 06/25*280 948  80    0*280 954  80    0*280 960  80    0*281 965  80    0*
00015 06/26*282 970  70    0*283 976  60    0*284 983  60    0*286 989  50    0*
00020 06/27*290 994  50    0*295 998  40    0*3001000  40    0*3051001  40    0*
00025 06/28*3101002  40    0*  0   0   0    0*  0   0   0    0*  0   0   0    0*
00030 HRBTX1
```

The header (beginning with 00005) and trailer (beginning with 00030) records are single rows. The header has eight fields. The first field is the line number in intervals of five and padded with leading zeros. The second is the start day for the cyclone in MM/DD/YYYY format. The third is `M= 4`, indicating four data records to follow before the trailer record. The fourth field is a number indicating the cyclone sequence for the season; here, 1 indicates the first cyclone of 1851. The fifth field, beginning with `SNBR=`, is the cyclone number over all cyclones and all seasons. The sixth field is the cyclone name. Cyclones were named beginning in 1950. The seventh field indicates whether the cyclone hit the United States, with `XING=1` indicating that it did and `XING=0` indicating that it did not. A hit is defined as the center of the cyclone crossed the coast on the continental United States as a tropical storm or hurricane. The final field indicates the Saffir–Simpson hurricane scale (1–5) impact in the United States based on the estimated maximum sustained winds at the coast. The value 0 was used to indicate U.S. tropical storm landfalls but has been deprecated.

The next four rows contain the data records. Each row has the same format. The first field is again the line number. The second field is the cyclone day in MM/DD format. The next 16 fields are divided into four blocks of four fields each. The first block is the 00 UTC record, and the next three blocks are in 6-hour increments (6, 12, and 18 UTC). Each block is the same and begins with a code indicating the stage of the cyclone, tropical cyclone `*`, subtropical cyclone `S`, extratropical low `E`, wave `W`, and remanent low `L`. The three digits immediately to the right of the stage code is the latitude of the center position in tenths of degree north (`280` is 28.0°N) and the next four digits are the longitude in tenths of a degree west (`948` is 94.8°W) followed by a space. The third set of three digits is the maximum sustained (1 minute) surface

[1] From `www.nhc.noaa.gov/pastall.shtml#hurdat`, August 2011.

(10 m) wind speed in knots. These are estimated to the nearest 10 kt for cyclones before to 1886 and to 5 kt afterward. The final four digits after another space is the central surface pressure of the cyclone in millibars if available. If not, the field is given a zero. Central pressures are available for all cyclones after 1978.

The trailer has at least two fields. The first field is the line number. The second field is the maximum intensity of the cyclone as a code using `HR` for hurricane, `TS` for tropical storm, and `SS` for subtropical storm. If there are additional fields, they relate to landfall in the United States. The fields are given in groups of four with the first three indicating location by state and the last indicating the Saffir–Simpson scale based on wind speeds in the state. Two-letter state abbreviations are used, with the exception of Texas and Florida, which are further subdivided as follows: `ATX`, `BTX`, `CTX` for south, central, and north Texas, respectively, and `AFL`, `BFL`, `CFL`, and `DFL` for northwest, southwest, southeast, and northeast Florida, respectively. An `I` is used as a prefix in cases where a cyclone had hurricane-level impacts is in a noncoastal state.

6.1.2 Import

The HURDAT file (e.g., *tracks.txt*) is appended each year with the set of cyclones from the previous season. The latest version is available usually by late spring or early summer from `www.nhc.noaa.gov/pastall.shtml`. Additional modifications to older cyclones are made when newer information becomes available. After downloading the HURDAT file, we use a FORTRAN executable file for the Windows platform (**BT2flat.exe**) to create a flat file (*BTflat.csv*) listing the data records. The file is created by typing

```
BT2flat.exe tracks.txt > BTflat.csv
```

The resulting comma-separated flat file is read into R and the lines between the separate cyclone records removed by typing

```
> best = read.csv("BTflat.csv")
> best = best[!is.na(best[, 1]),]
```

Further adjustment are made to change the hours to ones, the longitude to degrees east, and the column name for the type of cyclone.

```
> best$hr = best$hr/100
> best$lon = -best$lon
> east = best$lon < -180
> best$lon[east] = 360 + best$lon[east]
> names(best)[12] = "Type"
```

The first six lines of the data frame are shown here (`head(best)`).

```
  SYear Sn        name   Yr Mo Da hr  lat   lon Wmax pmin Type
1  1851  1 NOT NAMED  1851  6 25  0 28.0 -94.8   80    0    *
2  1851  1 NOT NAMED  1851  6 25  6 28.0 -95.4   80    0    *
3  1851  1 NOT NAMED  1851  6 25 12 28.0 -96.0   80    0    *
```

```
4  1851  1 NOT NAMED  1851  6 25 18 28.1 -96.5   80    0    *
5  1851  1 NOT NAMED  1851  6 26  0 28.2 -97.0   70    0    *
6  1851  1 NOT NAMED  1851  6 26  6 28.3 -97.6   60    0    *
```

Note the 10-kt precision on the Wmax column. This is reduced to 5 kt from 1886 onward.

Cyclones in the data frame are identified by SYear and Sn. To make it easier to subset by cyclone you add a unique cyclone identifier as follows. First, use the paste function to create a character id string that combines the SYear and Sn columns. Second, table the number of cyclone records with each character id and save these as an integer vector (nrs). Third, create a structured vector indexing the number of cyclones beginning with the first one. Fourth, repeat the index by the number of records in each cyclone and save the result in a Sid vector.

```
> id = paste(best$SYear, format(best$Sn), sep = ":")
> nrs = as.vector(table(id))
> cycn = 1:length(nrs)
> Sid = rep(cycn, nrs[cycn])
```

Next create a column identifying the cyclone hours. This is needed to perform time interpolations. Begin by creating a character vector with strings identifying the year, month, day, and hour. Note that first you need to take care of years when cyclones crossed into a new calendar year. In the best-track file, the year remains the year of the season. The character vector is turned into a POSIXlt object with the strptime function (see Chapter 5) and the time zone argument set to GMT (UTC).

```
> yrs = best$Yr
> mos = best$Mo
> yrs[mos==1] = yrs[mos==1]+1
> dtc = paste(yrs, "-", mos, "-", best$Da, " ",
+   best$hr, ":00:00", sep="")
> dt = strptime(dtc, format="%Y-%m-%d %H:%M:%S",
+   tz="GMT")
```

Each cyclone record begins at 0, 6, 12, or 18 UTC. Retrieve those hours for each cyclone using the cumsum function and the number of cyclone records as an index. Offsets are needed for the first and last cyclones. Then subsample the time vector obtained here at the corresponding values of the index and populate those times for all cyclone records. Then, the cyclone hour is the time difference between the two vectors in units of hours and is saved as Shour.

```
> i0 = c(1, cumsum(nrs[-length(nrs)]) + 1)
> dt0 = dt[i0]
> dt1 = rep(dt0, nrs[cycn])
> Shour = as.vector(difftime(dt, dt1, units="hours"))
```

Finally, include the two new columns in the best data frame.

```
> best$Sid = Sid
> best$Shour = Shour
> dim(best)
[1] 41192    14
```

The best-track data provide information on 1,442 individual tropical cyclones over the period 1851–2010, inclusive. The data frame you created contains these data in 41,192 separate 6-hourly records each having 14 columns. You can output the data as a spreadsheet using the `write.table` function.

If you want to send the file to someone that uses R or load it into another R session, use the `save` function. This exports a binary file that is imported back using the `load` function.

```
> save(best, file="best.RData")
> load("best.RData")
```

Alternatively, you might be interested in the functions available in the **RNetCDF** and **ncdf** packages for exporting data in Network Common Data Form.

6.1.3 Intensification

You add value to these data by computing intensification (and decay) rates. The rate of change is a derivative value. Here you use the Savitzky–Golay smoothing filter (Savitzky and Golay, 1964) specifically designed for calculating derivatives. The filter preserves the maximum and minimum cyclone intensities. Moving averages dampen the extremes and derivatives estimated using finite differencing have larger errors.

The smoothed value of wind speed at a particular location is estimated using a local polynomial regression of degree three on a window of six values (including three locations before and two after). This gives a window width of 1.5 days. The daily intensification rate is the coefficient on the linear term of the regression divided by 0.25, since the data are given in quarter-day increments. A third-degree polynomial captures most of the fluctuation in cyclone intensity without overfitting and ensures consistency with the 5-kt precision of the raw wind speed.

The functions are available in **savgol.R**. Download the file from the book web site and source it. Then use the function `savgol.best` on your `best` data frame saving the results back in `best`.

```
> source("savgol.R")
> best = savgol.best(best)
```

The result is an appended data frame with two new columns, `WmaxS` and `DWmaxDt` giving the filtered estimates of wind speed and intensification, respectively. The filtered speeds have units of knots consistent with the best-track winds and the intensification rates are in knots per hour.

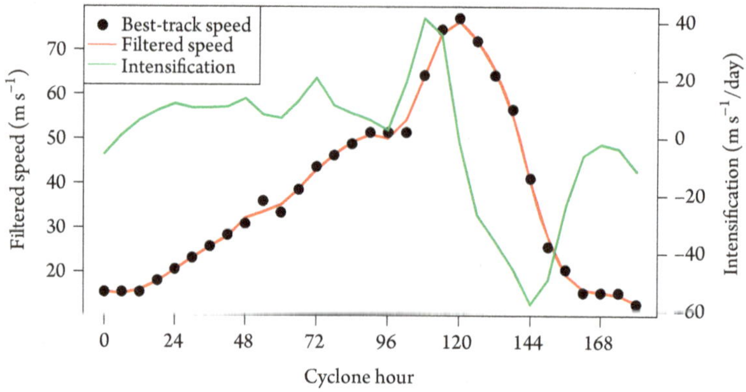

Figure 6.1 Hurricane Katrina data.

As a comparison of the filtered and raw wind speeds, you look at the results from Hurricane Katrina of 2005. To make the code easier to follow, first save the rows corresponding to this cyclone in a separate object.

```
> Kt = subset(best, SYear == 2005 &
+    name == "KATRINA    ")
```

Next, plot the raw wind speeds as points and then overlay the filtered winds as a red line.

```
> plot(Kt$Shour, Kt$Wmax, pch=16, xlab="Cyclone Hour",
+    ylab="Wind Speed (m/s)")
> lines(Kt$Shour, Kt$WmaxS, lwd=2, col="red")
```

The spaces in the name after `KATRINA` are important as the variable name is a fixed-length character vector. On average, the difference between the filtered and raw wind speeds is 0.75 m s^{-1}, which is below the roundoff of 2.5 m s^{-1} (5 kt) used in the best track. Over the entire set of data, the difference is less, averaging 0.49 m s^{-1}. Importantly, for estimating rates of change, the filtered wind speeds capture the maximum cyclone intensity.

Figure 6.1 shows filtered (red) and raw (circles) wind speeds and intensification rates (green) for Hurricane Katrina of 2005. Genesis occurred at 1800 UTC on Tuesday, August 23, 2005. The cyclone lasted for 180 h (7.5 days) dissipating at 600 UTC on Wednesday, August 31, 2005. The maximum best-track wind speed of 77.2 m s^{-1} occurs at cyclone hour 120. This value equals the smoothed wind speed at that same hour.

6.1.4 Interpolation

For each cyclone, the observations are 6 h apart. For spatial analysis and modeling, this can be too coarse as the average forward motion of hurricanes is about 6 m s^{-1}

(12 kt). You therefore fill in the data using interpolation to 1 h. You also add an indicator variable for whether the cyclone is over land.

The interpolation is done with splines. The spline preserves values at the regular 6-hour times and uses a piecewise polynomial to obtain values between these times. For the cyclone positions, the splines use spherical geometry. The functions are available in **interpolate.R**.

```
> source("interpolate.R")
> load("landGrid.RData")
> bi = Sys.time()
> best.interp = interpolate.tracks(best,
+   get.land=TRUE, createindex=TRUE)
> ei = Sys.time()
```

Additionally, a land mask is used to determine whether the cyclone is over land or water. Be patient, as the interpolation takes time to run. To see how long it takes, save the time (`Sys.time()`) before and after the interpolation and then take the difference in seconds.

```
> round(difftime(ei, bi, units="secs"), 1)
Time difference of 108 secs
```

The interpolation output is saved in the object `best.interp` as two lists: the data frame in a list called `data` and the index in a list called `index`. The index list is the record number by cyclone. The data frame has 239,948 rows and 30 columns. Additional columns include information related to the spherical coordinates and whether the position is over land as well as the cyclone's forward velocity (magnitude and direction). For instance, forward speed is in the column labeled `maguv` in units of kt. To obtain the average speed in meter per second over all cyclones, type

```
> mean(best.interp$data$maguv) * .5144
```

Finally, you add a day of year (`jd`) column giving the number of days since January 1 of each year. This is useful for examining intraseasonal activity (see Chapter 10). You use the function `ISOdate` from the **chron** package on the `ISOtime` column in the `best.interp$data` data frame. You first create a POSIXct object for the origin.

```
> x = best.interp$data
> start = ISOdate(x$Yr, 1, 1, 0)
> current = ISOdate(x$Yr, x$Mo, x$Da, x$hr)
> jd = as.numeric(current - start, unit="days")
> best.interp$data$jd = jd
> rm(x)
```

The hourly North Atlantic cyclone data prepared in the above manner are available in the file *best.use.RData*. The data include the best-track 6-hourly values plus the smoothed and interpolated values using the methods described here. The file is created by selecting specific columns from the interpolated data frame above. For example, type

```
> best.use = best.interp$data[, c("Sid", "Sn",
+    "SYear", "name", "Yr", "Mo", "Da", "hr", "lon",
+    "lat", "Wmax", "WmaxS", "DWmaxDt", "Type",
+    "Shour", "maguv", "diruv", "jd", "M")]
> save(best.use, file="best.use.RData")
```

You input these data as a data frame and list the first six lines by typing

```
> load("best.use.RData")
> head(best.use)
```

The `load` function imports an object saved as a compressed file with the `save` function. The object name in your workspace is the file name without the `.RData`.

Once the hurricane data are prepared in the manner described above, you can use functions to extract subsets of the data for particular applications. Here we consider a function to add regional information to the cyclone locations and another function to obtain the lifetime maximum intensity of each cyclone. These data sets are used throughout the book.

6.1.5 Regional Activity

Information about a cyclone's absolute location is available through the geographic coordinates (latitude and longitude). It is convenient to also have relative location information specifying, for instance, whether the cyclone is within a predefined area. Here your interest is near-coastal cyclones, so you consider three U.S. regions including the Gulf coast, Florida, and the East coast. The regions are shown in Figure 6.2. Boundaries are whole number parallels and meridians. The areas are large enough to capture enough cyclones, but not too large as to include many noncoastal strikes.

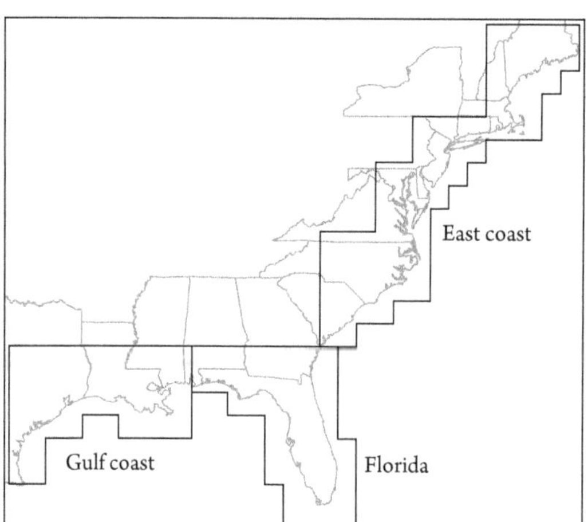

Figure 6.2 Coastal regions.

Relative location is coded as a logical vector indicating whether or not the cyclone is inside the region. The three near-coastal regions are nonoverlapping, and you create one vector for each region. But it is also of interest to know whether the cyclone was in either of these areas or none of them.

The functions are in **datasupport.R** package. They include `import.grid`, which inputs a text file defining the parallels and meridians of the near-coastal regions and adds a regional name with the Gulf coast defined as region one, Florida defined as region two, the East coast defined as region three, and the entire coast as region four.

```
> source("datasupport.R")
> grid = import.grid("gridboxes.txt")
> best.use = add.grid(data=best.use, grid=grid)
```

6.1.6 Lifetime Maximum Intensity

An important variable for understanding hurricane climate is lifetime maximum intensity. Lifetime refers to the time from genesis to dissipation and lifetime maximum refers to the highest wind speed during this lifetime. The intensity value and the location where the lifetime maximum occurs are of general interest.

Here you use the `get.max` function in the **getmax.R** package. To make it accessible to your workspace, type

```
> source("getmax.R")
```

To apply the function on the `best.use` data frame using the default options `idfield="Sid"` and `maxfield="Wmax"`, type

```
> LMI.df = get.max(best.use)
```

List the values in 10 columns of the first 6 rows of the data frame rounding numeric variables to one decimal place.

```
> round(head(LMI.df)[c(1, 5:9, 12, 16)], 1)
      Sid   Yr Mo Da hr   lon WmaxS maguv
3.4     1 1851  6 25 16 -96.3  79.6   4.4
15      2 1851  7  5 12 -97.6  80.0   0.0
17      3 1851  7 10 12 -60.0  50.0   0.0
47.2    4 1851  8 23  2 -86.5  98.9   6.5
76.2    5 1851  9 15  2 -73.5  50.0   0.0
89.2    6 1851 10 17  2 -76.5  59.0   5.7
```

The data frame `LMI.df` contains the same information as `best.use` except there is only one row per cyclone. Each row contains the lifetime maximum intensity and the corresponding location and other attribute information for the cyclone at the time the maximum was *first* achieved. If a cyclone is at its lifetime maximum intensity for more than 1 h, only the first hour information is saved. To subset the data frame of lifetime maximum intensities for cyclones of tropical storm intensity or stronger since 1967 exporting the data frame as a text file, type

```
> LMI.df = subset(LMI.df, Yr >= 1967 & WmaxS >= 34)
> write.table(LMI.df, file="LMI.txt")
```

6.1.7 Regional Maximum Intensity

Here your interest is the cyclone's maximum intensity only when it is within a specified region (e.g., near the coast). You create a set of data frames arranged as a list that includes the cyclone maximum within each of the regions defined in §6.1.5. You start by defining the first and last years of interest and create a structured list of those years, inclusive.

```
> firstYear = 1851
> lastYear = 2010
> sehur = firstYear:lastYear
```

These definitions make it easy for you to add additional of data as they become available or to focus your analysis on data only over the most recent years.

Next define a vector of region names and use the function `get.max.flags` (**datasupport.R**) to generate the set of data frames saved in the list object `max.regions`.

```
> Regions = c("Basin", "Gulf", "Florida", "East", "US")
> max.regions = get.max.flags(se=sehur, field="Wmax",
+    rnames=Regions)
```

You view the structure of the resulting list with the `str` function. Here you specify only the highest level of the list by setting the the argument `max.level` to one.

```
> str(max.regions, max.level=1)
List of 5
 $ Basin  :'data.frame': 1442 obs. of  23 variables:
 $ Gulf   :'data.frame': 246 obs. of  23 variables:
 $ Florida:'data.frame': 330 obs. of  23 variables:
 $ East   :'data.frame': 280 obs. of  23 variables:
 $ US     :'data.frame': 606 obs. of  23 variables:
```

The object contains a list of five data frames with names corresponding to the regions defined earlier. Each data frame has the same 23 columns of data defined in `best.use`, but the number of rows depends on the number of cyclones passing through the region. Note, the `Basin` data frame contains all cyclones.

To list the first six rows and several of the columns in the `Gulf` data frame, type

```
> head(max.regions$Gulf[c(1:7, 11)])
      Sid Sn SYear        name   Yr Mo Da  Wmax
3.4     1  1  1851 NOT NAMED  1851  6 25  80.8
130.4   7  1  1852 NOT NAMED  1852  8 26 100.6
353.4  20  1  1854 NOT NAMED  1854  6 26  71.8
389.4  23  4  1854 NOT NAMED  1854  9 18  90.9
443.3  29  5  1855 NOT NAMED  1855  9 16 111.1
456.3  30  1  1856 NOT NAMED  1856  8 10 132.1
```

You treat `max.regions$Gulf` as a regular data frame although it is part of a list. The output indicates that the sixth Gulf cyclone in the record is the 30th cyclone in the best-track record (`Sid` column) and the 1st cyclone of the 1856 season. It has a maximum intensity of 68 m s^{-1} while in the region. You export the data using the `save` function as before.

```
> save("max.regions", file="max.regions.Rdata")
```

6.1.8 Tracks by Location

Suppose you want to know only about hurricanes that have affected a particular location. Or those that have affected several locations (e.g., San Juan, Miami, and Kingston). Hurricanes specific to a location are extracted with functions in the **getTracks** package. To see how this works, load the *best.use.RData* data and install the source code.

```
> load("best.use.RData")
> source("getTracks.R")
```

The function is `get.tracks`. It takes as input the longitude and latitude of your location along with the search radius (nautical miles) and the number of cyclones and searches for tracks that are within this distance. It computes a score for each track with closer cyclones getting a higher score.

Here you use the function to find the five cyclones of at least tropical storm strength that have come closest to the Norfolk Naval Air Station (NGU) (76.28°W longitude and 36.93°N latitude) during the period 1900 through 2010. You save the location and a search radius of 100 nmi in a data frame. You also set the start and end years of your search and the number of cyclones before calling `get.tracks`.

```
> loc = data.frame(lon=-76.28, lat=36.93, R=100)
> se = c(1900, 2010); Ns = 5
> ngu = get.tracks(x=best.use, locations=loc, N=Ns,
+   se=se)
> names(ngu)
[1] "tracks"    "SidDist"   "N"         "locations"
```

The output contains a list with four objects. The objects `N` and `locations` are the input parameters. The object `SidDist` is the cyclone identifier for each cyclone captured by the input criteria listed from closest to farthest from NGU. The corresponding track attributes are given in the list object `tracks` with each component a data frame containing the individual cyclone attributes from `best.use`. The tracks are listed in order by increasing distance. For example, `ngu$SidDist[1]` is the distance of the closest track and `ngu$tracks[[1]]` is the data frame corresponding to this track.

You plot the tracks on a map reusing the code from Chapter 5. Here you use a gray scale on the track lines corresponding to a closeness ranking with darker lines indicating closer tracks.

```
> map("world", ylim=c(12, 60), xlim=c(-90, -50))
> points(ngu$location[1, 1], ngu$location[1, 2],
+   col="red", pch=19)
> for(i in Ns:1){
+ clr = gray((i - 1)/Ns)
+ Lo = ngu$tracks[[i]]$lon
+ La = ngu$tracks[[i]]$lat
+ n = length(Lo)
+ lines(Lo, La, lwd=2, col=clr)
+ arrows(Lo[n - 1], La[n - 1], Lo[n], La[n], lwd=2,
+    length=.1, col=clr)
+ }
> box()
```

The results are shown in Figure 6.3 for two locations: NGU only and for two locations: NGU and Roosevelt Naval Air Station in Puerto Rico (NRR). Darker tracks indicate closer cyclones. This application is useful for cyclone-relative hurricane climatologies (see Scheitlin et al. [2010]).

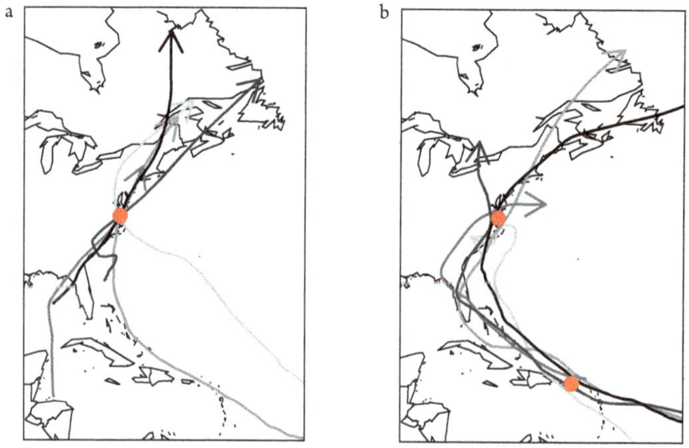

Figure 6.3 Five closest cyclones. (a) NGU and (b) NRR and NGU.

6.1.9 Attributes by Location

Location-specific hurricane attributes are needed for local surge and wind models. To extract these data, first determine the cyclone observations within a grid box centered on your location of interest. This is done using the `inside.lonlat` utility function (**getTracks.R**). Here your location is NGU from above.

```
> ins = inside.lonlat(best.use, lon=loc[1, 1],
+    lat = loc[1, 2], r = 100)
> length(ins)
[1] 239948
```

Your grid box size is determined by twice the value of argument `r` in units of nautical miles. The box is square as the distances are computed on a great-circle. The function returns a logical vector with length equal to the number of cyclone hours in `best.use`.

Next you subset the rows in `best.use` for values of `TRUE` in `ins`.

```
> ngu.use = best.use[ins, ]
```

Since your interest is cyclones of hurricane intensity, further subset using `WmaxS`.

```
> ngu.use = subset(ngu.use, WmaxS >= 64)
> length(unique(ngu.use$Sid))
[1] 54
```

There are 54 hurricanes passing through your grid box over the period of record. A similar subset is obtained using a latitude/longitude grid by typing

```
> d = 1.5
> lni = loc[1, 1]
> lti = loc[1, 2]
> ngu.use = subset(best.use, lat <= lti + d &
+    lat >= lti - d & lon <= lni + d &
+    lni >= lni - d & WmaxS >= 64)
```

Finally use the `get.max` function to select cyclone-specific attributes. For example, to determine the distribution of *minimum* translation speeds for all hurricanes in the grid and to put plot them as a histogram, type

```
> source("getmax.R")
> ngu.use$maguv = -ngu.use$maguv
> ngu.use1 = get.max(ngu.use, maxfield="maguv")
> speed = -ngu.use1$maguv * .5144
> hist(speed, las=1, xlab="Forward Speed (m/s)",
+        main="")
```

The results is shown in Figure 6.4. Notice that you take the additive inverse of the speed since your interest is in the minimum. A parametric distribution is fit or

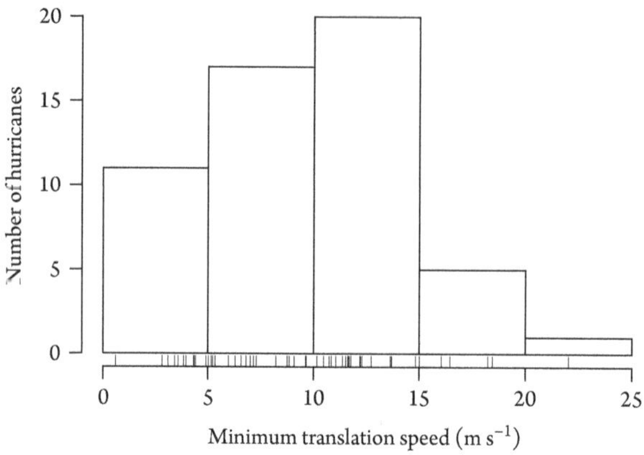

Figure 6.4 Minimum per hurricane forward speed for hurricanes near NGU.

resampling is used to generate inputs for hurricane surge and wind models. (see Chapter 13).

6.2 ANNUAL AGGREGATION

It is also useful to have hurricane data aggregated in time. Aggregation is often done annually since hurricane occurrence has a strong seasonal cycle (see Chapter 10). Annual aggregation makes it convenient to merge hurricane data with monthly climate variables.

6.2.1 Annual Cyclone Counts

Annual counts are the most frequently analyzed hurricane climate data. Here you aggregate counts by year for the entire basin and the near-coastal regions defined in §6.1.5. First, simplify the region names to a single letter using the `substring` function making an exception to the U.S. region by changing it back to `US`.

```
> load("max.regions.RData")
> names(max.regions) = substring(names(max.regions),
+   1, 1)
> names(max.regions)[names(max.regions)=="U"] = "US"
```

This allows you to add the Saffir–Simpson category as a suffix to the names.

The `make.counts.basin` function (**datasupport.R**) performs the annual aggregation of counts by category and region with the `max.regions` list of data frames as the input and a list of years specified with the `se` argument.

```
> source("datasupport.R")
> sehur = 1851:2010
> counts = make.counts.basin(max.regions, se=sehur,
+   single=TRUE)
> str(counts, list.len=5, vec.len=2)
```

```
'data.frame':   160 obs. of  31 variables:
 $ Year: int  1851 1852 1853 1854 1855 ...
 $ B.0 : int  6 5 8 5 5 ...
 $ B.1 : int  3 5 4 3 4 ...
 $ B.2 : int  1 2 3 2 3 ...
 $ B.3 : int  1 1 2 1 1 ...
  [list output truncated]
```

The result is a data frame with columns labeled $X.n$ for $n = 0, 1, \ldots, 5$, where X indicates the region. For example, the annual count of hurricanes at category one or higher affecting Florida is given in the column labeled `F.1`. The start year is 1851 and the end year is 2010.

Here you create a two-by-two matrix of plots showing hurricane counts by year for the basin, and the U.S., Gulf Coast, and Florida regions. The `with` function allows you to use the column names with the `plot` method.

```
> par(mfrow=c(2, 2))
> with(counts, plot(Year, B.1, type="h", xlab="Year",
+    ylab="Basin count"))
> with(counts, plot(Year, US.1, type="h", xlab="Year",
+    ylab="U.S. count"))
> with(counts, plot(Year, G.1, type="h", xlab="Year",
+    ylab="Gulf coast count"))
> with(counts, plot(Year, F.1, type="h", xlab="Year",
+    ylab="Florida count"))
```

The plots are shown in Figure 6.5. Regional hurricane counts indicate no long-term trend, but the basinwide counts show an active period beginning late in the twentieth century. Some of this variation is related to fluctuations in climate as examined in Chapter 7. Next the annually and regionally aggregated counts are merged with monthly and seasonal climate variables.

6.2.2 Environmental variables

The choice of climate variables is large. You narrow it down by considering what is known about hurricane climate. For example, it is well understood that ocean heat provides the fuel, a calm atmosphere provides a favorable setting, and the location and strength of the subtropical ridge provide the steering currents. Thus statistical models of hurricane counts should include covariates that index these climate variables including sea-surface temperature (SST), as an indicator of oceanic heat content, El Niño-Southern Oscillation (ENSO) as an indicator of vertical wind shear, and the North Atlantic Oscillation (NAO) as an indicator of steering flow. Variations in solar activity might also influence hurricane activity. We speculate that an increase in solar ultraviolet (UV) radiation during periods of strong solar activity might suppress tropical cyclone intensity as the temperature near the tropopause will warm

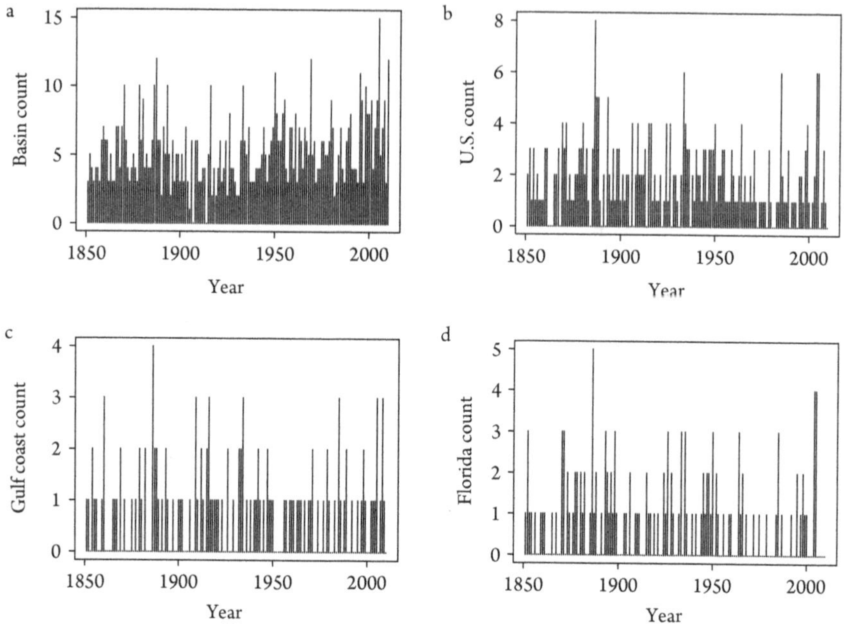

Figure 6.5 Hurricane counts. (a) Basin, (b) U.S., (c) Gulf Coast, and (d) Florida.

through absorption of radiation by ozone and modulated by dynamic effects in the stratosphere (Elsner and Jagger, 2008).

Thus you choose four climate variables including North Atlantic Ocean SST, the Southern Oscillation Index (SOI) as an indicator of ENSO, an index for the NAO, and sunspot numbers (SSN). Monthly values for these variables are obtained from the following sources.

- **SST**: The SST variable is an area-weighted average (°C) using values in 5° latitude–longitude grid boxes from the equator north to 70°N latitude and spanning the North Atlantic Ocean (Enfield et al. 2001).[2] Values in the grid boxes are from a global SST data set derived from the UK Met Office (Kaplan et al. 1998).
- **SOI**: The SOI is the contemporaneous difference in monthly sea-level pressures between Tahiti (T) in the South Pacific Ocean and Darwin (D) in Australia ($T - D$) (Trenberth, 1984).[3] The SOI is inversely correlated with equatorial eastern and central Pacific SST, so an El Niño warm event is associated with negative values of the SOI.
- **NAO**: The NAO is the fluctuation in contemporaneous sea-level pressure differences between the Azores and Iceland. An index value for the NAO is calculated as the difference in monthly normalized pressures at Gibraltar and over Iceland

[2] From `www.esrl.noaa.gov/psd/data/correlation/amon.us.long.data`, November 2011.

[3] From `www.cgd.ucar.edu/cas/catalog/climind/SOI.signal.annstd.ascii`, November 2011.

(Jones et al, 1997).[4] The NAO index indicates the strength and the position of the subtropical Azores/Bermuda High.

- **SSN**: The SSN variable are the Wolf sunspot numbers measuring the number of sunspots present on the surface of the sun. They are produced by the Solar Influences Data Analysis Center (SIDC) of World Data Center for the Sunspot Index at the Royal Observatory of Belgium and available from NOAA's National Geophysical Data Center.[5]

You combine the climate and solar variables by month (May through October) and season with the aggregate hurricane counts by year. You use the `useCov` (**datasupport.R**) function to input the data. The file must have a column indicating the year and 12 or 13 additional columns indicating the months and perhaps an annual average. The argument `miss` inputs the missing value code that is used in the file. The argument `ma` is for centering and scaling the values. The default is none; `"c"` centers, `"cs"` centers and scales, and `"l"` subtracts the values in the last row from values in each column. To accommodate using previous year's data for modeling current year's cyclone counts, the resulting data frame is augmented with columns corresponding to a 1-year shift of all months using the argument `last=TRUE`. Column names for the previous year's months are appended with a `.last`.

You input and organize all climate variables at once with the `readClimate` function. Copy and paste the code to your R session.

```
> readClimate = function(){
+   sst = readCov("data/amon.us.long.mean.txt",
+     header=FALSE, last=TRUE, miss=-99.990, ma="l",
+     extrayear=TRUE)
+   soi = readCov("data/soi_ncar.txt", last=TRUE,
+     miss=-99.9, extrayear=TRUE)
+   nao = readCov("data/nao_jones.txt", last=TRUE,
+     miss=c(-99.99, 9999), extrayear=TRUE)
+   ssn = readCov("data/sunspots.txt", header=TRUE,
+     last=TRUE, extrayear=TRUE)
+ return(list(sst=sst, soi=soi, nao=nao, ssn=ssn))
+ }
```

The list of data frames, one for each climate variable, is created by typing

```
> climate = readClimate()
> str(climate, max.level=1)
List of 4
 $ sst:'data.frame':    157 obs. of  25 variables:
 $ soi:'data.frame':    147 obs. of  25 variables:
```

[4] From `www.cru.uea.ac.uk/~timo/datapages/naoi.htm`, November 2011.
[5] From `ftp.ngdc.noaa.gov/STP/SOLAR_DATA/SUNSPOT_NUMBERS/INTERNATIONAL/monthly/MONTHLY`, November 2011.

```
 $ nao:'data.frame':     192 obs. of  25 variables:
 $ ssn:'data.frame':     162 obs. of  25 variables:
```

Each data frame has 25 columns (variables) corresponding to two sets of monthly values (current and previous year) plus a column of years. The number of rows (observations) in the data frames varies with the NAO being the longest, starting with the year 1821, although not all months in the earliest years have values. To list the first six rows and several of the columns in the nao data frame, type

```
> head(climate$nao[c(1:2, 21:23)])
    Yr Jan.last   Aug   Sep   Oct
1 1821       NA -0.14    NA    NA
2 1822       NA -0.19 -1.09 -2.00
3 1823       NA  2.90  0.67 -1.39
4 1824    -3.39 -0.08  0.19    NA
5 1825    -0.16  1.43 -0.95  1.98
6 1826    -0.23  2.72 -0.76  0.18
```

Note how climate$nao is treated as a data frame although it is part of a list.

The final step is to merge the climate data with the cyclone counts organized in §6.2.1. This is done by creating a single data frame of your climate variables. First, create a list of month names by climate variable. Here you consider only the months from May through October. You use August through October as a group for the SOI and SST variables, May and June as a group for the NAO variable, and September for the SSN variable.

```
> months = list(
+   soi=c("May", "Jun", "Jul", "Aug", "Sep", "Oct"),
+   sst=c("May", "Jun", "Jul", "Aug", "Sep", "Oct"),
+   ssn=c("May", "Jun", "Jul", "Aug", "Sep", "Oct"),
+   nao=c("May", "Jun", "Jul", "Aug", "Sep", "Oct"))
> monthsglm = list(
+   soi=c("Aug", "Sep", "Oct"),
+   sst=c("Aug", "Sep", "Oct"),
+   ssn="Sep",
+   nao=c("May","Jun"))
```

Next, use the make.cov (**datasupport.R**) function on the climate data frame, specifying the month list and the start and end years. Here you use the word "covariate" in the statistical sense to indicate a variable this is predictive of cyclone activity. In statistics, a covariate is also called an explanatory variable, an independent variable, or a predictor.

```
> covariates = cbind(make.cov(data=climate,
+   month=months, separate=TRUE, se=sehur),
+   make.cov(data=climate, month=monthsglm,
+   separate=FALSE, se=sehur)[-1])
```

The cbind function brings together the columns into a single data frame. The last six rows and a sequence of columns from the data frame are listed by typing,

```
> tail(covariates[seq(from=1, to=29, by=5)])
    Year soi.Sep sst.Aug ssn.Jul nao.Jun    soi
155 2005     1.4   0.622    40.1   -1.00  0.800
156 2006    -1.9   0.594    12.2   -0.41 -3.867
157 2007     0.4   0.245     9.7   -3.34  0.833
158 2008     4.6   0.361     0.8   -2.05  3.767
159 2009     1.0   0.345     3.2   -3.05 -2.033
160 2010     8.0   0.725    16.1   -2.40  6.200
```

The columns are labeled *X.m*, where *X* indicates the covariate (soi, sst, sun, and nao) and *m* indicates the month using a three-letter abbreviation with the first letter capitalized. Thus, for example, June values of the NAO index are in the column labeled nao.Jun. The hurricane-season-averaged covariate is also given in a column labeled without the month suffix. Season averages use August through October for SST and SOI, May and June for NAO, and September only for SSN.

As you did with the counts, here you create a two-by-two plot matrix showing the seasonal-averaged climate and solar variables by year (Fig. 6.6).

```
> par(mfrow=c(2, 2))
> with(covariates, plot(Year, sst, type="l",
+   xlab="Year", ylab="SST [C]"))
> with(covariates, plot(Year, nao, type="l",
+   xlab="Year", ylab="NAO [s.d.]"))
> with(covariates, plot(Year, soi, type="l",
+   xlab="Year", ylab="SOI [s.d.]"))
> with(covariates, plot(Year, ssn, type="l",
+   xlab="Year", ylab="Sunspot Count"))
```

The long-term warming trend in SST is quite pronounced as is the cooling trend during the 1960s and 1970s. The NAO values show large year-to-year variations and a tendency for negative values during the early part of the twenty-first century. The SOI values also show large interannual variations. Sunspot numbers show a pronounced periodicity near 11 years (solar cycle) related to changes in solar dynamics.

Finally, you use the merge function to combine the counts and covariates data frames, merging on the variable Year which appears in both.

```
> annual = merge(counts, covariates, by="Year")
> save(annual, file="annual.RData")
```

The result is a single data frame with 160 rows and 59 columns. The rows correspond to separate years and the columns include the cyclone counts by Saffir–Simpson scale and the monthly covariates defined here. The data frame is exported to the file *annual.RData*.

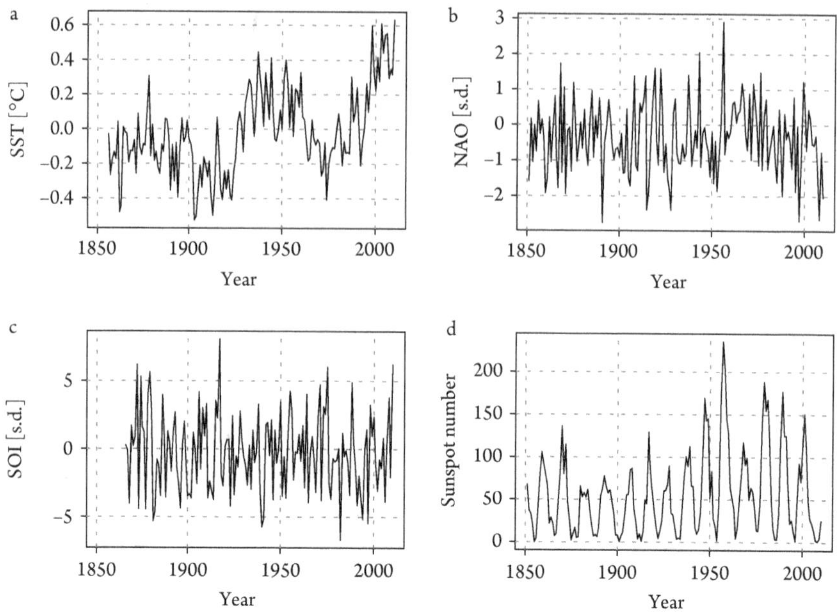

Figure 6.6 Climate variables. (a) SST, (b) NAO, (c) SOI, and (d) sunspots.

6.3 COASTAL COUNTY WINDS

6.3.1 Description

Other hurricane climate data sets are available. County wind data are compiled in Jarrell et al. (1992) from reports on hurricane experience levels for coastal counties from Texas to Maine. The data are coded by Saffir–Simpson category and are available as an Excel™spreadsheet.[6]

The file consists of one row for each year and one column for each county. A cell contains a number if a tropical cyclone affected the county in a given year, otherwise, it is left blank. The number is the Saffir–Simpson intensity scale. For example, a county with a value of 2 indicates that category-two scale wind speeds were likely experienced at least somewhere in the county. If the number is inside parentheses, then the county received an indirect hit and the highest winds were likely at least one category weaker. Cells having multiple entries, separated by commas, indicate that the county was affected by more than one hurricane during that year.

The data set is originally constructed as follows. First, a Saffir–Simpson category is assigned to the hurricane at landfall based on central pressure and wind intensity estimates in the best-track data set. Some subjectivity enters the assignment particularly with hurricanes during earlier years of the twentieth century and with storms moving inland over a sparsely populated area. Thus there is some ambiguity about the category for hurricanes with intensities near the category cutoff. The category is

[6] From `www.aoml.noaa.gov/hrd/hurdat/Data_Storm.html`, November 2011.

Table 6.1 Data symbols and interpretation. The symbol is from Appendix C of Jarrell et al. (1992).

Symbol	*Saffir–Simpson Range*	*Wind Speed Range ($m\,s^{-1}$)*
(1)	[0,1)	33–42
1	[1,2)	33–42
(2)	[1,2)	33–42
2	[2,3)	42–50
(3)	[1,3)	33–50
3	[3,4)	50–58
(4)	[1,4)	33–58
4	[4,5)	58–69
(5)	[1,5)	33–69
5	[5,∞)	69–1000

sometimes adjusted based on storm surge estimates, in which case the central pressure may not agree with the scale assignment. Beginning with the 1996 hurricane season, scale assignments are based solely on maximum winds.

Second, a determination is made about which coastal counties received direct and indirect hits. A direct hit is defined as the innermost core regions, or "eye," moving over the county. Each hurricane is judged individually based on the available data, but a general rule of thumb is applied in cases of greater uncertainty. That is, a county is regarded as receiving a direct hit when all or part of a county falls within a distance d to the left of a storm's landfall and a distance 2d to the right (with respect to an observer at sea looking toward shore), where d is the radius to maximum winds defined as the distance from the cyclone's center to the circumference of maximum winds around the center.

The determination of an indirect hit is based on a hurricane's strength and size and on the configuration of the coastline. In general, it is determined that the counties on either side of the direct-hit zone that received hurricane force winds or tides of at least 1–2 m above normal are considered an indirect hit. Subjectivity is also necessary here because of coastline geography and uncertainty about the hurricane's exact path.

Table 6.1 lists the possible cell entries for a given hurricane and our interpretations of the symbol in terms of the Saffir–Simpson category and wind speed range. The first column is the symbol used in Appendix C of Jarrell et al. (1992). The second column is the corresponding Saffir–Simpson scale range likely experienced somewhere in the county. The third column is the interpreted maximum sustained (1 min) near-surface (10 m) wind speed range ($m\,s^{-1}$).

The data are incomplete in the sense that you have a range of wind speeds rather than a single estimate. In statistics, the data are called "interval censored." Note that (1) is the same as 1, because they both indicate a cyclone with at least hurricane-force winds.

6.3.2 Counts and Magnitudes

The raw data need to be organized. First, remove characters that code for information not used. This includes codes such as 'W','E','*' and '_'. Second, convert all combinations of multiple-hit years, say (1, 2), to (1), (2) for parsing. Once parsed, all table cells consist of character strings that are either blank or contain cyclone hit information separated by commas. Finally, input the cleaned data to R with the first column used for row names by typing

```
> cd = read.csv("HS.csv", row.names=1, header=TRUE)
```

Remove the state row and save as a separate vector by typing

```
> states = cd[1,]
> cdf = cd[-1,]
> cdf[c(1, 11), 1:4]
     CAMERON WILLACY KENEDY KLEBERG
1900
1910     (2)     (2)      2       2
```

Rows 1 and 11 are printed so that you can see the data frame structure. In 1900, these four southern Texas counties were not affected by a hurricane (blank row), but in 1910, Kenedy and Kleberg counties had a direct hit by a category-two hurricane that was also felt indirectly in Cameron and Willacy counties.

Next, you convert this data frame into a matrix object containing event counts and a list object for event magnitudes. First set up a definition table to convert the category data to wind speeds. Note the order is (1), ..., (5), 1, ..., 5. The column names `time` and `time2` are required for use with `Surv` function to create a censored data type.

```
> wt = data.frame(
+   time = c(rep(33, 6), 42, 50, 58, 69),
+   time2 = c(42, 42, 50, 58, 69, 42, 50, 58, 69,
+   1000))
> rownames(wt) = c(paste("(", 1:5, ")", sep=""),
+   paste(1:5))
```

Next, expand the data frame into a matrix. Each entry of the matrix is a character string vector. The character string is a zero vector for counties without a hurricane for a given year. For counties with a hurricane, the string contains symbols as shown in Table 6.1, one for each hurricane. This is done using `apply` and the `strsplit` function as follows:

```
> pd = apply(cdf, c(1, 2), function(x)
+   unlist(strsplit(gsub(" ", "", x), ",")))
```

Next, extract a matrix of counts and generate a list of events one for each county along with a list of years required for matching with the covariate data. Note that the year is extracted from the names of the elements.

```
> counts = apply(pd, c(1, 2), function(x)
+   length(x[[1]]))
> events = lapply(apply(pd, 2, unlist),
+   function(x)
+   data.frame(Year=as.numeric(substr(names(x), 1, 4)),
+   Events=x, stringsAsFactors=FALSE))
```

Finally, convert events to wind speed categories. You do this using the `Surv` function from the **survival** package (Therneau, 2012) as follows:

```
> require(survival)
> winds = lapply(events, function(x)
+   data.frame(Year=x$Year,
+   W = do.call("Surv", c(wt[x$Events, ],
+   list(type="interval2")))))
```

The object `winds` is a list of the counties with each list containing a data frame. To extract the data frame from county 57 corresponding to Miami-Dade, Florida, type

```
> miami = winds[[57]]
> class(miami)
[1] "data.frame"
> head(miami)
  Year        W
1 1904 [33, 42]
2 1906 [33, 42]
3 1906 [50, 58]
4 1909 [50, 58]
5 1926 [58, 69]
6 1926 [33, 50]
```

The data frame contains a numerical year variable and a categorical survival variable. The survival variable has three components indicating the minimum and maximum Saffir–Simpson category.

You will use the `winds` and `counts` objects in Chapter 8 to create a probability model for winds exceeding threshold intensity levels. Here you export the objects as separate files using the `save` function so you can read them back using the `load` function.

```
> save(winds, file="catwinds.RData")
> save(counts, file="catcounts.RData")
```

The saved files are binary (8-bit characters) to ensure that they transfer without converting end-of-line markers.

6.4 NETCDF FILES

Climate data, such as monthly SST grids, are organized as arrays and stored in netCDF files. NetCDF (Network Common Data Form) is a set of software libraries and data formats from the Unidata community that support the creation, access, and sharing of data arrays. The National Center for Atmospheric Research (NCAR) uses netCDF files to store large data sets. The **ncdf** package (Pierce, 2011) provides functions for working with netCDF files in R. Install the package by typing

```
> require(ncdf)
```

You also might want to check out the functions available in **RNetCDF** for processing netCDF files.

Here your interest is with NOAA's extended reconstructed SST version 3b data set for the North Atlantic Ocean.[7] The data are provided by the NOAA/OAR/ESRL PSD in Boulder, Colorado. The data are available in file *sstna.nc* for the domain bounded by the equator and 70°N latitude and 100°W and 10°E longitude for the set of months starting with January 1854 through November 2009.

First, use the function `open.ncdf` to input the SST data.

```
> nc = open.ncdf("sstna.nc")
```

Next, convert the `nc` object of class ncdf into a three-dimension array and print the array's dimensions.

```
> sstvar = nc$var$sst
> ncdata = get.var.ncdf(nc, sstvar)
> dim(ncdata)
[1]   56   36 1871
> object.size(ncdata)
30175696 bytes
```

The file contains 3,771,936 monthly SST values distributed across 56 longitudes, 36 latitudes, and 1,871 months.

Additional work is needed before analysis can begin. First, extract the array dimensions as vector coordinates of longitudes, latitudes, and time. Then change the longitudes to negative west of the prime meridian and reverse the latitudes to increase from south to north. Also convert the time coordinate to a POSIX time (see Chapter 5) using January 1, 1800 (the year). as the origin.

```
> vals = lapply(nc$var$sst$dim, function(x)
+   as.vector(x$vals))
> vals[[1]] = (vals[[1]] + 180) %% 360 - 180
> vals[[2]] = rev(vals[[2]])
> timedate = as.POSIXlt(86400 * vals[[3]],
```

[7] From www.esrl.noaa.gov/psd/data/gridded/data.noaa.ersst.html, November 2011.

```
+    origin=ISOdatetime(1800, 1, 1, 0, 0, 0, tz="GMT"),
+    tz="GMT")
> timecolumn = paste("Y", 1900 + timedate$year, "M",
+    formatC(as.integer(timedate$mo + 1), 1, flag="0"),
+    sep="")
> names(vals) = sapply(nc$var$sst$dim, "[", "name")
> vals = vals[1:2]
```

Note that the double percent symbol is the modulo operator, which finds the remainder of a division of the number to the left of the symbol by the number to the right.

Next, coerce the array into a data frame with one column per time period and assign column names.

```
> ncdata1 = ncdata[, (dim(ncdata)[2]:1), ]
> dims = dim(ncdata1)
> dim(ncdata1) = c(dims[1] * dims[2], dims[3])
> colnames(ncdata1) = timecolumn
> ncdata1 = as.data.frame(ncdata1)
> ncdataframe = cbind(expand.grid(vals), ncdata1)
```

Then find missing values at nonland locations and save the resulting data frame.

```
> misbyrow = apply(ncdataframe, 1, function(x)
+    sum(is.na(x)))
> ncdataframe = ncdataframe[misbyrow==0, ]
> save("ncdataframe", file="ncdataframe.RData")
```

Finally, create a subset data frame with only July 2005 SST values on your latitude–longitude grid by typing

```
> sst = ncdataframe[paste("Y2005", "M",
+    formatC(7, 1, flag="0"), sep="")]
> names(sst) = "SST"
> sst$lon = ncdataframe$lon
> sst$lat = ncdataframe$lat
> write.table(sst, file="sstJuly2005.txt")
```

These data are used in Chapters 7 and 9.

This chapter showed how to extract hurricane data sets from raw data files. We began by showing how to create a spreadsheet-friendly flat file from the available best-tracks. We showed how to add value to these data by smoothing, interpolating, and computing derivative variables. We also showed how to parse the data regionally and locally and to create a subset based on lifetime maximum intensity. Next, we demonstrated how to aggregate the data annually and insert relevant environmental variables. We then examined a coastal county wind data set and showed how to work with NetCDF files.

Part II of this book focuses on using these data to analyze and model hurricane activity. We begin, in Chapter 7 with models for hurricane frequency.

PART TWO

Models and Methods

7

FREQUENCY MODELS

"Statistics is the grammar of science."
—Karl Pearson

Here in Part II, we focus on statistical models for understanding and predicting hurricane climate. This chapter shows you how to model hurricane occurrence. This is done using the annual count of hurricanes making landfall in the United States. We also consider the occurrence of hurricanes across the basin and by origin.

We begin with exploratory analysis and then show you how to model counts with Poisson regression. Issues of model fit, interpretation, and prediction are considered in turn. The topic of how to assess forecast skill is examined including how to perform cross-validation. Alternatives to the Poisson regression model are considered. Logistic regression and receiver operating characteristics (ROCS) are also covered.

7.1 COUNTS

You use the data set *US.txt* which contains a list of tropical cyclone counts by year (see Chapter 2). The counts indicate the number of hurricanes hitting in the United States (excluding Hawaii). Input the data, save them as a data frame object, and print out the first six lines by typing

```
> H = read.table("US.txt", header=TRUE)
> head(H)
  Year All MUS G FL E
1 1851   1   1 0  1 0
2 1852   3   1 1  2 0
3 1853   0   0 0  0 0
4 1854   2   1 1  0 1
```

```
5 1855   1   1 1   0 0
6 1856   2   1 1   1 0
```

The columns include year `Year`, number of U.S. hurricanes `All`, number of major U.S. hurricanes `MUS`, number of U.S. Gulf coast hurricanes `G`, number of Florida hurricanes `FL`, and number of East coast hurricanes `E`. Save the total number of years in the record as `n` and the average number hurricanes per year as `rate`.

```
> n = length(H$Year); rate = mean(H$All)
> n; rate
[1] 160
[1] 1.69
```

The average number of U.S. hurricanes is 1.69 per year over these 160 years.

First plot a time series and a distribution of the annual counts. Together, the two plots provide a nice summary of the information in your data relevant to any modeling effort.

```
> par(las=1)
> layout(matrix(c(1, 2), 1, 2, byrow=TRUE),
+    widths=c(3/5, 2/5))
> plot(H$Year, H$All, type="h", xlab="Year",
+    ylab="Hurricane Count")
> grid()
> mtext("a", side=3, line=1, adj=0, cex=1.1)
> barplot(table(H$All), xlab="Hurricane Count",
+    ylab="Number of Years", main="")
> mtext("b", side=3, line=1, adj=0, cex=1.1)
```

The `layout` function divides the plot page into rows and columns as specified in the matrix function (first argument). The column widths are specified using the `width` argument. The plot symbol is a vertical bar (`type="h"`). The tic labels on the vertical axis are presented in whole numbers consistent with count data.

Figure 7.1 shows the time series and distribution of annual hurricanes over the 160-year period. There is a total of 271 hurricanes. The year-to-year variability and the distribution of counts appear to be consistent with a random process. There are 34 years without a hurricane and one year (1886) with seven hurricanes. The number of years with a particular hurricane count provides a histogram.

7.1.1 Poisson Process

The shape of the histogram suggests that a Poisson distribution is a good description for these data. The density function of the Poisson distribution shows that the

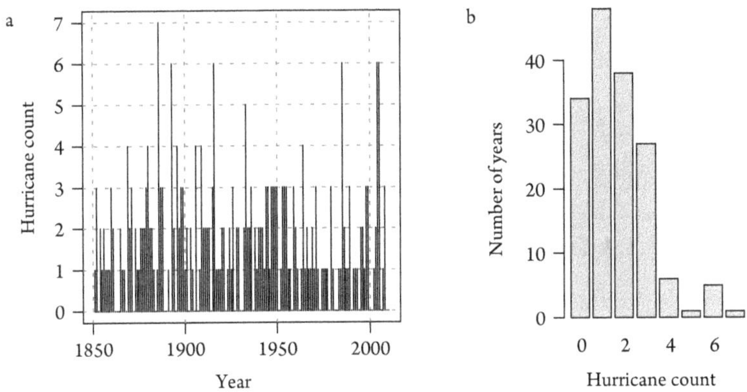

Figure 7.1 Annual hurricane occurrence. (a) Time series and (b) distribution.

probability p of obtaining a count x when the mean count (rate) is λ is given by

$$p(x) = \frac{e^{-\lambda}\lambda^{x}}{x!} \tag{7.1}$$

where e is the exponential function and ! is the factorial symbol. The equation indicates that probability of no events is $p(0) = e^{-\lambda}$.

With λ =1.69 hurricanes per year, the probability of no hurricanes in a random year is

```
> exp(-rate)
[1] 0.184
```

This implies that the probability of at least one hurricane is 0.82 or 82 percent.

Using the `dpois` function, you can determine the probability for any number of hurricanes. For example, to determine the probability of observing exactly one hurricane when the rate is 1.69 hurricanes per year, type

```
> dpois(x=1, lambda=rate)
[1] 0.311
```

Or the probability of five hurricanes expressed as a percentage is

```
> dpois(5, rate) * 100
[1] 2.14
```

Recall that you can leave off the argument names in the function if the argument values are placed in the order. The argument order can be found by placing a question mark in front of the function name and leaving off the parentheses. This brings up the function's help page (see Chapter 2).

To answer the question, *What is the probability of two or fewer hurricanes?* you use the cumulative probability function `ppois` as follows:

```
> ppois(q=2, lambda=rate)
[1] 0.759
```

To answer the question, *What is the probability of more than two hurricanes?* you add the argument `lower.tail=FALSE`.

```
> ppois(q=2, lambda=rate, lower.tail=FALSE)
[1] 0.241
```

7.1.2 Inhomogeneous Poisson Process

The Poisson distribution has the property that the variance is equal to the mean. Thus data that can be described with a Poisson distribution has a variance to mean ratio close to one. You compute this ratio with your data by typing

```
> round(var(H$All)/rate, 2)
[1] 1.24
```

This says that the variance of hurricane counts is 24 percent larger than the mean. Is this unusual for a Poisson distribution?

You check by performing a Monte Carlo (MC) simulation experiment. A MC simulation relies on repeated random sampling from a distribution. Compare it to a bootstrap resampling procedure, which relies on repeated random sampling from a set of data (see Section 3.9.3). A single random sample of size n from a Poisson distribution with a rate equal to 1.5 is obtained by typing

```
> rpois(n=5, lambda=1.5)
[1] 0 1 0 2 1
```

Here you repeat this $m = 1{,}000$ times and let n be the number of years in your hurricane record and λ be the rate. For each sample, you compute the ratio of the variance to the mean.

```
> set.seed(3042)
> ratio = numeric()
> m = 1000
> for (i in 1:m){
+ h = rpois(n=n, lambda=rate)
+ ratio[i] = var(h)/mean(h)
+ }
```

The vector `ratio` contains 1,000 values of the ratio. To help answer the *Is this unusual?* question, you determine the proportion of ratios greater than 1.24:

```
> sum(ratio > var(H$All)/rate)/m
[1] 0.028
```

Only 2.8 percent of the ratios are larger, so the answer from your MC experiment is "yes," and variability in hurricane counts is higher than you would expect from a Poisson distribution with a constant rate.

This might indicate that the rate varies with time. Although you can compute a long-term average, some years have a higher expected rate than others. The variation in the rate is due to things such as El Niño. So you expect more variance (extra dispersion) in counts relative to a constant rate (homogeneous Poisson) distribution. This is the rationale behind seasonal forecasts. The variation in the annual rate is not obvious from looking at the variation in counts. Even with a constant rate, the counts will vary.

You modify your MC simulation using the gamma distribution for the rate and then examine the ratio of variance to the mean from a set of Poisson counts with the variable rate. The gamma distribution describes the variability in the rate using the shape and scale parameters. The mean of the gamma distribution is the shape times the scale. You specify the shape to be 5.6 and the scale to be 0.3 so that the product matches closely the long-term average count. You could choose other values that produce the same average.

Now your simulation first generates 1,000 random gamma values, and then for each gamma, 160 years of hurricane counts are generated.

```
> ratio = numeric(); set.seed(3042); m = 1000
> for (i in 1:m){
+    h = rpois(n=n, lambda=rgamma(m, shape=5.6,
+    scale=.3))
+    ratio[i] = var(h)/mean(h)
+ }
> sum(ratio > var(H$All)/rate)/m
[1] 0.616
```

In this case, we find that 61.6 percent of the ratios are larger than the observed rate, so we conclude that the observed hurricane counts are more consistent with a variable-rate (inhomogeneous) Poisson model.

These examples demonstrate an important use of statistics: to simulate data that have the same characteristics as your observations. Figure 7.2 shows a plot of the observed hurricane record over the 160-year period together with plots from three simulated records of the same length and having the same overdispersed Poisson variation as the observed record. As shown earlier, such simulated records provide a way to test hypotheses about natural variability in hurricane climate. Summary characteristics of a 100 years of hurricanes at a coastal location may be of little value, but running a sediment transport model at that location with a much larger number of simulated hurricane counts will provide an assessment of the uncertainty in sediment movement resulting from variation in hurricane frequency.

7.2 ENVIRONMENTAL VARIABLES

The parameter of interest is the annual hurricane rate. Given the rate, you have a probability distribution for any possible hurricane count. You noted that the observed

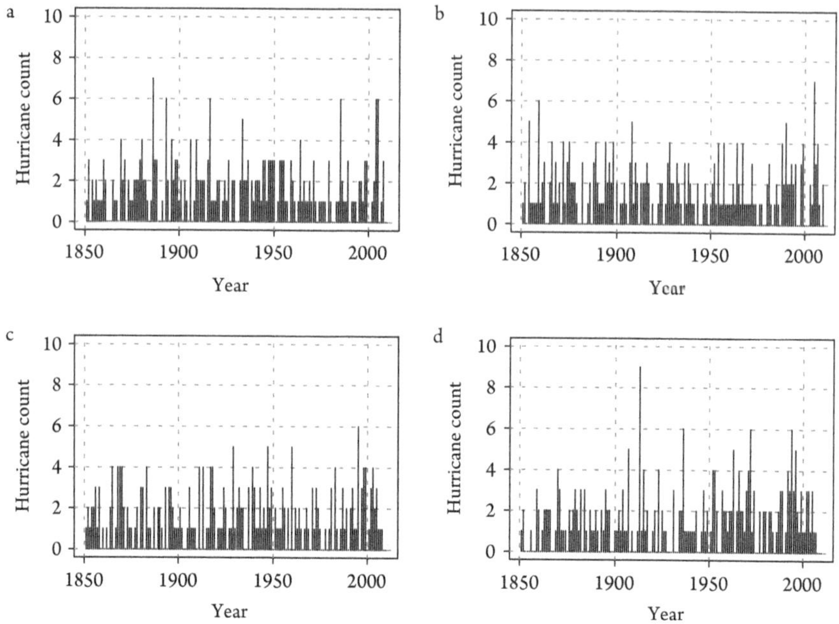

Figure 7.2 Hurricane occurrence using (a) observed and (b–d) simulated counts.

counts are consistent with a Poisson distribution having a variable rate. But where does this variability come from?

On the annual time scale to a first-order, high ocean heat content and cold upper air temperature provide the fuel for a hurricane, a calm atmosphere allows a hurricane to intensify, and the position and strength of the subtropical high pressure steers a hurricane that does form. Thus, hurricane activity responds to changes in climate conditions including sea-surface temperature (SST) as an indicator of oceanic heat content, sunspot number as an indicator of upper air temperature, El Niño-Southern Oscillation (ENSO) as indicator of wind shear, and the North Atlantic Oscillation as an indicator of steering flow.

SST provides an indication of the thermodynamic environment, as do sunspots. An increase in solar UV radiation during periods of strong solar activity will have a suppressing effect on tropical cyclone intensity as the air above the hurricane will warm through absorption of radiation by ozone. ENSO is characterized by basin-scale fluctuations in sea-level pressure (SLP) across the equatorial Pacific Ocean. The Southern Oscillation Index (SOI) is defined as the normalized SLP difference between Tahiti and Darwin. The SOI is strongly anticorrelated with equatorial Pacific SSTs so that an El Niño warming event is associated with negative SOI values.

The NAO is characterized by fluctuations in SLP differences. Monthly values are an indicator of the strength and/or position of the subtropical Bermuda High. The relationship might result from a teleconnection between the mid-latitudes and tropics whereby a below-normal NAO during the spring leads to dry conditions over the continents and to greater summer–fall middle tropospheric ridging (enhancing the dry conditions). Ridging over the eastern and western sides of the North Atlantic basin tends to keep the middle tropospheric trough, responsible for hurricane recurvature, farther to the north during the peak of the season (Elsner and Jagger, 2006). The data sets containing the environmental variables are described in Chapter 6, where you also plotted them (see Figure 6.6). With the exception of the NAO index, the monthly values are averages over the 3 months of August through October. The NAO index is averaged over the 2 months of May and June.

7.3 BIVARIATE RELATIONSHIPS

Consider the relationship between hurricane frequency and one of the environmental variables (covariate). Scatter plots are not very useful as there are many covariate values for a given count. It is better to plot a summary of the covariate distribution for each count.

The five-number summary provides information about the median, the range, and the quartile values of a distribution (see Chapter 5). So, for example, you compare the five-number summary of the NAO during years with no hurricanes and during years with three hurricanes by typing

```
> load("annual.RData")
> nao0 = annual$nao[H$All == 0]
> nao3 = annual$nao[H$All == 3]
> fivenum(nao0); fivenum(nao3)
[1] -2.030 -0.765 -0.165  0.600  1.725
[1] -2.655 -1.315 -0.685 -0.085  1.210
```

For each quantile of the five-number summary, the NAO value is lower when there are three hurricanes compared to when there are no hurricanes. A plot showing the five-number summary values for all years and all covariates is shown in Figure 7.3. See Figure 5.2 for an explanation of this plot. Note that the covariate is plotted on the horizontal axis in a scatter plot, so you draw the lines horizontal.

The plots show that the SOI and NAO are likely important in statistically explaining hurricane counts as there appears to be a systematic variation in counts across the range of values. The variation is less clear with SST and sunspot number. The bivariate relationships do not necessarily tell the entire story. A covariate might be important in explaining the residual variability so all variables might be significant in a multivariate model (see Chapter 3).

7.4 POISSON REGRESSION

The model of choice for count data is Poisson regression. Poisson regression assumes that the response variable can be described by a Poisson distribution with the

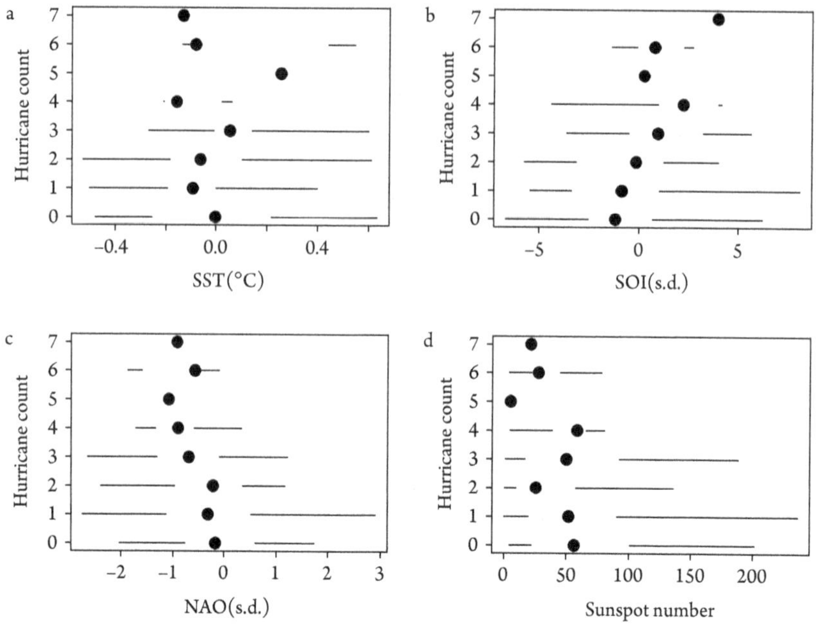

Figure 7.3 Bivariate relationships between covariates and hurricane counts.

logarithm of the expected response modeled with a linear combination of explanatory variables. It is an example of a generalized linear model (GLM).

7.4.1 Limitation of Linear Regression

The linear regression model described in Chapter 3 is not appropriate for count data. To illustrate, here you regress U.S. hurricane counts on the four explanatory variables (covariates) described earlier. You then use the model to make predictions specifying the SOI and NAO at three standard deviation departures from the average, a large sunspot number, and an average SST value.

To make things a bit simpler, you first create a data frame by typing

```
> df = data.frame(All=H$All, SOI=annual$soi,
+   NAO=annual$nao, SST=annual$sst, SSN=annual$ssn)
> df = df[-(1:15), ]
```

Here the data frame object `df` has columns with labels `All`, `SOI`, `NAO`, `SST`, and `SSN` corresponding to the response variable U.S. hurricane counts and the four explanatory variables. You remove the first 15 years because of missing SOI values.

You then create a linear regression model object using the `lm` function specifying the response and covariates accordingly.

```
> lrm = lm(All ~ SOI + NAO + SST + SSN, data=df)
```

Your model is saved in `lrm`.

Next you use the `predict` method on the model object together with specific explanatory values specified using the `newdata` argument. The names must match those used in your model object, and each explanatory variable must have a value.

```
> predict(lrm, newdata=data.frame(SOI=-3, NAO=3,
+     SST=0, SSN=250))
      1
-0.318
```

The prediction results in a negative number that is not a count. It indicates that the climate conditions are unfavorable for hurricanes, but the number from the linear regression model has no physical meaning.

7.4.2 Poisson Regression Equation

A Poisson regression model that specifies the logarithm of the annual hurricane rate is an alternative to linear regression. The assumption is that the hurricanes are independent in the sense that the arrival of one hurricane will not make another one more or less likely, but the rate of hurricanes varies from year to year due to the covariates.

The Poisson regression model is expressed as

$$\log(\lambda) = \beta_0 + \beta_1 x_1 + \ldots + \beta_p x_p \tag{7.2}$$

Here there are p covariates (indicated by the x_i's) and $p + 1$ parameters (β_i's). The model uses the logarithm of the rate as the response variable, but it is linear in the regression structure. It is not the same as a linear regression on the logarithm of counts. The model coefficients are determined by the *method of maximum likelihood.*

It is important to understand that with a Poisson regression you cannot explain all the variation in the observed counts; there will be unexplainable variation due to the stochastic nature of the process. Thus even if the model precisely predicts the rate of hurricanes, the set of predicted counts will have a degree of variability that cannot be reduced by the model (aleatory uncertainty). Consequently, if you think that most of the variability in the counts can be explained, then Poisson regression is not the appropriate model.

7.4.3 Method of Maximum Likelihood

Given the set of parameters as a vector β and a vector of explanatory variables x, the mean of the predicted Poisson distribution is given by

$$E(Y|x) = e^{\beta' x} \tag{7.3}$$

and thus, the Poisson distribution's probability density function is given by

$$p(y|x;\beta) = \frac{e^{y(\beta' x)} e^{-e^{\beta' x}}}{y!} \tag{7.4}$$

Suppose that you are given a data set consisting of n vectors $x_i \in \mathbb{R}^{n+1}$, $i = 1,\ldots,n$, along with a set of n values $y_1,\ldots,y_n \in \mathbb{R}$. Then, for a given set of parameters β, the probability of attaining this particular set of data is given by

$$p(y_1,\ldots,y_n|x_1,\ldots,x_n;\beta) = \prod_{i=1}^{n} \frac{e^{y_i(\beta' x_i)} e^{-e^{\beta' x_i}}}{y_i!}. \tag{7.5}$$

By the method of maximum likelihood, you wish to find the set of parameters β that makes this probability as large as possible. To do this, the equation is first rewritten as a likelihood function in terms of β.

$$L(\beta|X,Y) = \prod_{i=1}^{n} \frac{e^{y_i(\beta' x_i)} e^{-e^{\beta' x_i}}}{y_i!}. \tag{7.6}$$

Note that the expression on the right-hand side of the equation has not changed. By taking logarithms, the equation is easier to work with. The log-likelihood equation is given by

$$\ell(\beta|X,Y) = \log L(\beta|X,Y) = \sum_{i=1}^{n} \left(y_i(\beta' x_i) - e^{\beta' x_i} - \log(y_i!) \right). \tag{7.7}$$

Notice that the β's only appear in the first two terms of the summation. Therefore, given that you are interested only in finding the best value for β, you can drop the $y_i!$ and write

$$\ell(\beta|X,Y) = \sum_{i=1}^{n} \left(y_i(\beta' x_i) - e^{\beta' x_i} \right). \tag{7.8}$$

There is no closed form for the parameters that maximize this equation. However, the negative log-likelihood, $-\ell(\beta|X,Y)$, is a convex function, and so standard optimization or gradient ascent techniques can be applied to find the value of β, for which the probability is maximum.

7.4.4 Model Fit

The method of maximum likelihood is employed in the `glm` function to determine the model coefficients. The Poisson regression model is a type of generalized linear model (GLM) in which the logarithm of the annual rate is a linear function of the covariates (predictors).

To fit a Poisson regression model to U.S. hurricanes and save the model as an object, type

```
> prm = glm(All ~ SOI + NAO + SST + SSN, data=df,
+     family="poisson")
```

The model formula is identical to what you used to fit the linear regression model. The formula is read as "U.S. hurricane counts are modeled as a function of

Table 7.1 Coefficients of the Poisson regression model.

	Estimate	*Std. error*	*z value*	*Pr*(> |*z*|)
(Intercept)	0.5953	0.1033	5.76	0.0000
SOI	0.0619	0.0213	2.90	0.0037
NAO	−0.1666	0.0644	−2.59	0.0097
SST	0.2290	0.2553	0.90	0.3698
SSN	−0.0023	0.0014	−1.68	0.0928

SOI, NAO, SST, and SSN." Differences from the linear model fitting include the use of the `glm` function and the argument specifying `family="poisson"`.

You examine the model coefficients by typing

```
> summary(prm)
```

The model coefficients and the associated statistics are shown in Table 7.1. As anticipated from the bivariate relationships, the SOI and SST variables are positively related to the rate of U.S. hurricanes, and the NAO and sunspot number are negatively related. You can see that the coefficient on SST is positive but not statistically significant. Both the SOI and NAO have coefficients that provide convincing evidence against the null hypothesis, while the coefficient on the SSN provides suggestive but inconclusive evidence against the null.

Statistical significance is based on a null hypothesis that the coefficient is zero (Chapter 3). The ratio of the estimated coefficient to its standard error (z-value) has an approximate standard normal distribution assuming the null is true. The probability of finding a z-value this extreme or more is your p-value. The smaller the p-value, the less support there is for the null hypothesis given your data and model.

You use the `plotmo` function from the **plotmo** package (Milborrow, 2011b) to graph the model's response when varying one covariate while holding the other covariates constant at their median values.

```
> require(plotmo)
> plotmo(prm)
```

The results are shown in Figure 7.4. As SOI and SST increase so does the hurricane rate. By contrast the rate decreases with increasing NAO and SSN. The curvature arises from the fact that the covariates are related to the counts through the logarithm of the rate. The confidence bands are based on pointwise ± 2 standard errors. Note that the relatively large width for SOI values above 5 s.d. and for NAO values below −2 s.d.

7.4.5 Interpretation

Interpretation of the Poisson regression coefficients is different than the interpretation of the linear regression coefficients explained in Chapter 3. For example, the

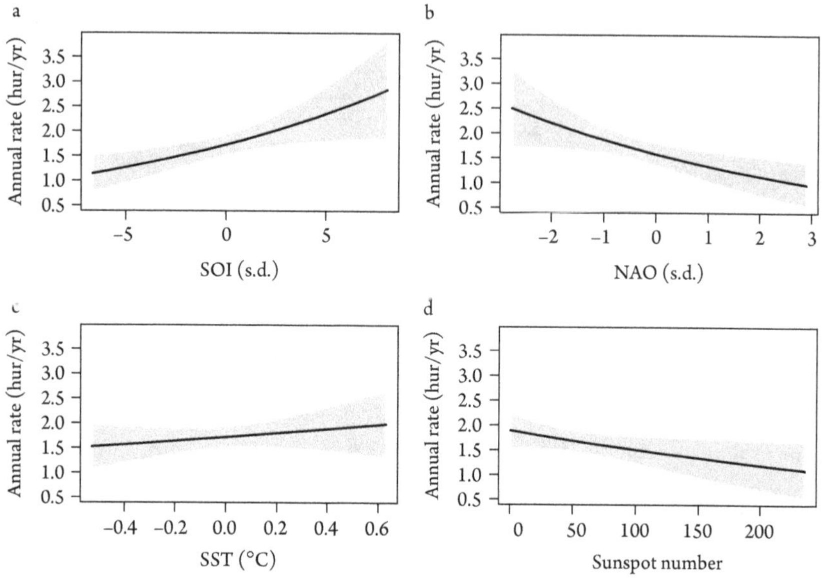

Figure 7.4 Dependence of hurricane rate on covariates in a Poisson regression.

coefficient value on the SOI indicates that for every one standard deviation (s.d.) increase in the SOI, the difference in the logarithm of hurricane rates is 0.062. Because there are other covariates, you must add "given that the other covariates in the model are held constant."

But what does the difference in the logarithm mean? Note that

$$\log A - \log B = \log\left(\frac{A}{B}\right) \tag{7.9}$$

so exponentiating the SOI coefficient value provides a ratio of the hurricane rates for a unit change in the SOI. You do this by typing

```
> exp(summary(prm)$coefficients[2, 1])
[1] 1.06
```

and find that for every one s.d. increase in SOI, the hurricane rate increases by a factor of 1.06 or 6 percent. Similarly, since the NAO coefficient value is -0.167, you find that for every one s.d. increase in the NAO, the hurricane rate decreases by a factor of 15 percent.

7.5 MODEL PREDICTIONS

Given the model coefficients obtained by the `glm` function and saved as a model object, you make predictions using the `predict` method. For comparison, you predict the rate of hurricanes given the same coefficient values used earlier for the linear regression model.

```
> predict(prm, newdata=data.frame(SOI=-3, NAO=3,
+   SST=0, SSN=250), type="response")
    1
0.513
```

The argument `type="response"` gives the prediction in terms of the mean response (hurricane rate). By default the argument is `type="link"`, which results in a prediction in terms of the link function (here the logarithm of the mean response). Recall that the linear regression model gave a prediction that was physically unrealistic. Here, the predicted value indicates a small hurricane rate as you would expect given the covariate values, but the rate is a realistic nonnegative number.

The predicted rate together with Eq. 7.1 provides a probability for each possible count. To see this you create two bar plots, one for a forecast of hurricanes under unfavorable conditions and another for a forecast of hurricanes under favorable conditions. First you save the predicted rate for the specified values of the covariates. You then create a vector of counts from zero to six that is used as the set of quantiles for the `dpois` function and as the names argument in the `barplot` function. The plotting parameters are set using the `par` function. To make it easier to compare the probability distributions, limits on the vertical axis (`ylim`) are also set.

```
> fav = predict(prm, newdata=data.frame(SOI=2, NAO=-2,
+   SST=0, SSN=50), type="response")
> ufa = predict(prm, newdata=data.frame(SOI=-2, NAO=2,
+   SST=0, SSN=200), type="response")
> h = 0:6
> par(mfrow=c(1, 2), las=1)
> barplot(dpois(x=h, lambda=ufa), ylim=c(0, .5),
+   names.arg=h, xlab="Number of Hurricanes",
+   ylab="Probability")
> mtext("a", side=3, line=1, adj=0, cex=1.1)
> barplot(dpois(x=h,lambda=fav), ylim=c(0,.5),
+   names.arg=h, xlab="Number of Hurricanes",
+   ylab="Probability")
> mtext("b", side=3, line=1, adj=0, cex=1.1)
```

The result is shown in Figure 7.5. The forecast probability of two or more hurricanes is 72 percent in years with favorable conditions but decreases to 16 percent in years with unfavorable conditions. The probability of no hurricanes during years with favorable conditions is 8 percent, which compares with a probability of 48 percent during years with unfavorable conditions. The model translates climate swings to changes in landfall probabilities.

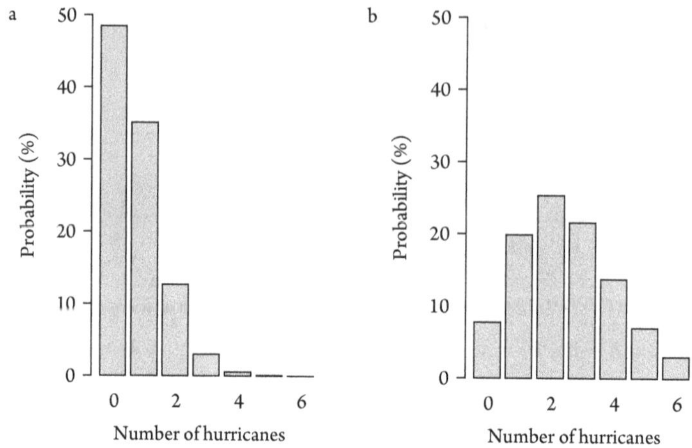

Figure 7.5 Forecast probabilities for (a) unfavorable and (b) favorable conditions.

7.6 FORECAST SKILL

7.6.1 Metrics

Forecast skill refers to how well predictions match observations. There are several ways to quantify this match. Here you consider three of the most common: the mean absolute error (MAE), the mean squared error (MSE), and the correlation coefficient (r).

Let λ_i be the predicted rate for year i and o_i be the corresponding observed count for that year. Then the three measures of skill over n years are defined by

$$\text{MAE} = \frac{1}{n}\sum_{i=1}^{n} |\lambda_i - o_i| \tag{7.10}$$

$$\text{MSE} = \frac{1}{n}\sum_{i=1}^{n} (\lambda_i - o_i)^2 \tag{7.11}$$

$$\text{r} = \frac{\sum(\lambda_i - \bar{\lambda})(o_i - \bar{o})}{\sqrt{\sum(\lambda_i - \bar{\lambda})^2 \sum(o_i - \bar{o})^2}} \tag{7.12}$$

You obtain the predicted rate for each year in the record (λ_i) by typing

```
> prm = glm(All ~ SOI + NAO + SSN, data=df,
+     family="poisson")
> pr = predict(prm, type="response")
```

You first create a new model removing the insignificant SST covariate. Since each prediction is made for a year with a known hurricane count, it is referred to as a "hindcast." The word "forecast" is reserved for a prediction made for a year where the hurricane count is unknown (typically in the future).

The Poisson regression hindcasts are given in terms of rates while the observations are counts. So instead of using a rate you use the probability distribution of observing

Table 7.2 Forecast skill (in sample). 'Useful' is the percentage skill above climatology.

	Poisson	*Climatology*	*Useful*
MAE	1.08	1.16	7.04
MSE	1.93	2.18	11.24
r	0.34		
MAEp	1.44	1.50	4.08
MSEp	3.67	3.92	6.26

$j = 0, 1, \ldots, \infty$ hurricanes. A probabilistic form of the earlier formulae are

$$\text{MAEp} = \frac{1}{n}\sum_{i=1}^{n}\sum_{j=0}^{\infty} \text{dpois}(j, \lambda_i) \cdot \left|j - o_i\right| \tag{7.13}$$

$$\text{MSEp} = \frac{1}{n}\sum_{i=1}^{n}\sum_{j=0}^{\infty} \text{dpois}(j, \lambda_i) \cdot (j - o_i)^2 \tag{7.14}$$

$$= \text{MSE} + \bar{\lambda} \tag{7.15}$$

where dpois(j, λ_i) is the probability of j hurricanes in the ith year given the predicted rate λ_i, and o_i is the observed count. There is no probabilistic form to the correlation as a measure of forecast skill.

A prediction model is deemed useful if the skill level exceeds the level of a naive reference model. The percentage above the skill obtained from a naive model is referred to useful skill. The naive model is typically climatology. To obtain the climatological rate, you type

```
> clim = glm(All ~ 1, data=df, family="poisson")
> cr = predict(clim, type="response")
```

Note that the only difference from your earlier model specification is that the term to the right of the tilde is 1. The model predicts the mean hurricane rate over the period of record. The value is the same for each year.

Table 7.2 shows skill metrics for your U.S. hurricane model and the percentage of useful skill relative to climatology. Correlation is undefined for a forecast of climatology. The useful skill level is between 4.1 and 11.2 percent. Although not high, it represents a significant improvement.

7.6.2 Cross-Validation

The earlier procedure results in an in-sample assessment of forecast skill. All years of data are used to estimate a single set of model coefficients with the model subsequently used to hindcast each year's hurricane activity. But how well will your model perform when making predictions of future activity? This question is best answered with an out-of-sample assessment of skill. An out-of-sample assessment (1) excludes a single year of observations, (2) determines the MLE coefficients of the Poisson regression model using observations from the remaining years, and (3) uses the model to predict the hurricane count for the year left out. This is done n times, removing each year's data successively. The above skill metrics are then used on these out-of-sample predictions. The procedure is called "crossvalidation," and where single cases are left out, it is called leave-one-out cross validation (LOOCV).

To perform LOOCV on your Poisson regression model, you loop over all years using the index `i`. Within the loop, you determine the model using all years except `i` (`df[-i, ]` in the `data` argument). You then make a prediction only for the year you left out (`newdata=df[i, ]`). Note that your climatology model is cross validated as well.

```
> j = 0:15; n = length(df$All)
> prx = numeric()
> crx = numeric()
> for(i in 1:n){
+ prm = glm(All ~ SOI + NAO + SSN,
+   data=df[-i, ], family="poisson")
+ clm = glm(All ~ 1, data=df[-i, ], family="poisson")
+ prx[i] = predict(prm, newdata=df[i, ], type="r")
+ crx[i] = predict(clm, newdata=df[i, ], type="r")
+ }
```

Skill assessment is done in the same way as for in-sample assessment. The results of the cross-validation assessment of model skill are give in Table 7.3. Out-of-sample skill levels are lower. This is an estimate of the average skill the model will have when making actual forecasts. You should show out-of-sample skill if you intend to use your model to predict the future.

The difference in percentage of usefulness between in-sample and out-of-sample skill is a measure of the overfit in your model. Overfit arises when your model

Table 7.3 Forecast skill (out of sample). 'Useful' in the percentage skill above climatology.

	Poisson	*Climatology*	*Useful*
MAE	1.11	1.16	4.91
MSE	2.05	2.21	7.03
r	0.26		
MAEp	1.46	1.51	2.81
MSEp	3.80	3.95	3.78

interprets random fluctuations as signal. Cross-validation helps you protect yourself against being fooled by this type of randomness.

7.7 NONLINEAR REGRESSION STRUCTURE

Poisson regression specifies a linear relationship between your covariates and the logarithm of the hurricane rate. This linearity can be restrictive if the influence of a covariate changes over the range of covariate valves. Multivariate adaptive regression splines (MARS) is a form of regression introduced by Friedman (1991) that allows for nonlinearity in the covariates.

MARS builds models of the form

$$\hat{f}(x) = \sum_{i=1}^{k} c_i B_i(x) \tag{7.16}$$

where $B_i(x)$ is a basis function and c_i is a constant for a given basis. The model is thus a weighted sum of the k basis functions. A basis function takes one of three forms: either a constant representing the intercept term, a *hinge* function of the form $\max(0, x - a)$, where a is a constant representing the *knot* for the hinge function, or a product of two or more hinge functions to allow the basis function the ability to handle interaction between two or more covariates. A hinge function is zero for part of its range, so it partitions your multivariate data into disjoint regions.

The `earth` function in the **earth** package (Milborrow, 2011a) provides functionality for MARS. The syntax is the same as other models in R. Here you create a MARS model for your hurricane counts and environmental covariates by typing

```
> require(earth)
> mars = earth(All ~ SOI + NAO + SST + SSN, data=df,
+    glm=list(family="poisson"))
```

A summary method on the model object indicates that only the SOI and NAO are selected as important in explaining hurricane rates. The correlation between the predicted rate and the observed counts is obtained by taking the square root of the R-squared value.

```
> sqrt(mars$rsq)
[1] 0.469
```

This value exceeds the correlation from your Poisson regression by 39.7 percent, suggesting that MARS might be a better prediction model.

The partial dependence plot (Fig. 7.6) shows the hinge functions for the two selected covariates. The knots on the SOI are located at about -2 and $+4$ s.d. There is no relationship between the SOI and hurricane counts for the lowest values of SOI, and there is a sharp inverse relationship for the largest values. Caution is advised against overinterpretation as the graph describes only the SOI–hurricane relationship for median NAO values. The single knot on the NAO indicates no relationship between the NAO and hurricane counts for values less than about -0.5 s.d. Again,

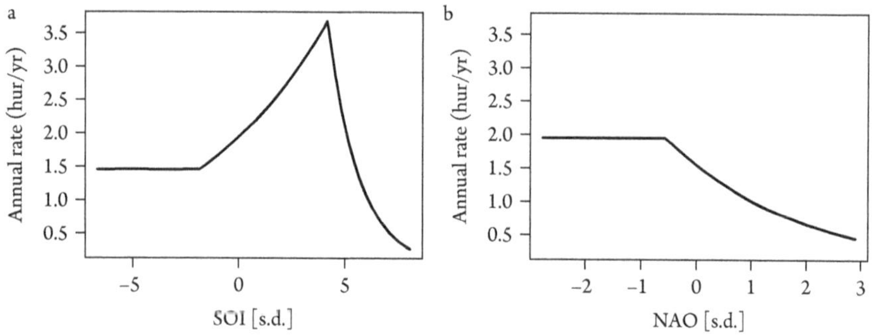

Figure 7.6 Dependence of hurricane rate on covariates in a MARS model.

this applies only at median SOI values. There are no standard errors from which to obtain confidence bands.

Before making a forecast, you cross validate your model to get an out-of-sample correlation between the observed counts and predicted rates. You do this by specifying the number of cross-validations with the `nfold` argument[1]. The function `earth` builds `nfold` cross-validated models. For each fold, it builds a model with in-sample data and then uses the model to compute the R-squared value from predictions made on the out-of-sample data. For instance, with `nfold=2`, the number of years in the in-sample and out-of-sample is roughly the same. The choice of years is chosen randomly, so you set a seed to allow replication of your results.

Here you set `nfold` at 5 as a compromise between having enough data to build the model and make predictions with it. You stabilize the variance further by specifying `ncross` to allow 40 different `nfold` cross-validations.

```
> set.seed(3042)
> marsCV = earth(All ~ SOI + NAO, data=df, nfold=5,
+   ncross=40, glm=list(family="poisson"))
```

The R-squared results are saved in your `marsCV` object in column `cv.rsq.tab`. The last row gives the mean R-squared value that provides an estimate of the average skill your model will have when it is used to make actual forecasts. The square root of that value is the correlation, obtained by typing

```
> rn = dim(marsCV$cv.rsq.tab)[1]
> mars.r = sqrt(marsCV$cv.rsq.tab[rn, ][1])
> mars.r
  All
0.291
```

The mean correlation is 11 percent higher than the correlation from the HOOCV of your Poisson regression model (see Table 7.3). This is still an improvement but below that estimated from your in-sample skill.

[1] Cross validation is done if the argument is greater than one.

7.8 ZERO-INFLATED COUNT MODEL

The Poisson regression model is a good place to start when working with count data, but it might not be good enough when the counts are overdispersed or when there are a large number of zeros. Consider the question of whether your Poisson regression of hurricane counts is adequate. You examine model adequacy using the residual deviance. The residual deviance is -2 times the log-likelihood ratio of a model without covariates compared to your model with covariates.

The residual deviance along with the residual degrees of freedom is available from the summary method on your `glm` object.

```
> prm = glm(All ~ SOI + NAO + SSN, data=df,
+    family="poisson")
> s = summary(prm)
> rd = s$deviance
> dof = s$df.residual
```

Under the null hypothesis that your model is adequate, the residual deviance has a χ^2 distribution with degrees of freedom equal to the residual degrees of freedom. Here the situation is reversed from the normal situation in which the null hypothesis is the opposite of what you hope for. To obtain the p-value for a test of model adequacy, type

```
> pchisq(rd, dof, lower.tail=FALSE)
[1] 0.0255
```

The residual deviance is 175.61 on 141 degrees of freedom resulting in a p-value of 0.0255. Thus there is evidence that something is missing.

The problem may be that hurricanes tend to arrive in clusters even after taking into account the covariates that influence hurricane rates from year to year. This clustering produces overdispersion in observed counts. You will examine this possibility and what to do about it in Chapter 11.

Another problem could be that the count data have too many zeros. This is typical when there are two processes at work: one determining whether there is at least one event and the other determining how many events. An example is the occurrence of cloud-to-ground lightning strikes. There will be many more hours with no strikes due to convective processes that are different than processes that produce one or more strikes. These kinds of data can be handled with zero-inflated models.

Zero-inflated count models are mixture models combining a point mass at zero and a count distribution. This leaves you with two sources of zeros: one from the point mass distribution and the other from the count distribution. Usually the count model is a Poisson regression and the point-mass model is a binomial regression.

The `zeroinfl` function in the **pscl** package (Zeileis et al., 2008) can be used to fit a zero-inflated model using the method of maximum likelihood. The formula describes the count data model, i.e., `y ~ x1 + x2` specifying a count data regression, where all zero counts have the same probability of belonging to the zero-inflation component. This is equivalent to the model `y ~ x1 + x2 | 1`, making it explicit that the zero-inflated model has only an intercept. Additional predictors can be added so that not all zeros have the same probability of belonging to the point mass component or to the count component. A typical formula is `y ~ x1 + x2 | z1 + z2`. The covariates in the zero and the count component can be overlapping (or identical).

For example, to model your U.S. hurricane counts where the count model uses all four covariates and where the binomial model uses only the SST variable, type

```
> require(pscl)
> zim = zeroinfl(All ~ SOI + NAO + SST + SSN | SST,
+    data=df)
```

The model syntax includes a vertical bar to separate the covariates between the two model components. The returned object is of class `zeroinfl` and is similar to the object of class `glm`. The object contains a list of the coefficients and terms for each model component. To summarize the model object, type

```
> summary(zim)
Call:
zeroinfl(formula = All ~ SOI + NAO + SST + SSN |
    SST, data = df)

Pearson residuals:
   Min      1Q Median      3Q     Max
-1.492 -0.774 -0.127  0.583  3.061

Count model coefficients (poisson with log link):
            Estimate Std. Error z value Pr(>|z|)
(Intercept)  0.64087    0.10593    6.05  1.5e-09
SOI          0.06920    0.02236    3.09   0.0020
NAO         -0.16928    0.06554   -2.58   0.0098
SST          0.57178    0.28547    2.00   0.0452
SSN         -0.00239    0.00143   -1.67   0.0942

Zero-inflation model coefficients (binomial with
   logit link):
            Estimate Std. Error z value Pr(>|z|)
(Intercept)    -4.44       2.03   -2.19    0.029
SST             6.80       3.89    1.75    0.080

Number of iterations in BFGS optimization: 16
Log-likelihood: -231 on 7 Df
```

Results show that the four covariates have a statistically significant influence on the number of U.S. hurricanes and that SST has a significant relationship with whether or not there will be at least one hurricane. But note that the sign on the SST coefficient is positive in the zero-inflation component indicating that higher SST is associated with *more* years *without* hurricanes. The sign is also positive in the count component indicating that, given at least one hurricane, higher SST is associated with a higher probability of two or more hurricanes.

A cross-validation exercise indicates that this the zero-inflated model performs slightly worse than the Poisson model. The useful skill as measured by the mean absolute error is 3.7 percent above climatology for your zero-inflated model compared with 4.9 percent above climatology for your Poisson model.

7.9 MACHINE LEARNING

You can remove the Poisson assumption all together by employing a machine-learning algorithm that searches your data to find patterns related to the annual counts (data mining). A regression tree is a type of machine learning algorithm that outputs a series of decisions with each decision leading to a value of the response or to another decision. If the value of the NAO is less than −1 s.d., for example, then the response is two hurricanes. If it is greater, then is the SOI greater than 0.5 s.d. and so on. A single tree will capture the relationships between annual counts and your predictors.

To see how this works, import the annual hurricane data and subset on years since 1950. Create a data frame containing only the basin-wide hurricane counts and SOI and SST as the two predictors.

```
> load("annual.RData")
> dat = subset(annual, Year >= 1950)
> df = data.frame(H=dat$B.1, SOI=dat$soi, SST=dat$sst)
```

Then using the `tree` function from the **tree** package (Ripley, 2011), type

```
> require(tree)
> rt = tree(H ~ SOI + SST, data=df)
```

The model syntax is the same, with the response variable to the left of the tilde and the covariates to the right. To plot the regression tree, type

```
> plot(rt); text(rt)
```

Instead of interpreting the parameter values from a table of coefficients, you interpret a regression tree from an upside-down tree diagram. You start at the top. The first branch is a split on your SST variable at a value of 0.33. The split is a rule. Is the value of SST less than 0.33°C? If yes, branch to the left; if no, branch to the right. All splits work this way. Following on the right, the next split is on SOI. If SOI is greater than 0.12 s.d., then the mean value of all years under these conditions is 10.8 hur/yr. This is the end of the branch (leaf). You check this by typing

```
> mean(df$H[df$SST >= .33 & df$SOI > .12])
[1] 10.8
```

The model is fit using binary recursive partitioning. Splits are made along coordinate axes of SST and SOI so that on any branch, a split is chosen that maximally distinguishes the hurricane counts. Splitting continues until the variables cannot be split or there are too few years (less than 6 by default). Here, SST is the variable explaining the most variation in the counts so it gets selected first. Again, the value at which the split occurs is based on maximizing the difference in counts between the two subsets. The tree has five branches.

In general the key questions are as follows: which variables are best to use and which value gives the best split? The choice of variables is similar to the forward selection procedure of stepwise regression (Chapter 3). A prediction is made by determining which leaf is reached based on the values of your predictors. To determine the mean number of hurricanes when SOI is −2 s.d. and SST is 0.2°C, you use the `predict` method and type

```
> predict(rt, data.frame(SOI=-2, SST=.2))
   1
7.35
```

The predicted value depends on the tree and the tree depends on what years are used to grow it. For example, regrow the tree by leaving the last year out and make a prediction using the same two predictor values.

```
> rt2 = tree(H ~ SOI + SST, data=df[-61, ])
> predict(rt2, data.frame(SOI=-2, SST=.2))
   1
5.71
```

Results are different. Which prediction do you choose? Forecast sensitivity occurs with all statistical models, but it is more acute in models that contain a large number of parameters. Each branch in a regression tree is a parameter, so with your two predictors the model has five parameters.

A random forest algorithm attempts to sidestep the question of prediction choice by making predictions from many trees (Breiman, 2001). It creates a sample from the set of all years and grows a tree using data only from the sampled years. It then repeats the sampling and grows another tree. Each tree gives a prediction and the mean is taken. The function `randomForest` in the **randomForest** package provides a random forest algorithm. For example, type

```
> require(randomForest)
> rf = randomForest(H ~ SOI + SST, data=df)
```

By default, the algorithm grows 500 trees. To make a prediction type,

```
> predict(rf, data.frame(SOI=-2, SST=.2))
   1
4.91
```

Regression trees and random forest algorithms tend to overfit your data, especially when searching over a large set of *potential* predictors in noisy climate data. Overfitting produces small in-sample error, but large out-of-sample error. Again, a cross-validation exercise is needed if you want to claim the algorithm has superior predictive skill. Cross-validation removes the noise specific to each year's set of observations and estimates how well the random forest algorithm finds skillful prediction rules when this coincident information is unavailable. For example, *Does the random forest algorithm provide better prediction skill than a Poisson regression?* To answer this question, you arrange a HOOCV as follows:

```
> n = length(df$H)
> rfx = numeric(n)
> prx = numeric(n)
> for(i in 1:n){
+   rfm = randomForest(H ~ SOI + SST, data=df[-i, ])
+   prm = glm(H ~ SOI + SST, data=df[-i, ],
+             family="poisson")
+   new = df[i, ]
+   rfx[i] = predict(rfm, newdata=new)
+   prx[i] = predict(prm, newdata=new,
+                    type="response")
+ }
```

The out-of-sample mean-squared prediction error is computed by typing

```
> mean((df$H - prx)^2); mean((df$H - rfx)^2)
[1] 5.07
[1] 5.36
```

Results indicate that the Poisson regression performs slightly better than the random forest algorithm in this case although the difference is not statistically significant. The correlation between the actual and predicted value is 0.539 for the Poisson model and 0.502 for the random forest algorithm.

Figure 7.7 shows the bivariate influence of SST and SOI on hurricane counts using the random forest algorithm and Poisson regression. Hurricane counts increase with SST and SOI, but for high values of SOI, the influence of SST is stronger. Similarly for high values of SST, the influence of the SOI is more pronounced. The random forest is able to capture nonlinearities and thresholds but at the expense of interpreting some noise as signal, as seen by the relatively high count with SOI values near −3 s.d. and SST values near −0.1°C.

7.10 LOGISTIC REGRESSION

Some of our research in the 1990s focused on the climatology of hurricanes from nontropical origins (Elsner et al., 1996; Kimberlain and Elsner, 1998). We analyzed available information from each North Atlantic hurricane since 1944 to determine

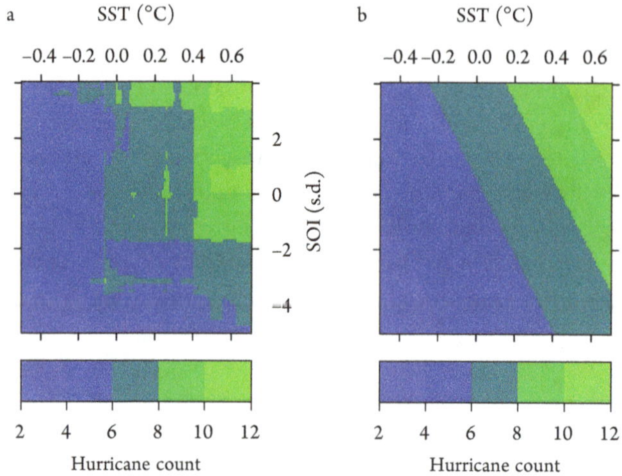

Figure 7.7 Hurricane predictions. (a) Random forest and (b) Poisson regression.

whether we could discern middle-latitude influences on development. We classified hurricanes into tropical and baroclinic based on primary origin and development mechanisms. Here you would like to have a model that predicts a hurricane's group membership based simply on where the hurricane originated.

Logistic regression is the model of choice when your response variable is dichotomous. Many phenomena can be studied in this way. An event either occurs or it does not. The focus is to predict the occurrence of the event. A hurricane forecaster is keen about whether an area of disturbance will develop into a cyclone, given the present atmospheric conditions.

Logistic regression is a generalization of the linear regression model, where the response variable does not have a normal distribution and the regression structure is linear in the covariates. Like Poisson regression, the model coefficients are determined using the method of maximum likelihood.

The mean of a binary variable is a percentage. For example, generate 10 random binary values and compute the mean by typing

```
> set.seed(123)
> x = rbinom(n=10, size=1, prob=.5)
> x
 [1] 0 1 0 1 1 0 1 1 1 0
> mean(x)
[1] 0.6
```

Think of the 1s as heads and 0s as tails from 10 flips of a fair coin (`prob=.5`). You find 6 heads in 10 flips. The mean number is the percentage of heads in the sample. The percentage of a particular outcome can be interpreted as a probability so it is denoted as π. The logistic regression model specifies how π is related to a set of explanatory variables.

7.10.1 Exploratory Analysis

You input the hurricane data by typing

```
> bh = read.csv("bi.csv", header=TRUE)
> table(bh$Type)
  0   1   3
187  77  73
```

The type as determined in Elsner et al. (1996) is given by the variable `Type` with 0 indicating tropical-only, 1 indicating baroclinic influences, and 3 indicating baroclinic initiation. The typing was done subjectively using all the available synoptic information about each hurricane. While the majority of hurricanes form over warm ocean waters of the deep tropics ('tropical-only') some are aided in their formation by interactions with midlatitude jet stream winds ('baroclinically induced'). The stronger, tropical-only hurricanes develop farther south and primarily occur in August and September. The generally weaker, baroclinically induced hurricanes occur throughout the season.

First combine the baroclinic types into a single group and add this column to the data frame.

```
> bh$tb = as.integer(bh$Type != 0)
> table(bh$tb)
  0   1
187 150
```

With this grouping, there are 187 tropical and 150 baroclinic hurricanes in the record. Thus you can state that a hurricane drawn at random from this set of cyclones has about a 55 percent chance of being tropical-only.

Your interest is to improve on this climatological model by adding a covariate. Here you consider the latitude at which the cyclone first reaches hurricane strength. As an exploratory step, you create box plots of the latitudes grouped by hurricane type.

```
> boxplot(bh$FirstLat ~ bh$tb, horizontal=TRUE,
+   notch=TRUE, yaxt="n", boxwex=.4,
+    xlab="Latitude of Hurricane Formation")
> axis(2, at=c(1, 2), labels=c("Tropical",
+    "Baroclinic"))
```

Here you make the boxes horizontal (Fig. 7.8) because latitude is your explanatory variable. With the argument `notch` switched on (`notch=TRUE`), notches are drawn on the box sides. The vertical dash inside the box is the median latitude. Notches extend to $\pm 1.58\times$ IQR/$\sqrt{n}$, where IQR is the interquartile range (see Chapter 2) and n is the sample size. If the notches of two box plots do not overlap, this provides evidence that the two medians are statistically different (Chambers et al., 1983).

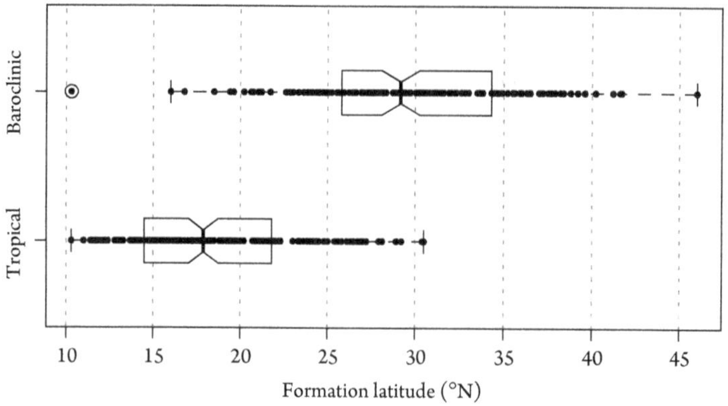

Figure 7.8 Genesis latitude by hurricane type.

The median formation latitude for the set of tropical hurricanes is 17.9°N, and for the set of baroclinic hurricanes, it is farther north at 29.1°N. This makes physical sense as cyclones farther south are less likely to have influence from middle-latitude baroclinic disturbances. The relatively small overlap between the two sets of latitudes strengthens your hypothesis that formation latitude will improve on a climatological model for hurricane type.

7.10.2 Logit and Logistic Functions

Linear regression is not the appropriate model for binary data. It violates the assumption of equal variance and normality of residuals resulting in invalid standard errors and erroneous hypothesis tests. In its place, you use a generalized linear model as you did earlier with the count data.

However, instead of using the logarithm as the link between the response and the covariates as you did in the Poisson regression model, here you use the *logit* function. The logit of a number π between 0 and 1 is

$$\text{logit}(\pi) = \log\left(\frac{\pi}{1-\pi}\right) = \log(\pi) - \log(1-\pi) \tag{7.17}$$

If π is a probability then $\pi/(1-\pi)$ is the corresponding *odds*, and the logit of the probability is the logarithm of the odds. Odds are expressed as for:against (read: for to against) something happening. So the odds of a hurricane strike that is posted at 1:4 means there is a 20 percent change that a strike will occur.

The logistic regression model is expressed statistically as

$$\text{logit}(\pi) = \beta_0 + \beta_1 x_1 + \ldots + \beta_p x_p \tag{7.18}$$

where π is the mean. There are p covariates (x_i's) and $p+1$ parameters (β_i's).

To convert logit(π) to π (probability of occurrence), you use the *logistic* function (inverse of the logit function) given as

$$\text{logistic}(\alpha) = \frac{1}{1+\exp(-\alpha)} = \frac{\exp(\alpha)}{1+\exp(\alpha)} \qquad (7.19)$$

7.10.3 Fit and Interpretation

To fit a logistic regression model to hurricane type with latitude as the covariate saving the model as an object, type

```
> lorm = glm(tb ~ FirstLat, data=bh,
+    family="binomial")
```

The function specification is similar to Poisson regression, but here the family is binomial instead of poisson. The formula is read as "hurricane type is modeled as a function of formation latitude."

The model coefficients are determined by the method of maximum likelihood in the `glm` function. To produce a table of the coefficients, you type

```
> summary(lorm)$coefficients
              Estimate Std. Error z value Pr(>|z|)
(Intercept)    -9.083     0.9615   -9.45 3.50e-21
FirstLat        0.373     0.0395    9.45 3.49e-21
```

The estimated coefficients are listed by row. The coefficient for the intercept is the log odds of a baroclinic hurricane at the equator. In other words, the odds of being baroclinic when the latitude is zero is $\exp(-9.0826) = 0.000114$. These odds are very low, but that makes sense since no hurricanes form at the equator.

Interest is on the coefficient of the formation latitude variable indicated by the row labeled `FirstLat`. The value is 0.373. Before fitting the model, you anticipate the formation latitude coefficient to have a positive sign. Why? Because baroclinic (tropical)-type hurricanes are coded as 1 (0) in your data set and the box plots show that as formation latitude increases, the chance that a hurricane has baroclinic influences increases. Note that if your response values are character strings (e.g., "to" and "be") rather than coded as 0s and 1s, things will still work, but R will assign 0s and 1s based on alphabetical order and this will affect how you make sense of the coefficient's sign.

The magnitude of the coefficient is interpreted as for every degree increase in formation latitude, the log odds increases by a constant 0.373 units, on average. This is not very informative. By taking the exponent of the coefficient value, the interpretation is in terms of an odds ratio.

```
> exp(summary(lorm)$coefficients[2])
[1] 1.45
```

Thus, for every degree increase in formation latitude the odds ratio increases on average by a constant factor of 1.45 (or 45 percent). This 45 percent increase does not depend on the value of latitude. That is, logistic regression is linear in the odds ratio. The interpretation is valid only over the range of latitudes in your data and physically meaningless for latitudes outside the range where hurricanes occur.

The table of coefficients includes standard errors and p-values. Statistical significance is based on a null hypothesis that the coefficient is zero. The ratio of the estimated coefficient to its standard error (z-value) has an approximate standard normal distribution assuming that the null is true. The probability of finding a z-value this extreme or more is the p-value. The smaller the p-value, the less support there is for the null hypothesis given the data and the model. The lack of support for the null allows you accept the model.

Also a confidence interval on the estimated coefficient is obtained by typing

```
> confint(lorm)[2, ]
 2.5 % 97.5 %
 0.301  0.456
```

This is interpreted to mean that although your best estimate for the log odds of a baroclinic hurricane given latitude is 0.373, there is a 95 percent chance that the interval between 0.301 and 0.456 will cover the true log odds.

7.10.4 Prediction

Predictions help you understand your model. As you did previously, you use the `predict` method on the model object. To predict the probability that a hurricane picked at random from your data will be baroclinic given that its formation latitude is 20°N latitude, you type

```
> predict(lorm, newdata=data.frame(FirstLat=20),
+     type="response")
    1
0.164
```

Thus, the probability of a baroclinic hurricane forming at this low latitude is 16.4 percent on average.

To create a plot of predictions across a range of latitudes, first prepare a vector of latitudes. The vector spans the latitudes in your data set. You specify an increment of 0.1° so the resulting prediction curve is smooth. You then use the predict method with `se.fit` equal to true and save the average prediction and the predictions corresponding to $\pm 1.96\times$ the standard error.

```
> lats = seq(min(bh$FirstLat), max(bh$FirstLat), .1)
> probs = predict(lorm,
+   newdata=data.frame(FirstLat=lats),
+   type="response", se.fit=TRUE)
> pm = probs$fit
```

```
> pu = probs$fit + probs$se.fit * 1.96
> pl = probs$fit - probs$se.fit * 1.96
```

Finally, you plot the data points at 0 and 1, as you did earlier, with the bar plot and add the predicted values using the `lines` function.

```
> plot(bh$FirstLat, bh$tb, pch=19, cex=.4,
+   ylab="Probability",
+   xlab="Formation Latitude (N)")
> grid()
> lines(lats, pm, lwd=2)
> lines(lats, pu, lwd=2, col="red")
> lines(lats, pl, lwd=2, col="red")
```

Results are shown in Figure 7.9. Tropical-only and baroclinically enhanced hurricane formation latitudes are shown along the $y = 0$ and $y = 100$ lines, respectively. The gray band is the 95 percent pointwise confidence interval. Model predictions make sense. The probability of a baroclinically enhanced hurricane is less than 20 percent at latitudes south of 20°N. However, by latitude 24°N, the probability exceeds 50 percent and by latitude 31°N the probability exceeds 90 percent. Note that although the odds ratio is constant, the probability is a nonlinear function of latitude.

7.10.5 Fit and Adequacy

Output from a summary method applied to your model object (`summary(lorm)`) prints statistics of model fit including null and deviance residuals and the AIC (see Chapter 3). These are shown below the table of coefficients. One measure of model fit is the significance of the overall model.

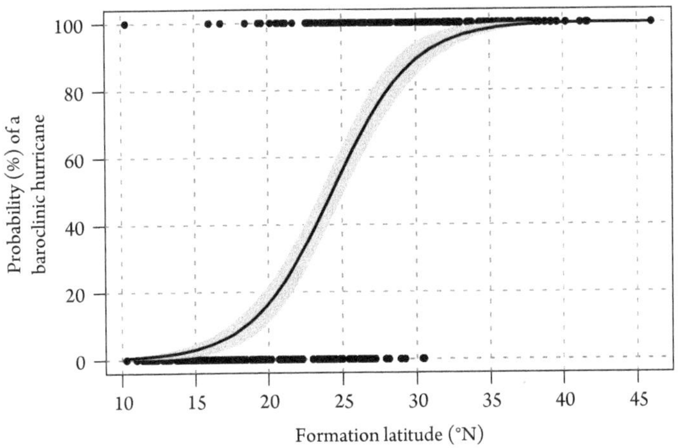

Figure 7.9 Logistic regression model for hurricane type.

This test asks whether the model with latitude fits significantly better than a model with an intercept only. An intercept-only model is called a "null" model (no covariates). The test statistic is the difference between the residual deviance for the model with and without latitude. The test statistic has a χ-squared distribution with degrees of freedom equal to the differences in degrees of freedom between the latitude model and the null model (i.e., the number of predictors in the model, here just one).

To find the difference in deviance between the two models (i.e., the test statistic) along with the difference in degrees of freedom, type

```
> dd = lorm$null.deviance - lorm$deviance
> ddof = lorm$df.null - lorm$df.residual
> dd; ddof
[1] 231
[1] 1
```

Then the *p*-value as evidence in support of the null model is obtained by typing

```
> 1 - pchisq(q=dd, df=ddof)
[1] 0
```

This leads you to reject the null hypothesis in favor of the model that includes latitude as a covariate.

A model can fit well but still be inadequate if it is missing an important predictor or if the relationship has a different form. Model adequacy is examined with the residual deviance statistic. The test is performed under the null hypothesis that the model is adequate (see §7.8). Under this hypothesis, the residual deviance has a χ-squared distribution with residual degrees of freedom. Thus, to test the model for adequacy, you type

```
> pchisq(q=lorm$deviance, df=lorm$df.residual)
[1] 4.24e-06
```

The small *p*-value indicates that the model is not adequate. So while formation latitude is a statistically significant predictor of baroclinic hurricanes, the model can be improved.

To try and improve things, you add another variable to the model. Here you create a new model adding the latitude at which maximum intensity first occurred (`MaxLat`) and examine the table of coefficients.

```
> lorm2 = glm(tb ~ FirstLat + MaxLat, data=bh,
+    family="binomial")
> summary(lorm2)$coefficients
            Estimate Std. Error z value Pr(>|z|)
(Intercept)   -8.560     0.9770   -8.76 1.93e-18
FirstLat       0.504     0.0662    7.62 2.50e-14
MaxLat        -0.134     0.0482   -2.77 5.57e-03
```

Although the latitude at maximum intensity is also statistically significant, something is wrong. The sign on the coefficient is negative indicating that baroclinic hurricanes are more likely if maximum latitude occurs farther south. This lacks physical sense, and it indicates a problem with the model.

The problem arises because of the high correlation between your two explanatory variables (see Chapter 3). You check the correlation between the two variables by typing

```
> cor(bh$FirstLat, bh$MaxLat)
[1] 0.855
```

The correlation exceeds 0.6, so it is best to remove one of your variables. You go back to your one-predictor model, but this time you use maximum latitude. You again check the model for statistical significance and adequacy and find both.

```
> lorm3 = glm(tb ~ MaxLat, data=bh, family="binomial")
> summary(lorm3)$coefficients
            Estimate Std. Error z value Pr(>|z|)
(Intercept)   -5.965     0.6760   -8.82 1.10e-18
MaxLat         0.207     0.0236    8.78 1.58e-18
> pchisq(q=lorm3$deviance, df=lorm3$df.residual)
[1] 0.543
```

Thus, you settle on a final model that includes the latitude at maximum intensity as the sole predictor for hurricane type.

7.10.6 Receiver Operating Characteristics

Your model predicts the probability that a hurricane has baroclinic influences given its latitude at lifetime maximum intensity. To make a decision from this forecast, you need to choose a threshold probability. For example, if the probability exceeds 0.5, then you predict a baroclinic hurricane.

Given your set of hindcast probabilities, one for each hurricane, and a threshold probability, you can create a two-by-two table of observed versus hindcast frequencies indicating how many times you correctly forecast baroclinic and tropical hurricanes with your model. Here you do this using the `table` function on the logical vector of your predictions together with the vector of observed hurricane types.

```
> tab = table(predict(lorm3, type="response") > .5,
+    bh$tb)
> dimnames(tab) = list(Predicted=c("True",
+    "False"), Observed=c("BE", "TO"))
> tab
```

```
          Observed
Predicted  BE  TO
    True  147  51
    False  40  99
```

Note that you use the `dimnames` function on the table object to get the row and column names. The results show that of the 150 tropical-only hurricanes, 99 are predicted as such by the model using the threshold of 0.5 probability. Similarly, of the 187 baroclinic hurricanes, 147 are predicted as such by the model.

In this binary setup, 147 is the number of true positives, 40 is the number of false negatives, 51 is the number of false positives, and 99 is the number of true negatives. The *sensitivity* of your classification scheme is defined as the true-positive proportion given as 147/(147+40) = 79%. The *specificity* is defined as the true-negative proportion given as 99/(40+99) = 71 %.

Note that the values for sensitivity and specificity depend on your choice of threshold. For example, by increasing the threshold to 0.7, the sensitivity changes to 94 percent and the specificity to 90 percent.

The false-positive proportion is one minus the specificity. As you allow for more false positives, you can increase the sensitivity of your model. In the limit, if your model predicts all hurricanes to be BE, then it will be perfectly sensitive (it makes no mistakes in predicting every baroclinic hurricane as baroclinic), but it will not be specific enough to be useful.

A graph of the sensitivity versus the false-positive rate (1−specificity) as the decision threshold is varied is called an ROC curve. Here you use the code in **roc.R** (Peter DeWitt) to generate the ROC curves for your logistic model. Source the code by typing

```
> source("roc.R")
```

The `roc` function uses the model formula together with a set of training (testing) and validation data to generate ROC output. Here you use the `sample` function to take 168 random samples from the set of integers representing the sequence of hurricanes. The corresponding set of hurricanes is used for training and the remaining hurricanes are used for validation.

```
> set.seed(23456)
> idx = sample(dim(bh)[1], trunc(.5*dim(bh)[1]))
> out = roc(formula(lorm), data = bh[idx, ],
+   testing.data=bh[-idx, ], show.auc=FALSE)
```

The output is a list of three elements with the first two containing the area under the ROC curves for the test and validation data, respectively. To plot the curves, type

```
> out$plot
```

Figure 7.10 is made using `ggplot` (see Chapter 5) and shows the two ROC curves corresponding to the training and validation data sets. The area under the curve is an

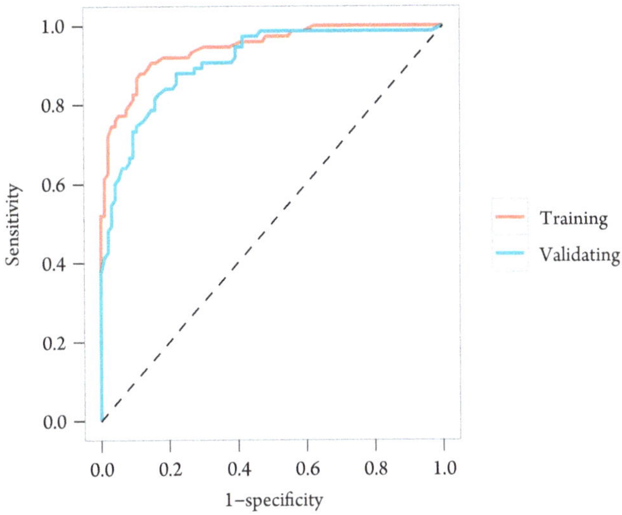

Figure 7.10 ROC curves for the logistic regression model of hurricane type.

indication of the amount of model skill with the diagonal line (dotted) indicating no skill. To list the areas, type

```
> round(out$testing.auc, 3)
[1] 0.944
> round(out$validation.auc, 3)
[1] 0.905
```

Both values are close to 1 indicating a skillful model. In general, you expect the validation area to be less than the testing area.

You interpret the ROC curve as follows. Looking at the graph, if you allow a false-positive proportion of 20 percent (0.2 on the horizontal axis), then you can expect to correctly identify 84 percent of the future BE hurricanes. Since you are interested in future hurricanes, you use the validation curve. Note that if you want to perform better than that, say correctly identifying 95 percent of future BE hurricanes, then you need to accept a false-positive rate of about 40 percent.

This chapter showed how to build models for the occurrence of hurricanes. We began by modeling the annual counts of U.S. hurricanes with a Poisson regression and using environmental variables as covariates. We showed how to make predictions with the model and interpret the coefficients. We showed how to assess forecast skill including how to run a cross-validation exercise. We then showed how to include nonlinear terms in the regression with multivariate adaptive regression splines. We also took a look at a zero-inflated and a random forest model. We finished with an examination of logistic regression for predicting hurricane type. We showed how to interpret the coefficients, make predictions, and evaluate the ROCs when a decision threshold is imposed.

8

INTENSITY MODELS

"We must think about what our models mean, regardless of fit, or we will promulgate nonsense."
—Leland Wilkinson

Strong hurricanes, such as Camille in 1969, Andrew in 1992, and Katrina in 2005, cause catastrophic damage. It is important to have an estimate of when the next big one will occur. You also want to know what influences the strongest hurricanes and whether they are getting stronger as the earth warms.

This chapter shows you how to model hurricane intensity. The data are basinwide lifetime highest intensities for individual tropical cyclones over the North Atlantic and county-level hurricane wind intervals. We begin by considering trends using the method of quantile regression and then examine extreme-value models for estimating return periods. We also look at modeling cyclone winds when the values are given by category, and use Miami-Dade County as an example.

8.1 LIFETIME HIGHEST INTENSITY

Here you consider cyclones above tropical storm intensity ($\geq$ 17 m s^{-1}) during the period 1967–2010, inclusive. The period is long enough to see changes but not too long that it includes intensity estimates before satellite observations. We use "intensity" and "strength" synonymously to mean the fastest wind inside the cyclone.

8.1.1 Exploratory Analysis

Consider the set of events defined by the location and wind speed at which a tropical cyclone first reaches its lifetime maximum intensity (see Chapter 5). The data are in the file *LMI.txt*. Import and list the values in 10 columns of the first 6 rows of the data frame by typing

```
> LMI.df = read.table("LMI.txt", header=TRUE)
> round(head(LMI.df)[c(1, 5:9, 12, 16)], 1)
        Sid   Yr Mo Da hr   lon WmaxS maguv
26637.5 941 1967  9  3 17 -52.2  70.5  27.5
26703.4 942 1967  9 20 10 -97.1 136.2   8.0
26747.2 943 1967  9 13  2 -51.0  94.5   4.2
26807.2 944 1967  9 13 20 -65.0  74.3   3.8
26849.5 945 1967  9 28 23 -56.9  47.3   9.0
26867   946 1967 10  3  0 -93.7  69.0   5.6
```

The data set is described in Chapter 6. Here your interest is the smoothed intensity estimate at the time of lifetime maximum (`WmaxS`).

First, convert the wind speeds from the operational units of knots to the SI units of meter per second.

```
> LMI.df$WmaxS = LMI.df$WmaxS * .5144
```

Next, determine the quartiles (0.25 and 0.75 quantiles) of the wind speed distribution. The quartiles divide the cumulative distribution function (CDF) into three equal-sized subsets.

```
> quantile(LMI.df$WmaxS, c(.25, .75))
 25%  75%
25.5 46.0
```

You find that 25 percent of the cyclones have a lifetime maximum wind speed less than 26 m s^{-1} and 75 percent have a maximum wind speed less than 46 m s^{-1}, so that 50 percent of all cyclones have a maximum wind speed between 26 and 46 m s^{-1} (interquartile range–IQR). Similarly, the quartiles (deciles) divide the sample of storm intensities into four (10) groups with equal proportions of the sample in each group. The quantiles, or percentiles, refer to the general case.

The CDF gives the empirical probability of observing a value in the record less than a given wind speed maximum. The quantile function is the inverse of the CDF allowing you to determine the wind speed for specified quantiles.

Thus, given a sample of maximum wind speeds $w_1, \ldots, w_n$, the τth sample quantile is the τth quantile of the corresponding empirical CDF. Formally, let W be a random maximum storm intensity, then the kth "q"-quantile is defined as the value "w" such that

$$p(W \leq w) \geq \tau \text{ and } p(W \geq w) \geq 1 - \tau \qquad (8.1)$$

where $\tau = \frac{k}{n}$.

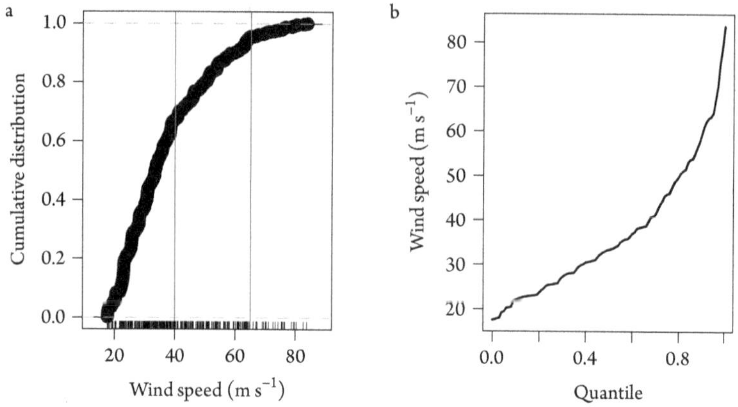

Figure 8.1 Fastest cyclone wind. (a) Cumulative distribution and (b) quantile.

Figure 8.1 shows the cumulative distribution and quantile functions for the 500 tropical cyclone intensities in the data frame. The CDF appears to have three distinct regions, delineated by the vertical lines. The function is nearly a straight line for intensities less than 40 m s^{-1} and greater than 65 m s^{-1}.

Is there a trend in cyclone intensities? Start with a plot of your data. By specifying the first argument in the `boxplot` function as a model formula, you create a sequence of conditional box plots. For example, to create a series of wind speed box plot conditional on year, type

```
> boxplot(LMI.df$WmaxS ~ as.factor(LMI.df$SYear))
```

Note that the conditioning variable must be specified as a factor. The graph is useful for examining the distribution of your wind speed data over time.

Recall from Chapter 5 that you created a series of box plots of the SOI by month that minimized the amount of redundant ink. Here you reuse this code, modifying it a bit, to create a series of wind speed box plots by year. Begin by creating a vector of years and saving the length of the vector as a numeric object.

```
> yrs = 1967:2010
> n = length(yrs)
```

Next create the plot frame without the data and without the horizontal axis tic labels. Then add a label to the vertical axis.

```
> plot(c(1967, 2010), c(15, 85), type="n", xaxt="n",
+     bty="n", xlab="",
+     ylab="Lifetime maximum wind speed (m/s)")
> axis(1, at=yrs, labels=yrs, cex=.5)
```

The function `fivenum` lists the minimum, first quartile, median, third quartile, and maximum value, in that order, so to obtain the median value from a vector of values called x, you type `fivenum(x)[3]`. You loop over each year indexed by `i`

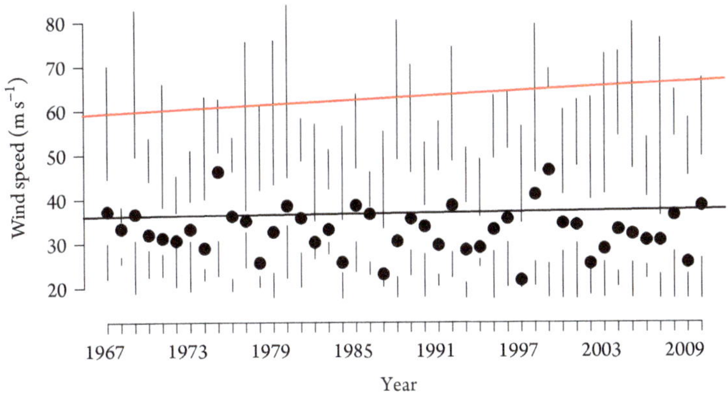

Figure 8.2 Lifetime highest wind speeds by year.

and plot the median wind speed value for that year as a point using the `points` function. In the same loop, you create vertical lines connecting the minimum with the first quartile and the third quartile with the maximum using the `lines` function.

```
> for(i in 1:n){
+ fn = fivenum(LMI.df$WmaxS[LMI.df$SYear == yrs[i]])
+ points(yrs[i], fn[3], pch=19)
+ lines(c(yrs[i], yrs[i]), c(fn[1], fn[2]))
+ lines(c(yrs[i], yrs[i]), c(fn[4], fn[5]))
+ }
```

Note that the subset operator `[` is used to obtain wind speed values by year.

The results are shown in Figure 8.2. Here we added the least-squares regression line about the annual mean lifetime highest wind speed (black line) and the least-squares regression line about the annual lifetime highest wind speed (red). Although there is only a shallow upward trend in average cyclone intensities, there is a substantial trend in the intensities of the strongest cyclones.

The theory of maximum potential intensity, which relates intensity to ocean heat, refers to a theoretical limit given proper environmental conditions (Emanuel, 1988). So the upward trend in the observed lifetime maximum intensity is consistent with what you expect given the observed increases in ocean temperature.

It is informative then to explore the relationship of lifetime highest wind speed to ocean temperature. In Chapter 2, you imported the monthly North Atlantic sea-surface temperature (SST) data by typing

```
> SST = read.table("SST.txt", header=TRUE)
```

Here you subset the data using years since 1967 and only keep June values.

```
> lg = SST$Year >= 1967
> sst.df = data.frame(Yr=SST$Year[lg],
+    sst=SST$Jun[lg])
```

Next, merge your SST data frame with your cyclone intensity data. This is done using the `merge` function. Merge is performed on the shared column name `Yr` as specified with the `by` argument.

```
> lmisst.df = merge(LMI.df, sst.df, by="Yr")
> head(lmisst.df[c("Yr", "WmaxS", "sst")])
    Yr WmaxS sst
1 1967  36.3  21
2 1967  70.1  21
3 1967  48.6  21
4 1967  38.2  21
5 1967  24.3  21
6 1967  35.5  21
```

Note that since there are more instances of `Yr` in the intensity data frame (one for each cyclone), the June SST values in the SST data frame get duplicated for each instance. Thus all cyclones for a particular year get the same SST value.

You are interested in regressing cyclone intensity on SST, but the SST values are continuous rather than discrete. So you first create SST intervals. This is done with the `cut` function.

```
> brk = quantile(lmisst.df$sst, prob=seq(0, 1, .2))
> sst.i = cut(lmisst.df$sst, brk, include.lowest=TRUE)
```

Your cuts divide the SST values into five equal quantiles (pentiles). The intervals represent categories of much below (MB) normal, below (B) normal, normal (N), above (A) normal, and much above (MA) normal. The choice of quantiles is a compromise between having enough years for a given range of SSTs and having enough quantiles to assess differences.

You repeat this procedure for your SOI data. You create a merged data frame and cut the SOI values into pentads.

```
> SOI = read.table("SOI.txt", header=TRUE)
> lg = SOI$Year >= 1967
> soi.df = data.frame(Yr=SOI$Year[lg],
+   soi=SOI$Sep[lg])
> lmisoi.df = merge(LMI.df, soi.df, by="Yr")
> brk = quantile(lmisoi.df$soi, prob=seq(0, 1, .2))
> soi.i = cut(lmisoi.df$soi, brk, include.lowest=TRUE)
```

Finally, you create a series of box plots corresponding to the SST intervals. This time you use the `boxplot` function as described earlier. Begin by creating a character vector of horizontal axis labels corresponding to the SST intervals and, to simplify the code, save the wind speeds as a vector.

```
> xlabs = c("MB", "B", "N", "A", "MA")
> W = lmisst.df$WmaxS
```

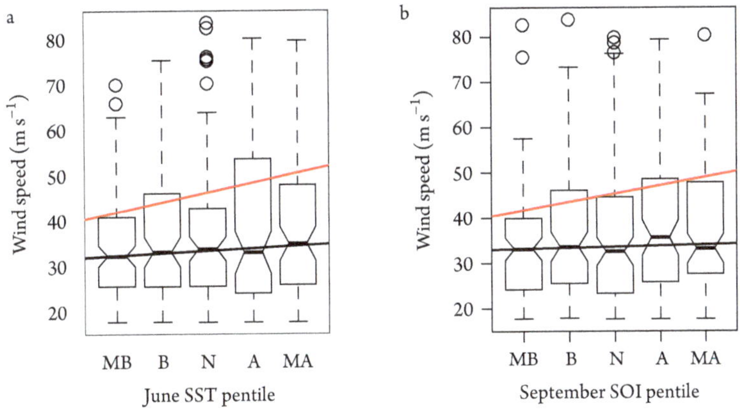

Figure 8.3 Lifetime highest intensity by (a) June SST and (b) September SOI.

You then save the output from a call to the `boxplot` function, making sure to turn off the plotting option.

```
> y = boxplot(W ~ sst.i, plot=FALSE)
```

Initiate the plot again and add regression lines through the medians and third quartile values using the saved statistics of the box plot and regressing on the sequence from one to five.

```
> boxplot(W ~ sst.i, notch=TRUE, names=xlabs,
+   xlab="June SST Quantiles",
+   ylab="Lifetime Maximum Wind Speed (m/s)")
> x = 1:5
> abline(lm(y$stats[3, ] ~ x), lwd=2)
> abline(lm(y$stats[4, ] ~ x), col="red", lwd=2)
```

The results are shown in Figure 8.3. Here we repeat the code using the September SOI covariate and create two box plots. The first pentad is the lowest 20 percent of all values. The upper and lower limits of the boxes represent the first and third quartiles of cyclone intensity. The median for each group is represented by the horizontal bar in the middle of each box. Notches on the box sides represent an estimated confidence interval about the median. The full range of the observed intensities in each group is represented by the horizontal bars at the end of the dashed whiskers. In cases in which the whiskers extend more than one and half times the interquartile range, they are truncated and the remaining outlying points are indicated by open circles. The red line is the best-fit line through the upper quartile and the black line is through the medians.

The box plot summarizes the distribution of maximum storm intensity by pentiles of the covariate. The graphs show an increase in upper quantiles of cyclone intensity with increasing SST and SOI. As SST increases, so does the intensity of the strongest cyclones. Also as SOI increases (toward more La Niña-like conditions), so does the intensity of the cyclones. Results from your exploratory analysis give you a reason to continue your investigation.

The next step is to model these data. The box plots provide evidence that a model for the mean will not capture the relationships as the trends are larger for higher quantiles. So instead of linear regression, you use quantile regression.

8.1.2 Quantile Regression

The quantile function and the conditional box plots shown above are useful for exploratory analysis. They are adequate for describing and comparing univariate distributions. However, since you are interested in modeling the relationship between a response variable (intensity) and the covariates (SST and SOI), it is necessary to introduce a regression model for the quantile function. Quantile regression extends the ordinary least-squares regression model to conditional quantiles of the response variable. Although you used linear regression on the conditional quantiles in the exploratory plots, this is not the same as quantile regression on the covariates.

Quantile regression allows you to examine cyclone intensity without the need to consider discrete levels of the covariate. Ordinary regression model specifies how the mean changes with changes in the covariates while the quantile regression model specifies how the quantile changes with changes in the covariates. Quantile regression relies on empirical quantiles, but uses parameters to assess the relationship between the quantile and the covariates.

The quantile regression model with two covariates is given by

$$\hat{\mu}(\tau) = \hat{\beta}_0(\tau) + \hat{\beta}_1(\tau)x_1 + \hat{\beta}_2(\tau)x_2 \tag{8.2}$$

where $\hat{\mu}(\tau)$ is the predicted conditional quantile of tropical cyclone intensity (W) and where the $\hat{\beta}_i$'s are obtained by minimizing the piecewise linear least absolute deviation function given by

$$\frac{1-\tau}{n} \sum_{w_i<q_i} |w_i - q_i| + \frac{\tau}{n} \sum_{w_i>q_i} |w_i - q_i| \tag{8.3}$$

for a given τ, where q_i is the predicted τ quantile corresponding to observation i ($\hat{\mu}_i(\tau)$).

The value of a simple trend analysis (involving only one variable—usually time) is limited by the fact that other explanatory variables also might be trending. In the context of hurricane intensity, it is well known that the ENSO cycle can alter the frequency and intensity of hurricane activity on the seasonal time scale. A trend over time in hurricane intensity could reflect a change in this cycle. Thus it is important to look at the trend after controlling for this factor. Here we show the trend as a function of Atlantic SST after controlling for the ENSO cycle. Thus we answer the question of whether the data support the contention that the increasing trend in the intensity of the strongest hurricanes is related to an increase in ocean warmth conditional on ENSO.

Here we use the **quantreg** package for performing quantile regression developed by Roger Koenker. Load the package and print a BibTeX citation.

```
> require(quantreg)
> x = citation(package="quantreg")
> toBibtex(x)
@Manual{,
  title = {quantreg: Quantile Regression},
  author = {Roger Koenker},
  year = {2011},
  note = {R package version 4.76},
  url = {http://CRAN.R-project.org/package=quantreg},
}
```

Begin with median regression. Here τ is set to 0.5 and is specified with the `tau` argument. The function `rq` performs the regression. The syntax for the model formula is the same as before. The output is assigned to the object `qrm`.

```
> Year = lmisst.df$Yr
> W = lmisst.df$WmaxS
> SOI = lmisoi.df$soi
> SST = lmisst.df$sst
> qrm = rq(W ~ Year + SST + SOI, tau=.5)
```

Rather than least-squares or maximum likelihood, a simplex method is used to fit the regression. It is a variant of Barrodale and Robert's (1974) approach described in Koenker and d'Orey (1987). If your data set has more than a few thousand observations, it is recommended that you change the default by specifying `method="fn"`, which invokes the Frisch-Newton algorithm described in Portnoy and Koenker (1997).

You obtain a concise summary of the regression results by typing

```
> qrm
Call:
rq(formula = W ~ Year + SST + SOI, tau = 0.5)

Coefficients:
(Intercept)        Year         SST         SOI
    238.039      -0.221      11.009       0.827

Degrees of freedom: 500 total; 496 residual
```

The output shows the estimated coefficients and information about the degrees of freedom. You find that the median lifetime intensity decreases with year (negative trend) and increases with SST and the SOI.

To obtain more details, you type

```
> summary(qrm)
```

Table 8.1 Coefficients of the median regression model.

	Coefficients	*Lower bd*	*Upper bd*
(Intercept)	238.04	11.67	414.78
Year	−0.22	−0.33	−0.05
SST	11.01	0.73	16.99
SOI	0.83	−0.65	1.93

Table 8.1 gives the estimated coefficients and confidence intervals (95%) for these parameters. The confidence intervals are computed by the rank inversion method developed in Koenker (2005).

The confidence interval includes zero for Year and the SOI indicating these terms are not significant in explaining the median per cyclone intensity. However, the SST variable is significant and positive. The relationship indicates that for every 1°C increase in SST, the median intensity increases by 11 m s^{-1}.

But this seems too large (by an order of magnitude), given the box plot (Fig. 8.3) and the range of SST values.

```
> range(SST)
[1] 20.8 21.8
```

The problem arises from having other variables in the model. To see this, refit the regression model after removing the variables that are not significant.

```
> qrm2 = rq(W ~ SST, tau=.5)
> summary(qrm2)
Call: rq(formula = W ~ SST, tau = 0.5)

tau: [1] 0.5

Coefficients:
            coefficients lower bd upper bd
(Intercept)  -14.23       -141.11   63.12
SST            2.23         -1.44    8.18
```

Now the relationship indicates that for every 1°C increase in SST, the median intensity increases by 2.23 m s^{-1}. This amount is not statistically significant as you might have guessed from your exploratory plot.

The theory of maximum potential intensity relates a theoretically highest wind speed to ocean temperature so it is interesting to consider quantiles above the median. You repeat the modeling exercise using $\tau = 0.9$. Here again, you find year and SOI not significant, so you exclude them in your final model.

```
> summary(rq(W ~ SST, tau=.9), se="iid")
```

Table 8.2 Coefficients of the 90th percentile regression model.

	Value	*Std. Error*	*t Value*	Pr(>|t|)
(Intercept)	−307.30	68.69	−4.47	0.00
SST	17.16	3.22	5.33	0.00

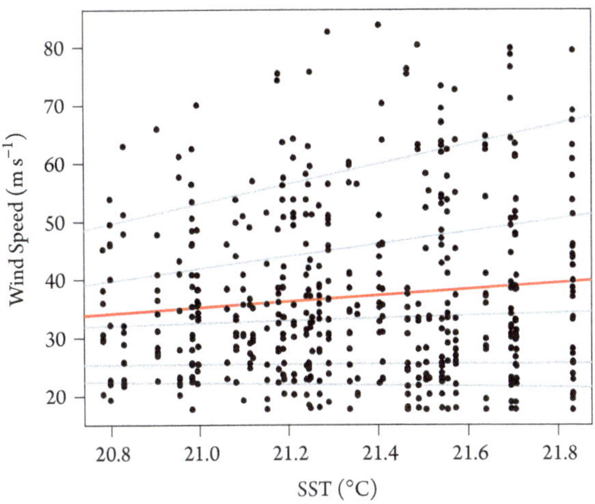

Figure 8.4 Quantile regressions of lifetime maximum intensity on SST.

Here instead of the rank-inversion CI, you obtain a more conventional table of coefficients (Table 8.2) that includes standard errors, t-statistics, and p-values using the `se="iid"` argument in the `summary` function.

As anticipated from theory and your exploratory data analysis, you see a statistically significant positive relationship between cyclone intensity and SST for the set of tropical cyclones within the top 10 percent of intensities. The estimated coefficient indicates that for every 1°C increase in SST, the upper percentile intensity increases by 17.2 m s^{-1}.

Other options exist for computing standard errors including a bootstrap approach (`se="boot"`; see Koenker [2005]), which produces a standard error in this case of 4.04 (difference of 26 percent). The larger standard error results in a significance level that is somewhat less, but the results still provide conclusive evidence of a climate change signal.

To visualize the intensity–SST relationship in more detail, you plot several quantile regression lines on a scatter plot. For reference, you include the least-squares regression line. The code is given here, and the results are shown in Figure 8.4. Note that you use `type="n"` in the plot function and use the `points` function to add them on top of the lines. The 0.1, 0.25, 0.5, 0.75, and 0.9 quantile regression lines are shown in gray and the least-squares regression line about the mean is shown in red. Trend lines are close to horizontal for the weaker tropical cyclones but have a significant upward slope for the stronger cyclones.

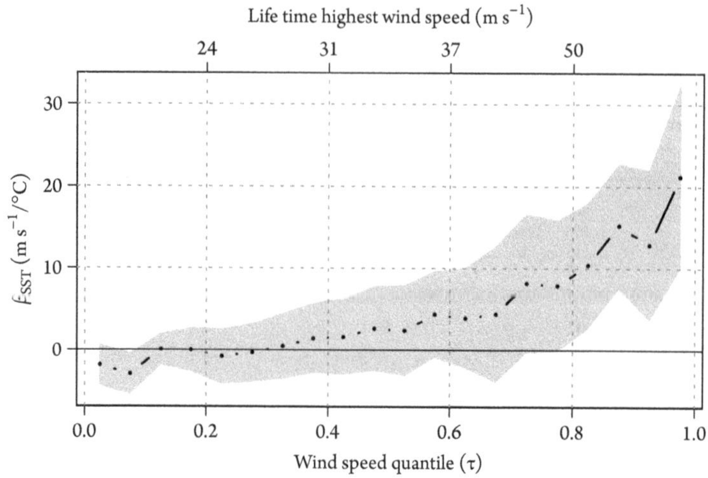

Figure 8.5 SST coefficient from a regression of LMI on SST and SOI.

To see all of the quantile regressions for a particular model, you specify a `tau=-1`. For example, save the quantile regressions of wind speed on SST and SOI in an object by typing

```
> model = rq(W ~ SST + SOI, tau=-1)
```

This will cause the `rq` function to find the entire sample path of the quantile process. The returned object is of class `rq.process`. You plot the regression coefficients for each variable in the model as a function of quantile by typing

```
> plot.rq.process(model)
```

The result for the SST variable is plotted in Figure 8.5. Values of τ range from 0.025 to 0.975 in intervals of 0.05. The 95 percent confidence band (gray) is based on a bootstrap method. The plot shows the rising trend of the most intense hurricanes as the ocean temperatures rise after accounting for the El Niño. The trends depart from zero for quantiles above about 0.4 and become significant for cyclones that exceed about 50 m s^{-1}. Additional capabilities for quantile modeling and inference are available in the **quantreg** package. Next, we consider a model for the most intense hurricanes.

8.2 FASTEST HURRICANE WINDS

Eighty percent of all hurricane damage is caused by less than 20 percent of the worst events (Jagger et al., 2008). The rarity of severe hurricanes implies that empirical models that estimate the probability of the next big one will be unreliable. Extreme value theory provides a statistical framework for modeling these rare wind events. Here you employ these models on hurricane wind speeds.

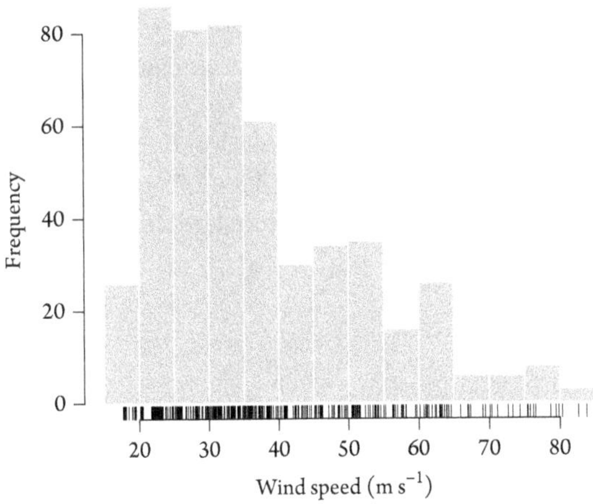

Figure 8.6 Histogram of lifetime highest wind speed.

8.2.1 Exploratory Analysis

To begin, you plot the lifetime maximum wind speeds for all North Atlantic tropical cyclones from the period 1967 to 2010 as a histogram. You use the same data set as in §8.1, where `W` is the vector of wind speeds.

```
> W = LMI.df$WmaxS
> hist(W, main="", las=1, col="gray", border="white",
+     xlab="Wind Speed (m/s)")
> rug(W)
```

The function uses 5 m s^{-1} intervals and the minimum intensity is 17.6 m s^{-1}.

Figure 8.6 shows a peak in the distribution between 20 and 40 m s^{-1} and a long right tail. Values in this tail are of interest. For a model of the fastest winds, you want to include enough of these high values that your parameter estimates are reliable (they do not change by much if you add or remove a few values). But you also want to be careful not to include too many to ensure that the values represent the strongest hurricanes.

8.2.2 Return Periods

How long can you expect to wait for the next big hurricane? A return period is the average recurrence interval. The recurrence interval is the time between successive hurricanes of a given intensity or stronger (events). Suppose you define an event as a hurricane with a threshold intensity of 75 m s^{-1}, then the annual return period is the inverse of the probability that such an event will be exceeded in any one year. Here "exceeded" refers to a hurricane with an intensity of at least 75 m s^{-1}.

For instance, a 10-year hurricane event has a $1/10 = 0.1$ or 10 percent chance of having an intensity exceeding a threshold level in any one year and a 50-year hurricane event has a 0.02 or 2 percent chance of having an intensity exceeding a higher threshold level in any one year. These are statistical statements. On average, a 10-year event will occur once every 10 years. The interpretation requires that for a year or set of years in which the event does not occur, the expected time until it occurs next remains 10 years, with the 10-year period resetting each year.

Note, there is a monotonic relationship between the intensity of the hurricane event (return level) and the return period. The return period for a 75-m s^{-1} return level must be longer than the return period for a 70-m s^{-1} return level. The empirical relationship is expressed as

$$\text{RP} = \frac{n+1}{m} \tag{8.4}$$

where n is the number of years in the record and m is the intensity rank of the event.[1]

You use this formula to estimate return periods for your set of hurricanes. First assign the record length and sort the lifetime maximum wind speeds in decreasing order. Then list the speeds of the six most intense hurricanes.

```
> n = length(1967:2010)
> Ws = sort(W, decreasing=TRUE)
> round(Ws, 1)[1:6]
[1] 83.8 82.6 80.3 79.8 79.4 78.7
```

Finally, compute the return period for these six events using the above formula, rounding to the nearest year.

```
> m = rev(rank(Ws))
> round((n + 1)/m, 0)[1:6]
[1] 45 22 15 11  9  8
```

Thus, an 83.8-m s^{-1} hurricane has a return period of 45 years and a 78.7-m s^{-1} hurricane has a return period of 8 years. Said another way, you can expect a hurricane of at least 80.3 m s^{-1} once every 45 years. The threshold wind speed for a given return period is called the return level.

Your goal here is a statistical model that provides a continuous estimate of the return level (threshold intensity) for a set of return periods. A model is more useful than a set of empirical estimates because it provides a smoothed return-level estimate for all return periods, and it allows you to estimate the return level for a return period longer than your data record.

The literature provides some examples of hurricane return periods. Rupp and Lander (1996) use the method of moments on annual peak winds over Guam to determine the parameters of an extreme value distribution leading to estimates of return periods for extreme typhoon winds. Heckert et al. (1998) use the peaks-over-threshold method and a reverse Weibull distribution to obtain return periods for extreme wind speeds at locations along the U.S. coastline.

Walshaw (2000) uses a Bayesian approach to jointly model extreme winds from tropical and nontropical systems. Jagger and Elsner (2006) use a maximum likelihood

[1] Sometimes $RP = n/(m - 0.5)$ is used instead.

and Bayesian approach to model tropical cyclone winds in the vicinity of the United States conditional on climate factors. In the former study, the Bayesian approach allows them to take advantage of information from nearby sites, and in the later study it allows them to take advantage of older, less reliable data.

Here you use functions in the **ismev** package (Coles and Stephenson, 2011) to fit an extreme-value model for hurricane winds using the method of maximum likelihood. We begin with some background material.

8.2.3 Extreme Value Theory

Extreme value theory is a branch of statistics. It concerns techniques and models for describing the rare event rather than the typical, or average, event. It is similar to the central limit theory in that it considers the limiting distributions of independent identically distributed (iid) random variables under an affine transformation.[2] According to the central limit theorem, the mean value of an iid random variable x converges to a normal distribution with mean 0 and variance 1 under the affine transformation $(\bar{x} - \mu)/\sqrt{n\sigma^2})$, where μ and σ are the mean and standard deviation of x, respectively.

Similarly, if the distribution of the maxima under some affine transformation converges, then it must converge to a member of the generalized extreme value (GEV) family (Embrechts et al., 1997). The maxima of most continuous random variables converge to a nondegenerate random variable. This asymptotic argument is used to motivate the use of extreme value models in the absence of empirical or physical evidence for assigning an extreme-level to a process. However, the argument does not hold for the maxima of discrete random variables including the Poisson and negative binomial. An excellent introduction to this topic is provided in Coles (2001).

Although by definition extreme values are scarce, an extreme-value model allows you to estimate return periods for hurricanes that are stronger than the strongest one in your data set. In fact, your goal is to quantify the statistical behavior of hurricanes extrapolated to unusually high levels. Extreme value theory provides models for this kind of extrapolation.

Given a set of observations from a continuous process, if you generate a sample from the set, take the maximum value from the sample, and repeat the procedure many times, you obtain a distribution that is different from that of the original (parent) distribution. For instance, if the observations are described by a normal distribution, the distribution of the maxima is described by a Gumbel distribution.

To see this, plot a density curve for the standard normal distribution and compare it to the density curve of the maxima from samples of size 100 taken from the same distribution. Here you generate 1,000 samples saving the maxima in the vector `m`.

```
> par(mfrow=c(1, 2))
> curve(dnorm(x), from=-4, to=4, ylab="Density")
> m = numeric()
```

[2] Linear transformation followed by a translation.

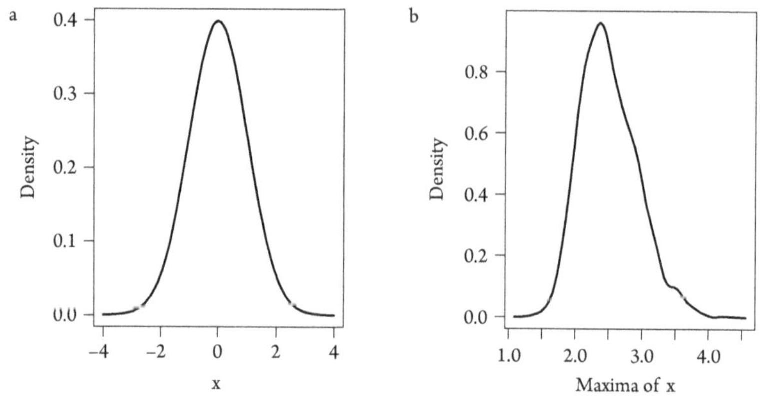

Figure 8.7 Density curves. (a) Standard normal and (b) maxima from samples of the standard normal.

```
> for(i in 1:1000) m[i] = max(rnorm(100))
> plot(density(m), xlab="Maxima of x", main="")
```

The results are shown in Figure 8.7. The maxima belong to a GEV distribution that is shifted to the right relative to the parent distribution and positively skewed. The three parameters of the GEV distribution are determined by values in the tail portion of the parent distribution.

8.2.4 Generalized Pareto Distribution

A GEV distribution fits the set of values consisting of the single strongest hurricane each year. Alternatively, consider the set of per-cyclone lifetime strongest winds in which you keep all values exceeding a given threshold level, say 60 m s^{-1}. Some years will contribute no values to your set and some years will contribute two or more.

A two-parameter generalized Pareto distribution (GPD) family describes this set of fast winds. The threshold choice is a compromise between retaining enough hurricanes to estimate the distribution parameters with sufficient precision but not too many that the intensities fail to be described by a GPD.

Specifically, given a threshold wind speed u, you model the exceedances, $W-u$, as samples from a GPD family so that for an individual hurricane with maximum winds W, the probability that W exceeds any value v given that it is above the threshold u is given by

$$p(W > v|W > u) = \begin{cases} \exp([-(v-u)]/\sigma) & \text{when } \xi = 0 \\ (1+\frac{\xi}{\sigma}[v-u])^{-1/\xi} & \text{otherwise} \end{cases} \tag{8.5}$$

where $\sigma > 0$ and $\sigma + \xi(v-u) \geq 0$. The parameters σ and ξ are scale and shape parameters of the GPD, respectively. Thus you can write $p(W > v|W > u) = \text{GPD}(v-u|,\sigma,\xi)$.

To illustrate, copy the following code to create a function called sGpd for the exceedance probability of a GPD.

```
> sGpd = function(w, u, sigma, xi){
+    d = (w - u) * (w > u)
+    sapply(xi, function(xi) if(xi==0) exp(-d/sigma)
+    else
+    ifelse(1 + xi/sigma * d < 0, 0,
+              (1 + xi/sigma * d)^(-1/xi)))
+     }
```

Given a threshold intensity u, the function computes the probability that a hurricane at this intensity or higher picked at random will have a maximum wind speed of at least W. The probability depends on the scale and shape parameters. For instance, given a scale of 10 and a shape of 0, the probability that a random hurricane will have a maximum wind speed of at least 70 m s^{-1} is obtained by typing

```
> sGpd(w=70, u=60, sigma=10, xi=0)
[1] 0.368
```

The scale parameter controls how fast the probability decays for values near the threshold. The decay is faster for smaller values of σ. The shape parameter controls the length of the tail. For negative values of ξ, the probability is zero beyond a certain intensity. With $\xi = 0$ the probability decay is exponential. For positive values of ξ, the tail is described as "heavy" or "fat," indicating a decay in the probabilities gentler than logarithmic. Figure 8.8 compares exceedance curves for different values of σ with $\xi = 0$ and for different values of ξ with $\sigma = 10$, keeping the threshold value at 60 m s^{-1}.

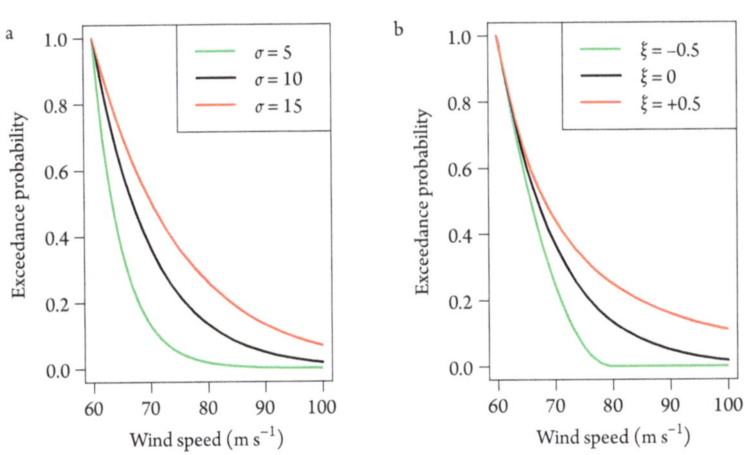

Figure 8.8 Exceedance curves for the generalized Pareto distribution. (a) Different σ's with $\xi = 0$ and (b) different ξ's with $\sigma = 10$.

8.2.5 Extreme Intensity Model

Given your set of lifetime maximum wind speeds in the object `W`, you use the `gpd.fit` function from the **ismev** package to find the scale and shape parameters of the GPD using the method of maximum likelihood. Here you set the threshold wind speed to 62 m s^{-1} again as a compromise between high enough to capture only the strongest hurricanes and low enough to have a sufficient number of wind speeds. The output is saved as a list object and printed to your screen.

```
> require(ismev)
> model = gpd.fit(W, threshold=62)
$threshold
[1] 62

$nexc
[1] 42

$conv
[1] 0

$nllh
[1] 124

$mle
[1]  9.832 -0.334

$rate
[1] 0.084

$se
[1] 2.794 0.244
```

This is a probability model that specifies the chance of a random hurricane obtaining any intensity value given that it has already reached the threshold intensity.

The function prints the threshold value, the number of extreme winds in the data set (`nexc`) as defined by the threshold, the negative log-likelihood value (`nllh`), the maximum-likelihood parameter estimates (`mle`), and the rate, which is the number of extreme winds divided by the total number of hurricane (per-hurricane rate).

You use your `sGpd` function to compute probabilities for a sequence of winds from the threshold value to 85 m s^{-1} in increments of 0.1 m s^{-1}.

```
> v = seq(63, 85, .1)
> p = sGpd(v, u=62, sigma=model$mle[1],
+    xi=model$mle[2])
```

You then use the plot method to graph them.

```
> plot(v, p, type="l", lwd=2, xlab="Wind Speed (m/s)",
+   ylab="p(W > v | W > 62)")
```

To turn the per-hurricane rate into an annual rate, you divide the number of extreme winds by the record length.

```
> rate = model$nexc/length(1967:2010)
> rate
[1] 0.955
```

Thus, the annual rate of hurricanes at this intensity or higher over the 44 years in the data set is 0.95 per year. Recall from the Poisson distribution that this implies a

```
> round((1 - ppois(0, rate)) * 100, 2)
[1] 61.5
```

percent chance that next year a hurricane will exceed this threshold.

8.2.6 Intensity and Frequency Model

The GPD describes hurricane intensities above a threshold wind speed. You know from Chapter 7 that the Poisson distribution describes the frequency of hurricanes above some intensity. Therefore you need to combine these two descriptions.

Let the annual number of hurricanes whose lifetime maximum intensity exceeds u have a Poisson distribution with mean rate λ_u. Then the average number of hurricanes with winds exceeding v (where $v \geq u$) is given by

$$\lambda_v = \lambda_u \times p(W > v|W > u) \tag{8.6}$$

This allows you to model hurricane occurrence separate from hurricane intensity. This is helpful because processes that govern hurricane frequently are not necessarily the same as the processes that govern hurricane intensity. Moreover, from a practical perspective, rather than a return rate per hurricane occurrence, this specification allows you to obtain an annual return rate on the extreme winds. This is more meaningful for the business of risk management and insurance.

Now, the probability that the highest lifetime maximum intensity in a given year will be less than v is

$$p(W\text{max} \leq v) = \exp(-\lambda_v) \tag{8.7}$$

$$= \exp[-\lambda_u \times \text{GPD}(v - u|\sigma, \xi)] \tag{8.8}$$

The return period RP is the inverse of the probability that $W\text{max}$ exceeds v, where v is called the return level. You compute the return period and create a return period plot using

```
> rp = 1/(1 - exp(-rate * p))
> plot(rp, v, type="l", lwd=2, log="x",
+   xlab="Return Period (yr)",
+   ylab="Return Level (m/s)")
```

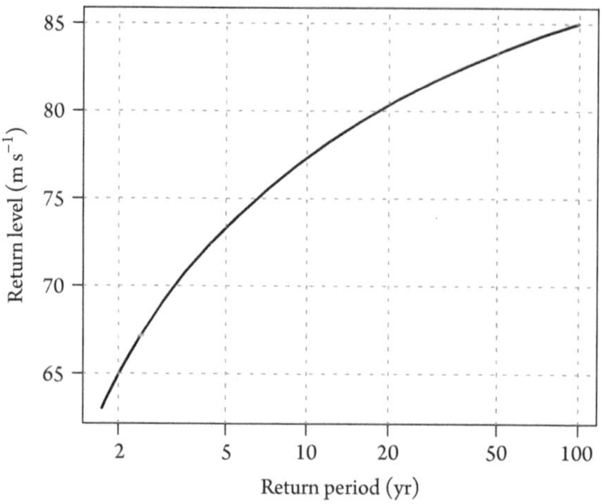

Figure 8.9 Return periods for the fastest hurricane winds.

Figure 8.9 shows the results. Return levels increase with increasing return period. The model estimates that an 81-m s^{-1} hurricane will occur on average once every 27 years and an 85-m s^{-1} hurricane will occur on average once every 100 years. However, based on the results in §8.1, these return periods might be getting shorter due to increasing ocean heat.

8.2.7 Confidence Intervals

You obtain confidence limits on the return period estimates shown in Figure 8.9 using a bootstrap approach (see Chapter 3). Suppose you are interested in the 95 percent CI on the return period of a 73-m s^{-1} hurricane. Your model tells you that the best estimate for the return period is 5 years.

To obtain the CI, you randomly sample your set of wind speeds with replacement to create a bootstrap replicate. You run your model on this replicate and get an estimate of the return period. You repeat this procedure 1,000 times each time generating a new return period estimate. You then treat the bootstrapped return periods as a distribution and find the lower and upper quantiles corresponding to the 0.025 and 0.975 probabilities.

To implement this procedure, you type

```
> thr = 62
> v = 73
> rps = numeric()
> m = 1000
> for(i in 1:m){
+   Wbs = sample(W, size=length(W), replace=TRUE)
+   modelbs = gpd.fit(Wbs, threshold=thr, show=FALSE)
+   ps = sGpd(v, u=thr, sigma=modelbs$mle[1],
+             xi=modelbs$mle[2])
```

```
+    rps[i] = 1/(1 - exp(-rate * ps))
+    }
> ci = round(quantile(rps, probs=c(.025, .975)))
```

The procedure provides a 95 percent CI of (3, 9) years about the estimated 5-year return period for a 73-m s^{-1} hurricane. You can estimate other CIs (e.g., 90%) by specifying the relevant percentiles in the `quantile` function.

8.2.8 Threshold Intensity

The GPD model requires a threshold intensity u. The choice is a trade-off between an intensity high enough that the positive residual values ($W - u \geq 0$) follow a GPD, but low enough that there are enough values to accurately estimate the GPD parameters.

For an arbitrary intensity level, you can compute the average of the positive residuals (excesses). For example, at an intensity of 60 m s^{-1}, the mean excess in units of m s^{-1} is

```
> mean(W[W >= 60] - 60)
[1] 8
```

By increasing the level, say to 70 m s^{-1}, the mean excess decreases to 6.23 m s^{-1}.

In this way, you compute a vector of mean excesses for a range of potential threshold intensities. The relationship between the mean excess and threshold is linear if the residuals follow a GPD. A plot of the mean excess across a range of intensity levels (mean residual life plot) helps you choose a threshold intensity.

The function `mrl.plot` is part of the **ismev** package that makes the plot for you. Type

```
> mrl.plot(W)
> grid()
```

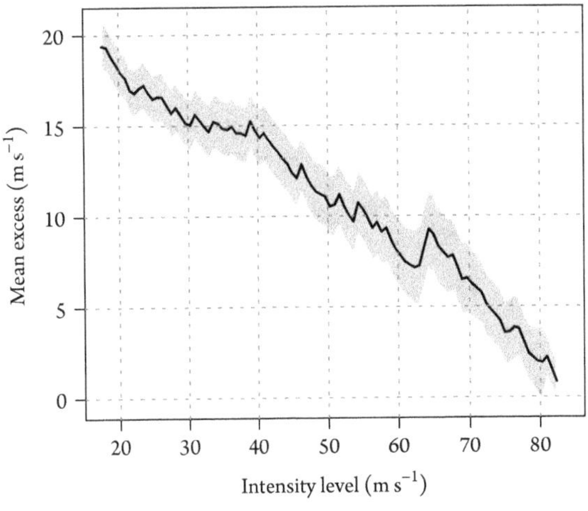

Figure 8.10 Mean excess as a function of threshold intensity.

The result is shown in Figure 8.10. There is a general decrease in the mean excess with increasing intensity levels. The 95 percent confidence band is shown in gray. The decrease is linear above an intensity value of 62 m s^{-1}, indicating that any threshold chosen above this results in a set of wind speeds that follow a GPD. To maximize the number of wind speed values for estimating the model parameters, the lowest such threshold is optimal.

Alternatively, you can proceed by trial and error. You calculate the parameters of the GPD for increasing thresholds and choose the minimum threshold at which the parameter values remain fixed.

8.3 CATEGORICAL WIND SPEEDS BY COUNTY

Hurricane wind speeds are often described with a Saffir–Simpson category. If possible you should avoid using categories for analysis and modeling. However, historical hurricane intensities data are sometimes provided only by category.

Here you model county-level categorical wind data. The data represent direct and indirect hurricane hits by Saffir—Simpson category. The data are described and organized in Chapter 6. The wind speed category and count data are saved in separate binary files. Make them available in your workspace by typing

```
> load("catwinds.RData")
> load("catcounts.RData")
```

The list of data frames is stored in the object `winds`. Lists are generic objects and can be of any type. To see the data frame for Cameron County, Texas (the first county in the list where the counties are numbered from 1 to 175 starting with south Texas), type

```
> winds[[1]]
  Year        W
1 1909 [42, 50]
2 1909 [33, 50]
3 1910 [33, 42]
4 1919 [33, 50]
5 1933 [42, 50]
6 1933 [50, 58]
7 1967 [50, 58]
8 1980 [50, 58]
9 2008 [33, 42]
```

The data frame contains a numerical year variable and a categorical survival variable. The survival variable has three components with the first two indicating the wind speed bounds of the cyclone. The bounds correspond to the Saffir—Simpson categories.

The data frame of corresponding hurricane counts is stored in the object `counts`. To see the first 10 years of counts from Cameron County, type

```
> counts[1:10, 1]
1900 1901 1902 1903 1904 1905 1906 1907 1908 1909
   0    0    0    0    0    0    0    0    0    2
```

There were no hurricanes in this part of Texas during the first nine years of the twentieth century, but there were two in 1909. The first eight county names in the data set are listed by typing

```
> colnames(counts)[1:8]
[1] "CAMERON"       "WILLACY"       "KENEDY"
[4] "KLEBERG"       "NUECES"        "SAN_PATRICIO"
[7] "ARANSAS"       "REFUGIO"
```

You use the two-parameter Weibull distribution to model the wind speed categories. The survival function ($S(w) = P(W > w)$) for the Weibull distribution ($W \sim \text{Weib}(b,a)$) is

$$S(w) = \exp\left(-\left(\frac{w}{b}\right)^{a}\right) \tag{8.9}$$

where a and b are the shape and scale parameters, respectively.

The Weibull distribution has the property that if $W \sim \text{Weib}(a,b)$, then a linear transformation of W results in a variable whose distribution is also Weibull (i.e., $kW \sim \text{Weib}(kb,a)$). Similarly, a power transformation results in a variable whose distribution is Weibull (i.e., $W^k \sim \text{Weib}(b^k, a/k)$).

However your data do not contain single wind speed values. Instead for a particular cyclone, the affected county has a lower and an upper wind speed bound. This is called censored data. You know that the wind speed is at least as strong as the lower bound and weaker than the upper bound.[3] In other words, W lies in an interval $[W_l, W_u]$, and the true wind speed follows a Weibull distribution. So instead of using the logarithm of the density function in the Weibull likelihood, you use the logarithm of the probability distribution function over the interval.

8.3.1 Marked Poisson Process

These data were originally modeled in Jagger et al. (2001). This model considered annual winds by keeping only the highest wind event for that year. That is, in a county that was hit by multiple hurricanes in a given year, only the strongest wind was used. Here you rework the analysis using a marked Poisson process meaning the wind events are independent and the number of events follows a Poisson distribution with a rate λ. The marks are the wind speed interval associated with the event. In this way, all events are included.

[3] Censored data attaches `.time1` and `.time2` to the bounds, but here the winds are from the same time.

You assume that the marks have a Weibull distribution with shape parameter a and scale parameter b. The scale parameter has units of wind speed in meter per second. Note that the mean exceedance wind speed is given by $\mu = b\Gamma(1 + 1/a)$ as can be seen by integrating the survival function (Eq. 8.9). The probability that the yearly maximum wind is less than or equal to w can be found by determining the probability of not seeing a wind of this speed.

Given the rate of events (λ) and the probability of an event exceeding w, the rate of events exceeding w is a thinned Poisson process with a rate given by

$$r(w) = \lambda \exp(-(w/b)^a) \tag{8.10}$$

So the probability of observing no events is $\exp(-r(w))$, and, the probability distribution of the yearly maximum winds is given by

$$F_{max}(w) = \exp(-\lambda \exp(-(w/b)^a)) \tag{8.11}$$

The return level (w) in years (n) associated with the return period is given by $1/n = 1 - F(w)$, the long run proportion of years with events exceeding w. Solving for w gives

$$w = b\left(\log\left(\frac{\lambda}{\log\left(\frac{n}{n-1}\right)}\right)\right)^{\frac{1}{a}} \tag{8.12}$$

which is approximately

$$w \approx b(\log(\lambda(n-.5)))^{\frac{1}{a}} \tag{8.13}$$

8.3.2 Return Levels

To help with the modeling, we packaged the functions Weibull survival (`sWeib`), distribution of maximum winds (`sWeibMax`), and return level (`rlWeibPois`) in the file **CountyWinds.R**. Use `source` to input these functions by typing

```
> source("CountyWinds.R")
```

To see how these functions work, suppose the annual hurricane rate for a county is $\lambda = 0.2$ and the Weibull survival parameters are a = 5 and b = 50 m s^{-1}. Then, to estimate the return level associated with a 100-year return period wind event, you type

```
> rlWeibPois(n=100, a=5, b=50, lambda=.2)
        100
[1,] 62.2
```

Thus, you can expect to see a hurricane wind event of magnitude 62.2 m s^{-1} in the county, on average, once every 100 years.

Note that since the event frequency is 1 in 5 years (0.2), the return period in years is given by 1/[1–exp(–0.2)] or

```
> round(1/(1 - exp(-.2)))
[1] 6
```

Note also that the Weibull distribution has support on the real number line to positive infinity. This means that there will be a nonzero probability of a wind exceeding any magnitude.

You can generate a series of return levels using the `rlWeibPois` function and the assigned parameters by typing

```
> rl = round(rlWeibPois(n=c(5, 10, 20) *
+   10^(rep(0:2, each=3)), a=5, b=50, lambda=.2), 1)
> rl
          5   10   20 50  100  200  500 1000 2000
[1,] NaN 45.7 53.2 59 62.2 64.9 67.9 69.8 71.5
```

Thus, on average, the county can expect to see a cyclone of 45.7 m s^{-1} once every 10 years. For a given return period, the return level scales linearly with the scale parameter b, but to a power of $1/a$ with the shape parameter. Note that the function returns an `NaN` (not a number) for the 5-year return level since it is below 33 m s^{-1} (minimum hurricane threshold).

8.3.3 Covariates

The earlier return-level computation assumes that all years have equal probability of events and equal probability of wind speed exceedances. This is a climatology model. You might be able to do better by conditioning on environmental factors. You include covariate effects by modeling the transformed parameters $\log\lambda$, $\log b$, and $\log a$ as linear functions of the covariates NAO, SST, SOI, and SSN (see Chapter 6).

For a given county, let L_i and U_i be the lower and upper bounds for each observation as given in the Table 6.1 and y_j be the yearly cyclone count. Furthermore, assume that $[\theta_\lambda, \theta_b, \theta_a]$ is a vector of model parameters associated with covariate matrices given as $\mathbf{X}_\lambda$, $\mathbf{X}_b$, and $\mathbf{X}_a$ of size $m \times p_\lambda, n \times p_a$, and $n \times p_b$, respectively.

The log-likelihood function of the process for a given county with n observations over m years is

$$
\begin{aligned}
\mathrm{LL}(\theta) = & \sum_{i=1}^{n} \log\left(\exp(-(L_i/b_i)^{a_i}) - \exp(-(U_i/b_i)^{a_i})\right) + \\
& \sum_{j=1}^{m} y_j \log(\lambda_i) - \lambda_i - \log(i!) \\
\log(a_i) = & \mathbf{X}_a[i,] \cdot \theta_a \\
\log(b_i) = & \mathbf{X}_{mu}[i,] \cdot \theta_b \\
\log(\lambda_i) = & \mathbf{X}_\lambda[i,] \cdot \theta_\lambda
\end{aligned}
$$

The log-likelihood separates into two parts: one for the counts and another for the wind speeds. This allows you to use maximum-likelihood estimation (MLE) for the count model parameters separate from the MLE for the wind speed model parameters. The count model is a generalized linear model, and you can use the `glm` function as you did in Chapter 7.

For the wind speeds, you can build the likelihood function (see Jagger et al. [2001]) or use a package. The advantage of the latter is greater functionality through the use of plot, summary, and predict methods. You can usually find an R package to do what you need using familiar methods. If not, you can write an extension to an existing package. If you write an extension, send it to the package maintainer so that your functions get added to future versions of the package.

The **gamlss** package together with the **gamlss.dist** package provides extensions to the `glm` function from the **stats** package and to the `gam` function from the **gam** package for generalized additive models. The **gamlss.cens** package allows you to fit parametric distributions to censored and interval data created using the `Surv` function in the **Survival** package for use with the **gamlss.dist** package. With this flexibility, you can estimate the parameters of the return-level model without writing code for the likelihood or its derivatives.

You make the packages available to your working directory by typing

```
> require(gamlss)
> require(gamlss.cens)
```

You are interested in estimating return levels at various return periods. You can do this using the MLE for the model parameters along with a set of covariates using `rlWeibPois` as described earlier. The covariate parameters have a degree of uncertainty due to finite sample size.

The return level parameters also have uncertainty. You propagate this uncertainty to your final return-level estimates in two ways. One way is to estimate the variance of the return level as a function of the parameter covariance matrix (delta method). Another way is to sample the parameters assuming that they have a normal distribution with a mean equal to the MLE estimate and with a variance—covariance matrix given by Σ, where Σ is a block diagonal matrix composed of a $p_\lambda \times p_\lambda$ covariance matrix from the count model and a $p_a + p_b \times p_a + p_b$ covariance matrix from the wind speed model. The parameters and the covariances are returned from the `vcov` function on the model object returned from `glm` and `gamlss`.

First, you generate samples of the transformed parameters and save them in separate vectors ($\log\lambda, \log b, \log a$). Then, you take the antilog of the inner product of the parameters and the corresponding set of predictions based on the covariates. You then pass these values and your desired return periods to `rlWeibPois` to obtain a return level for each return period of interest. Finally, you use the function `sampleParameters` provided in `CountyWinds.R` to sample the return levels for a given set of predictors and return periods.

8.3.4 Miami-Dade

The model can be applied to any county that has experienced more than a few hurricanes. Since not all counties have the same size, comparing wind probabilities across counties is not straightforward. By contrast, county-wide return levels are useful to local officials. You will model the county data with and without the covariates. As an example, here you model the categorical wind data for Florida's Miami-Dade County.

First, you extract the wind speed categories and the counts.

```
> miami.w = winds[[57]]
> Year = as.numeric(row.names(counts))
> H = counts[, 57]
> miami.c = data.frame(Year=Year, H=H)
```

Since you have two separate data sets, it is a good idea to see whether the cyclone counts match the winds by year and number. You do this by typing

```
> all(do.call("data.frame", rle(miami.w$Year))-
+   miami.c[miami.c$H > 0, c(2, 1)] == 0)
[1] TRUE
```

You first fit the counts to a Poisson distribution by typing

```
> fitc = glm(H ~ 1, data=miami.c, family="poisson")
```

Next, you fit the wind speed intervals to a Weibull distribution by typing

```
> WEIic = cens(WEI, type="interval", local=FALSE)
> fitw = gamlss(W ~ 1, data=miami.w, family=WEIic,
+   trace=FALSE)
```

Finally, you generate samples of return levels for a set of return periods by typing

```
> rp = c(5, 10, 20) * 10^(rep(0:2, each=3))
> rl = sampleParameters(R=1000, fitc=fitc,
+   fitw=fitw, n=rp)
```

You display the results with a series of box plots.

```
> boxplot(rl[, , ], xlab="Return Period (yr)",
+   ylab="Return Level (m/s)", main="Miami-Dade")
```

The results are shown in Figure 8.11, which also includes a plot using data from Galveston, Texas. The median return level is shown with a dot. Given the model and the data, a 50-year return period is a hurricane that produces winds of at least 60 m s^{-1} in the county. The return level increases with increasing return period. The uncertainty levels represent the upper and lower quartile values and the ends of the whiskers define the 95 percent confidence interval.

Andrew struck Miami in 1992 as a category-five hurricane with 70 m s^{-1} winds. Your model indicates that the most likely return period for a cyclone of this magnitude

```
The following object(s) are masked from 'H':

    Year
```

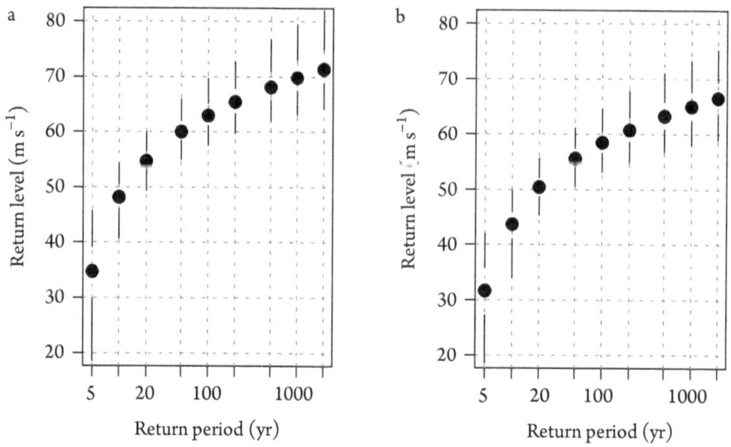

Figure 8.11 Return periods for winds in (a) Miami-Dade and (b) Galveston counties.

is 1,000 years, but it could be as short as 100 years. Return levels are higher at all return periods for Miami compared to Galveston. Miami is closer to the main tropical cyclone development region of the North Atlantic.

This chapter showed how to create models from cyclone intensity data. We began by considering the set of lifetime maximum wind speeds for basinwide cyclones and a quantile regression model for trends. We then showed how to model the fastest winds using models from extreme-value theory. The models estimate the return period of winds exceeding threshold intensities. We finished with a model for interval wind data that describes the hurricane experience at the county level. We demonstrated the model on data from Miami-Dade and Galveston Counties. A categorical model can be developed using tornado data, where intensities are estimated from damages given in intervals defined by the Fujita scale.

9

SPATIAL MODELS

"Design is a question of substance, not just form."
—Adriano Olivetti

In Chapter 7, annual counts were used to create rate models, and in Chapter 8, lifetime maximum winds were used to create intensity models. In this chapter, we show you how to use cyclone track data together with climate field data to create spatial models.

Spatial models make use of location information in data. Geographic coordinates locate the hurricane's center on the surface of the earth and wind speed provides an attribute. Spatial models make use of location separate from attributes. Given a common spatial framework, these models can accommodate climate data including indices (e.g., North Atlantic Oscillation) and fields (e.g., sea-surface temperature).

9.1 TRACK HEXAGONS

Here, we show you how to create a spatial framework for combining hurricane data with climate data. The method tessellates the basin with hexagons and populates them with local cyclone and climate information (Elsner et al., 2012).

9.1.1 Spatial Points Data Frame

In Chapter 5, you learned how to create a spatial data frame using functions from the **sp** package (Bivand et al., 2008). Let us review. Here you are interested in wind speeds along the entire track for all tropical storms and hurricanes during the 2005 North Atlantic season. You begin by creating a data frame from the *best.use.RData* file, where you subset on year and wind speed and convert the speed to meters per second.

```
> load("best.use.RData")
> W.df = subset(best.use, Yr==2005 & WmaxS >= 34
```

```
+   & Type=="*")
> W.df$WmaxS = W.df$WmaxS * .5144
```

The asterisk for `Type` indicates a tropical cyclone as opposed to a tropical wave or extratropical cyclone. The number of rows in your data frame is the total number of cyclone hours (3,010), and you save this by typing

```
> ch = nrow(W.df)
```

Next, assign the `lon` and `lat` columns as spatial coordinates using the `coordinates` function (**sp**). Finally, make a copy of your data frame, keeping only the spatial coordinates and the wind speed columns.

```
> require(sp)
> W.sdf = W.df[c("lon", "lat", "WmaxS")]
> coordinates(W.sdf) = c("lon", "lat")
> str(W.sdf, max.level=2, strict.width="cut")
Formal class 'SpatialPointsDataFrame' [package "sp"..
  ..@ data         :'data.frame': 3010 obs. of  1 var..
  ..@ coords.nrs : int [1:2] 1 2
  ..@ coords       : num [1:3010, 1:2] -83.9 -83.9 -8..
  .. ..- attr(*, "dimnames")=List of 2
  ..@ bbox         : num [1:2, 1:2] -100 11 -12.4 44.2
  .. ..- attr(*, "dimnames")=List of 2
  ..@ proj4string:Formal class 'CRS' [package "sp"]..
```

The result is a spatial points data frame with five slots. The data slot is a data frame and contains the attributes (here only wind speed). The coordinate (`coord`) slot contains the longitude and latitude columns from the original data frame and the coordinate numbers (here two spatial dimensions) are given in the `coords.nrs` slot. The bounding box (`bbox`) slot is a matrix containing the maximal extent of the hurricane positions as defined by the lower left and upper right longitude/latitude coordinates.

A summary of the information in your spatial points data frame is obtained by typing

```
> summary(W.sdf)
Object of class SpatialPointsDataFrame
Coordinates:
     min   max
lon -100 -12.4
lat   11  44.2
Is projected: NA
proj4string : [NA]
Number of points: 3010
```

```
Data attributes:
   Min. 1st Qu.  Median    Mean 3rd Qu.    Max.
   17.5    23.7    29.5    33.3    38.0    80.2
```

Again, there are 3,010 cyclone hours. The projection (`proj4string`) slot contains an `NA` character indicating that it has not yet been specified.

You give your spatial data frame a coordinate reference system using the `CRS` function. Here you specify a geographic reference (intersections of parallels and meridians) as a character string in an object called `ll_crs`, then use the `proj4string` function to generate a CRS object and assign it to your spatial data frame.

```
> ll = "+proj=longlat +datum=WGS84"
> proj4string(W.sdf) = CRS(ll)
```

Check this slot by typing

```
> slot(W.sdf, "proj4string")
CRS arguments:
 +proj=longlat +datum=WGS84 +ellps=WGS84
+towgs84=0,0,0
```

Next, you transform the geographic CRS into a Lambert conformal conic (LCC) planar projection using the parallels 30 and 60°N and a center longitude of 60°W. First save the reference system as a CRS object and then use the `spTransform` function from the **rgdal** package.

```
> lcc = "+proj=lcc +lat_1=60 +lat_2=30 +lon_0=-60"
> require(rgdal)
> W.sdf = spTransform(W.sdf, CRS(lcc))
> bbox(W.sdf)
        min     max
lon -3999983 3987430
lat  1493807 5521312
```

The coordinates are now planar rather than geographical, although the coordinate names remain the same (`lon` and `lat`). Projection of the hurricane locations to a plane makes it easy to perform distance calculations.

9.1.2 Hexagon Tessellation

Next, you construct a hexagon tessellation of the cyclone tracks. This is done by first situating points representing a staggered grid and then drawing hexagon boundaries around each grid point. You create the grid points inside the bounding box defined by your spatial points data frame using the `spsample` function. Here you specify the number of points, but the actual number will vary depending on the bounding box. You expand the bounding box by multiplying each coordinate value by 20 percent and use an offset vector to fix the position of the grid point, over the tracks.

```
> hpt = spsample(W.sdf, type="hexagonal", n=250,
+    bb=bbox(W.sdf) * 1.2, offset=c(1, -1))
```

The expansion and offset values require a bit of trial and error. The methods used in spatial sampling function assume that the geometry has planar coordinates, so your spatial data object should not have geographic coordinates.

Next, call the function to convert the points to hexagons. The more points, the smaller the area of each hexagon.

```
> hpg = HexPoints2SpatialPolygons(hpt)
```

The number of polygons generated and the area of each polygon is obtained by typing

```
> np = length(hpg@polygons)
> area = hpg@polygons[[1]]@area
> np; area
[1] 228
[1] 2.14e+11
```

This results in 228 equal-area hexagons. The length unit is meters. To convert the area from square meters to square kilometers, multiply by 10^{-6}. Thus the area of each hexagon is approximately 213,961 km^2.

9.1.3 Overlays

With your hexagons and cyclone locations having the same projection, you now obtain the maximum intensity and number of observations per hexagon. The function `over` combines points (or grids) and polygons by performing point-in-polygon operations on all point–polygon combinations. First you obtain a vector containing the hexagon identification number for each hourly cyclone observation by typing

```
> hexid = over(x=W.sdf, y=hpg)
```

The length of the vector is the number of hourly observations. Not all hexagons have cyclones, so you subset your spatial polygons object keeping only those that do.

```
> hpg = hpg[unique(hexid)]
```

Next, you create a data frame with a single column listing the maximum wind speed over the set of all cyclone observations in each hexagon.

```
> int = over(x=hpg, y=W.sdf, fn=max)
> colnames(int) = c("WmaxS")
> head(int)
```

```
      WmaxS
ID60   76.3
ID80   65.7
ID99   65.9
ID118  74.8
ID137  31.9
ID96   23.1
```

The result is a data frame with a single column representing the maximum wind speed value over all cyclone observations in each hexagon. The row names are the hexagon numbers prefixed with `ID`. Finally, you combine the spatial polygons with the maximum per hexagon wind speeds to create a spatial polygon data frame.

```
> hspdf = SpatialPolygonsDataFrame(hpg, int,
+    match.ID = TRUE)
```

You plot the hourly storm locations together with an overlay of your hexagons by typing

```
> plot(hspdf)
> plot(W.sdf, pch=20, cex=.3, add=TRUE)
```

Some hexagons contain many cyclone observations while others contain only a few. This difference might be important in modeling intensity, so you add total cyclone hours as a second attribute. First, replace the data in the spatial points data frame with an index of ones.

```
> W.sdf@data = data.frame(num=rep(1, ch))
```

Then, perform an overlay of the hexagons on the cyclone locations with the function argument (`fn`) set to `sum`. Check that the sum of the counts over all grids equals the total number of cyclone hours.

```
> co = over(x=hspdf, y=W.sdf, fn=sum)
> sum(co) == ch
[1] TRUE
```

Finally, add the counts to the spatial polygon data frame and list the first six rows of the data frame corresponding to the first six hexagons. Here, the rows of `co` correspond to those in the data slot of `hspdf` so there is no need to match the polygon IDs.

```
> hspdf$count = co[, 1]
> head(slot(hspdf, "data"))
      WmaxS count
ID60   76.3    95
ID80   65.7    50
ID99   65.9   104
ID118  74.8    50
```

```
ID137  31.9    13
ID96   23.1    45
```

9.1.4 Maps

You now have a spatial polygon data frame with each polygon as an equal-area hexagon on an LCC projection and two attributes (maximum wind speed and cyclone count) in the data slot. Choropleth maps of cyclone attributes are created using the `spplot` method introduced in Chapter 5.

Before mapping, you create a couple of lists that are used by the `sp.layout` argument. One specifies the set of hourly locations from the spatial points data frame and assigns values to plot arguments.

```
> l1 = list("sp.points", W.sdf, pch=20, col="gray",
+   cex=.3)
```

Another specifies the coastlines as a spatial lines object. Some additional work is needed.

Next acquire the **maps** and **maptools** packages. The first package contains the `map` function to generate an object of country borders and the second contains the `map2SpatialLines` conversion function. The conversion is made to geographic coordinates, which are then transformed to the LCC projection of the earlier spatial objects. You set geographic coordinate limits on the map domain to limit the amount of conversion and projection calculations.

```
> require(maps)
> require(maptools)
> cl = map("world", xlim=c(-120, 20),
+   ylim=c(-10, 70), plot=FALSE)
> clp = map2SpatialLines(cl, proj4string=CRS(ll))
> clp = spTransform(clp, CRS(lcc))
> l2 = list("sp.lines", clp, col="gray")
```

Depicting attribute levels on a map is done using a color ramp. A color ramp creates a character vector of color hex codes. The number of colors is specified by the argument in the color ramp function. Color ramp functions are available in the **colorRamps** package (Keitt, 2009). Acquire the package and assign 20 colors to the vector `cr` using the blue-to-yellow color ramp.

```
> require(colorRamps)
> cr = blue2yellow(20)
```

If the number of colors is less than the number of levels, the colors get recycled. A map of cyclone frequency is made by typing

```
> spplot(hspdf, "count", col="white",
+    col.regions=blue2yellow(20),
```

```
+     sp.layout=list(l1, l2),
+     colorkey=list(space="bottom"),
+     sub="Cyclone Hours")
```

Similarly, a map of the highest hurricane intensity is made by typing

```
> spplot(hspdf, "WmaxS", col="white",
+     col.regions=blue2red(20),
+     sp.layout=list(l2),
+     colorkey=list(space="bottom"),
+     sub="Highest Intensity (m/s)")
```

The results are shown in Figure 9.1. Areas along the southeastern coastline of the United States have the greatest number of cyclone hours while the region from the western Caribbean into the Gulf of Mexico has the highest hurricane intensities.

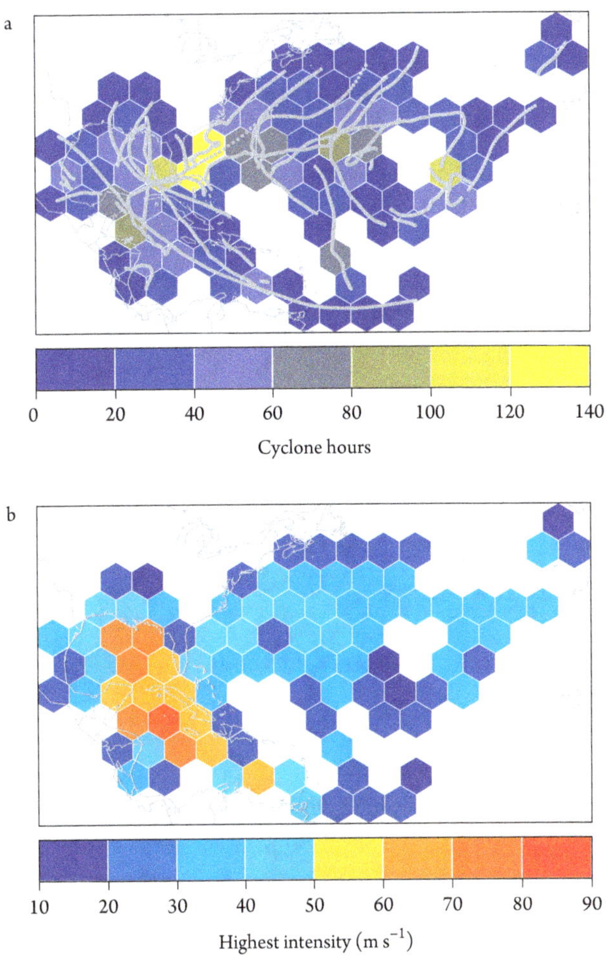

Figure 9.1 Cyclone frequency and intensity. (a) Hours and (b) highest intensity.

9.2 SST DATA

The effort needed to create track hexagons pays off nicely when you add regional climate data. Here you use sea-surface temperature (SST) from July 2005, as an example. July values indicate conditions occurring before the active part of the North Atlantic hurricane season.

In Chapter 6, you used functions in the **ncdf** package along with some additional code to extract a data frame of SST consisting of monthly values at the intersections of parallels and meridians at 2° intervals. We return to these data here. Input them by typing

```
> sst = read.table("JulySST2005.txt", header=TRUE)
> head(sst)
   SST  lon lat
1 24.1 -100   0
2 24.0  -98   0
3 23.8  -96   0
4 23.5  -94   0
5 23.4  -92   0
6 23.6  -90   0
```

The data are listed in the column labeled `SST`. You treat these values as point data because they have longitudes and latitudes although they are regional averages.

At 25°N latitude, the 2° spacing of the SST values covers an area of approximately the same size as the hexagon grids used in the previous section. The SST locations are converted to planar coordinates using the same LCC projection after the data frame is converted to a spatial data frame. Recall that you need to first assign a projection string to the `proj4string` slot.

```
> coordinates(sst) = c("lon", "lat")
> proj4string(sst) = CRS(ll)
> sst = spTransform(sst, CRS(lcc))
```

To examine the SST data as spatial points, type

```
> spplot(sst, "SST", col.regions=rev(heat.colors(20)))
```

This produces a plot of the spatial points data frame, where the SST attribute is specified as a character string in the second argument.

Your interest is the average SST within each hexagon. Again, use the `over` function to combine your SST values with the track polygons setting the argument `fn` to `mean`.

```
> ssta = over(x=hspdf, y=sst, fn=mean)
```

The result is a data frame with a single column representing the average over all SST values in each hexagon. The row names are the hexagon numbers prefixed with `ID`; however, hexagons completely over land have missing SST values.

Next, you add the averaged SSTs as an attribute to spatial polygon data frame and then remove hexagons without a value.

```
> hspdf$sst = ssta$SST
> hspdf = hspdf[!is.na(hspdf$sst),]
> str(slot(hspdf, "data"))
'data.frame':    88 obs. of  3 variables:
 $ WmaxS: num  76.3 65.7 65.9 74.8 23.1 ...
 $ count: num  95 50 104 50 45 58 56 26 40 24 ...
 $ sst  : num  29.2 29.4 29.8 30 29.5 ...
```

Finally, you create the July SST map corresponding to where the cyclones occurred during the season by typing

```
> spplot(hspdf, "sst", col="white",
+     col.regions=blue2red(20),
+     sp.layout=list(l1, l2),
+     colorkey=list(space="bottom"),
+     sub="Sea Surface Temperature (C)")
```

Results are shown in Figure 9.2. Ocean temperatures exceed 26°C over a wide swath of the North Atlantic extending into the Caribbean Sea and Gulf of Mexico. Coldest waters are noted off the coast of Nova Scotia and Newfoundland.

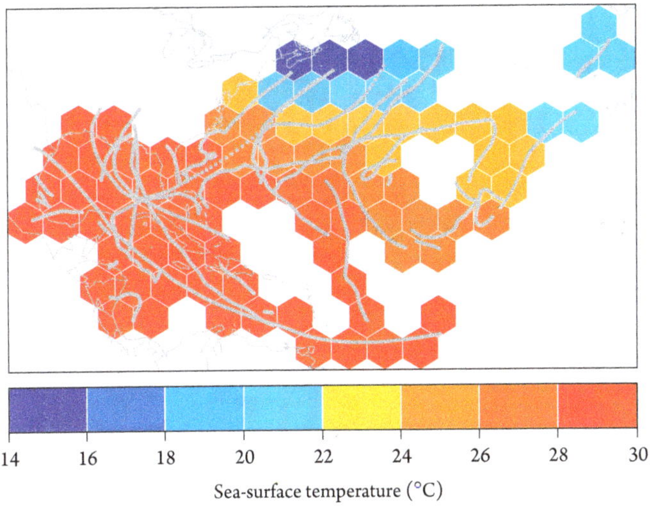

Figure 9.2 Sea-surface temperature in the cyclone hexagons.

9.3 SST AND INTENSITY

Analyzing the SST and cyclone data together is easy with your spatial polygon data frame. For instance, the average maximum wind speed in regions where the SST exceeds 28°C is obtained by typing

```
> mean(hspdf$WmaxS[hspdf$sst > 28])
[1] 45.1
```

Here you treat your spatial data frame as you would a regular data frame. Continuing, you create side-by-side box plots of wind speed conditional on the collocated values of SST (see Chapter 5) by typing

```
> boxplot(hspdf$WmaxS~hspdf$sst > 28)
```

Spatial information allows you to analyze relationships on a map. Figure 9.3 shows the hexagons colored by groups defined by a two-way table of cyclone intensity and SST using above and below median values. The median SST and intensity values calculated from your data are 28.2°C and 33.9 m s^{-1}, respectively. Red hexagons indicate regions of high intensity and relatively high ocean temperature and blue hexagons indicate regions of low intensity and relatively low ocean temperature. More interesting are regions of mismatch. Magenta hexagons show low intensity coupled with relatively high ocean temperature indicating "underperforming" cyclones (cyclones weaker than the thermodynamic potential of the environment). By contrast, cyan hexagons show high intensity coupled with relatively low temperature indicating "overperforming" cyclones (cyclones stronger than the thermodynamic potential of the environment).

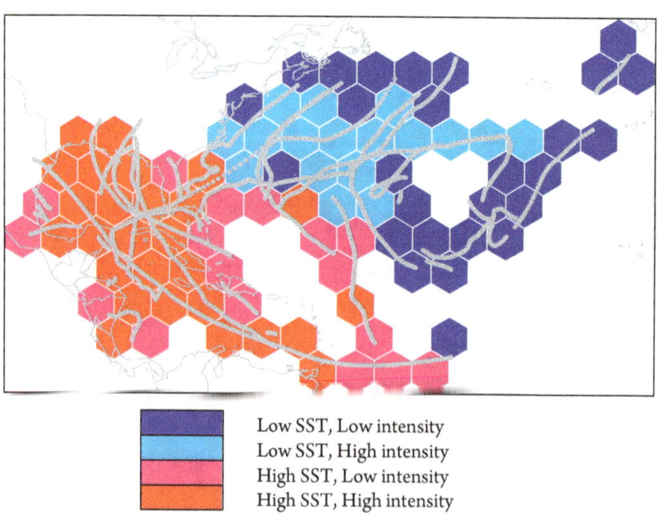

Figure 9.3 SST and cyclone intensity. Groups are based on median values.

Maps provide insight into the relationship between hurricanes and climate not accessible with basin-level analyses. Next, we show you how to model these spatial data. We begin with a look at spatial correlation and then use a spatial regression model to quantify the regional variation in intensity as a function of SST.

9.4 SPATIAL AUTOCORRELATION

Hurricane intensities in neighboring hexagons will tend to be more similar than intensities in hexagons farther away. Spatial autocorrelation quantifies the degree of similarity across geographic space. It is more complicated than temporal autocorrelation because map space has two dimensions and multiple directions.

9.4.1 Moran's I

A measure of spatial autocorrelation is Moran's I (Moran, 1950), defined as

$$I = \frac{m}{s}\frac{y^T Wy}{y^T y} \tag{9.1}$$

where m is the number of hexagons, y is the vector of values within each hexagon (e.g., cyclone intensities) where the values are deviations from the overall spatial mean, W is a weights matrix, s is the sum over all the weights, and the superscript T indicates the transpose operator.

Values of Moran's I range from -1 to $+1$ with a value of zero indicating a pattern with no spatial autocorrelation. Although not widely used in climate studies, de Beurs and Henebry (2008) use it to identify spatially coherent eco-regions and biomes related to the North Atlantic Oscillation (NAO).

To compute Moran's I, you need a weights matrix. The weights matrix is square with the number of rows and columns equal to the number of hexagons. Here the weight in row i, column j, of the matrix is assigned a zero unless hexagon i is contiguous with hexagon j. The **spdep** package (Bivand et al., 2011a) has functions for creating weights based on contiguity (and distance) neighbors. The process requires two steps.

First, you use the `poly2nb` function on your spatial polygon data frame to create a contiguity-based neighborhood list object.

```
> require(spdep)
> hexnb = poly2nb(hspdf)
```

A summary of the neighborhood list is obtained by typing

```
> hexnb
Neighbour list object:
Number of regions: 88
Number of nonzero links: 372
Percentage nonzero weights: 4.8
Average number of links: 4.23
```

The list is ordered by hexagon number, starting with the southwest-most hexagon. It has five neighbors: hexagon numbers 2 and 21. Hexagon numbers increase to the east and north. A hexagon has at most six contiguous neighbors. Hexagons at the borders have fewer neighbors. A graph of the hexagon connectivity, here defined by the first-order contiguity, is made by typing

```
> plot(hexnb, coordinates(hspdf))
> plot(hspdf, add=TRUE)
```

A summary method applied to the neighborhood list (`summary(hexnb)`) reveals the average number of neighbors and the distribution of connectivity among the hexagons.

You turn the neighborhood list object into a `listw` object using the `nb2listw` function that duplicates the neighborhood list and adds the weights. The `style` argument determines the weighting scheme. With the argument value set to `W`, the weights are the inverse of the number of neighbors. For instance, the six neighbors of a fully connected hexagon each get a weight of 1/6.

```
> wts = nb2listw(hexnb, style="W")
> summary(wts)
```

Now you are ready to compute the value of Moran's *I*. This is done using the `moran` function. The first argument is the per hexagon maximum intensity followed by the name of the `listw` object. Also needed is the number of hexagons and the global sum of the weights, which is obtained using the `Szero` function.

```
> m = length(hspdf$WmaxS)
> s = Szero(wts)
> moran(hspdf$WmaxS, wts, n=m, S0=s)
$I
[1] 0.564

$K
[1] 2.87
```

The function returns Moran's *I* and the sample kurtosis.[1] The value of 0.56 indicates fairly high spatial autocorrelation in cyclone intensity. This is expected, since the strength of a hurricane as it moves across the hexagons often does not vary by much and because stronger hurricanes tend to occur at lower latitudes.

9.4.2 Spatial Lag Variable

Insight into the Moran's *I* statistic is obtained by noting that it is the slope coefficient in a regression model of Wy on y, where Wy is the spatial lag variable (see Eq. 9.1).

[1] Kurtosis is a measure of the peakedness of the distribution. A normal distribution has a kurtosis of 3.

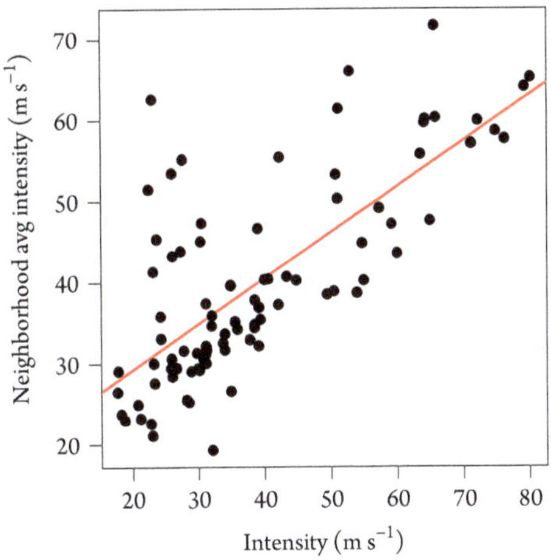

Figure 9.4 Moran's scatter plot for cyclone intensity.

Let `y` be the set of intensities in each hexagon, then you create a spatial lag intensity variable using the `lag.listw` function.

```
> y = hspdf$WmaxS
> Wy = lag.listw(wts, y)
```

Thus, for each intensity value in the vector object `y`, there is a corresponding value in the vector object `Wy`, representing the mean intensity over the neighboring hexagons. The neighborhood average does not include the value in `y`, so for a completely connected hexagon, the average is taken over the adjoining six neighboring values.

A scatter plot of the neighborhood average intensity versus the intensity in each hexagon shows the spatial autocorrelation relationship. The slope of a least-squares regression line through the points is the value of Moran's I. Use the following code below to create the scatter plot shown in Figure 9.4.

```
> par(las=1, pty="s")
> plot(y, Wy, pch=20, xlab="Intensity (m/s)",
+   ylab="Neighborhood Avg Intensity (m/s)")
> abline(lm(Wy ~ y), lwd=2, col="red")
```

The scatter plot (Moran's scatter plot) indicates a high level of spatial autocorrelation. High-intensity hexagons tend to be surrounded, on average, by hexagons with high intensity and vice versa as evidenced by the upward slope of the regression line.

9.4.3 Statistical Significance

The expected value of Moran's I under the hypothesis of no spatial autocorrelation is

$$E(I) = \frac{-1}{m-1} \tag{9.2}$$

where m is the number of hexagons. This allows you to test the significance of your sample Moran's I. The test is available in the `moran.test` function. The first argument is the vector of intensities and the second is the spatial weights matrix in the weights list form.

```
> moran.test(y, wts)
      Moran's I test under randomisation

data:  y
weights: wts

Moran I statistic standard deviate = 7.53,
p-value = 2.548e-14
alternative hypothesis: greater
sample estimates:
Moran I statistic        Expectation
          0.56410           -0.01149
         Variance
          0.00584
```

The output shows the standard deviate computed by taking the difference between the estimated I and its expected value under the null hypothesis of no autocorrelation. The difference is divided by the square root of the difference between the variance of I and the square of the mean of I. The p-value is the chance of observing a standard deviate this large or larger assuming that there is no spatial autocorrelation (the null hypothesis). The p-value is extremely small leading you to conclude that there is significant autocorrelation. The output also gives the value of Moran's I along with its expected value and variance.

By default, the variance of I is computed by randomizing the intensities across the hexagons. If the intensities have a normal distribution then a direct computation of the variance of I is made by adding the argument `random=FALSE`. Moran's I and the corresponding significance test are sensitive to the definition of neighbors and to the neighborhood weights, so conclusions should be stated as conditional on your definition of neighborhoods.

9.5 SPATIAL REGRESSION MODELS

Spatial regression models make use of spatial autocorrelation. If significant autocorrelation exists, spatial regression models have parameters that are more stable and statistical tests that are more reliable than nonspatial alternatives. For instance, confidence intervals on a regression slope will have the proper coverage probabilities and prediction errors will be smaller. Autocorrelation is included in a regression model by adding a spatial lag-variable or by including a spatially correlated error term (Anselin et al., 2006).

Spatial autocorrelation can also enter a model by having the relationship between the response and the explanatory variable vary across the domain. This is called geographically weighted regression (GWR) (Brunsdon et al., 1998, Fotheringham et al., 2000). GWR allows you to see where an explanatory variable contributes strongly to the relationship and where it contributes weakly. It is similar to a local linear regression.

For example, you compare a standard linear regression model of intensity on SST with a local linear regression of the same relationship using the `loess.smooth` function by typing

```
> x = hspdf$sst
> par(las=1, pty="s")
> plot(x, y, pch=20, xlab="SST (C)",
+   ylab="Intensity (m/s)")
> abline(lm(y ~ x), lwd=2)
> lines(loess.smooth(x, y, span=.85), lwd=2, col="red")
```

With the local regression, the relationship between intensity and SST changes for different values of SST (Fig. 9.5). For low values of SST, the relationship has a gentle slope and with high values of SST the relationship has a steeper slope. By contrast, the "global" linear regression results in a single moderate slope across all values of SST.

Fitting is done locally in the domain of SST. That is, for a point *s* along the SST axis, the fit is made using points in a neighborhood of *s* weighted by their distance to *s*. The size of the neighborhood is controlled by the `span` argument. With `span=.85`, the neighborhood includes 85 percent of the SST values.

With GWR, this localization is done in geographic space. For example, a regression of cyclone intensity on SST is performed using paired values of intensity and SST values across the 88 hexagons. For each hexagon, the weight associated with a paired

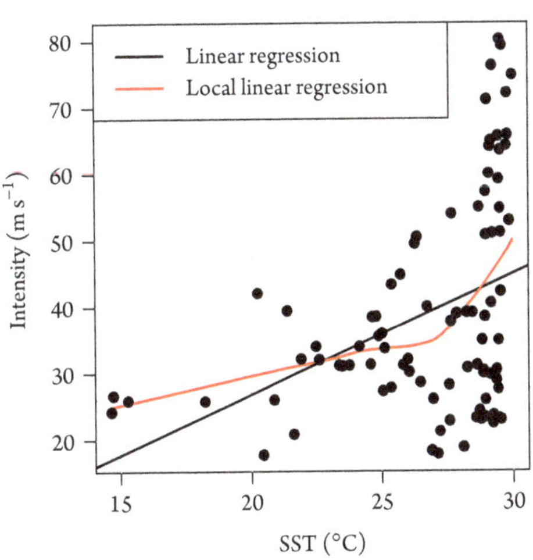

Figure 9.5 Linear and local linear regressions of cyclone intensity on SST.

value in another hexagon is inversely proportional to physical distance between the two hexagons. In this way, the relationship between intensity and SST is localized.

9.5.1 Linear Regression

The standard regression model consists of a vector y of response values and a matrix X containing the set of explanatory variables plus a row vector of 1s. The relationship is given by

$$y = X\beta + \varepsilon \tag{9.3}$$

where β is a vector of regression coefficients and $\varepsilon \sim N(0, \sigma^2 I)$ is a vector of independent and identically distributed residuals with variance σ^2. The maximum-likelihood estimate of β is given by

$$\hat{\beta} = (X^T X)^{-1} X^T y. \tag{9.4}$$

You begin with a linear regression of cyclone intensity on SST across the set of hexagons. Although your interest is the relationship between intensity and SST, you know that intensity within the hexagon will likely also depend on the number of cyclone hours. On average, a hexagon with a large number of hurricane hours will have a higher intensity. Thus, your regression model includes SST and cyclone hours as explanatory variables, in which case the SST coefficient from the regression is an estimate of the effect of SST on intensity after accounting for cyclone frequency.

You use the `lm` function to create a linear regression model. The `summary` method is used to obtain statistical information about the model.

```
> model = lm(WmaxS ~ sst + count, data=hspdf)
> summary(model)
Call:
lm(formula = WmaxS ~ sst + count, data = hspdf)

Residuals:
   Min     1Q Median     3Q    Max
-25.27 -10.03  -1.54   7.32  36.41

Coefficients:
            Estimate Std. Error t value Pr(>|t|)
(Intercept)  -0.9402    11.3521   -0.08   0.9342
sst           1.1981     0.4390    2.73   0.0077
count         0.2343     0.0547    4.28  4.9e-05

Residual standard error: 13.6 on 85 degrees of freedom
Multiple R-squared: 0.305,  Adjusted R-squared: 0.289
F-statistic: 18.7 on 2 and 85 DF,  p-value: 1.89e-07
```

The formula is specified as `WmaxS ~ sst + count` to indicate that the mean `WmaxS` is related to `sst` and `count`. Results show that SST and cyclone hours are important in statistically explaining cyclone intensity across the set of hexagons. The parameter on the SST variable is interpreted to mean that for every 1°C increase in sea-surface temperature, cyclone intensity goes up by 1.2 ms^{-1} after accounting for cyclone hours. The model explains 30.5 percent of the variation in intensity. These results represent the average relationship over the basin.

9.5.2 Geographically Weighted Regression

With GWR the SST parameter is replaced by a vector of parameters one for each hexagon. The relationship between the response vector and the explanatory variables is given by

$$y = X\beta(g) + \varepsilon \tag{9.5}$$

where g is a vector of geographic locations—here the set of hexagons with cyclone intensities—and

$$\hat{\beta}(g) = (X^T W X)^{-1} X^T W y \tag{9.6}$$

where W is a weights matrix given by

$$W = \exp(-D^2/h^2) \tag{9.7}$$

where D is a matrix of pairwise distances between the hexagons and h is the bandwidth. The elements of the weights matrix, w_{ij}, are proportional to the influence hexagons j have on hexagons i. Weights are determined by an inverse-distance function (kernel) so that values in nearby hexagons have greater influence on the local relationship between x and y compared with values in hexagons farther away.

The bandwidth controls the amount of smoothing. It is chosen as a trade-off between variance and bias. A bandwidth too narrow (steep gradients on the kernel) results in large variations in the parameter estimates (large variance). A bandwidth too wide leads to a large bias as the parameter estimates are influenced by processes that do not represent local conditions.

Functions for selecting a bandwidth and running GWR are available in the **spgwr** package (Bivand et al., 2011b). First, acquire the package and select the bandwidth using the `gwr.sel` function. The first argument is the model formula and the second is the data frame. The data frame can be a spatial points or spatial polygon data frame.

```
> require(spgwr)
> bw = gwr.sel(WmaxS ~ sst + count, data=hspdf)
> bw * .001
```

The procedure is an iterative optimization with improvements made based on previous values of the bandwidth. Values for the bandwidth and the cross-validated skill score are printed. After several iterations, no improvement in the score occurs. The bandwidth has dimensions of length representing a spatial distance. The units of the LCC projection in the spatial polygon data frame is meters, so to convert the bandwidth to kilometers, you multiply by 10^{-3}.

GWR is performed using the gwr function. The first two arguments are the same as in the function to select the bandwidth. The value of the bandwidth is supplied with the bandwidth argument.

```
> model = gwr(WmaxS ~ sst + count, data=hspdf,
+   bandwidth=bw)
```

The output is saved in an object of class gwr. To print a brief summary of the output, type

```
> model
Call:
gwr(formula = WmaxS ~ sst + count, data = hspdf,
   bandwidth = bw)
Kernel function: gwr.Gauss
Fixed bandwidth: 644341
Summary of GWR coefficient estimates:
                   Min.   1st Qu.    Median   3rd Qu.
X.Intercept. -8.12e+02 -3.21e+02 -1.20e+01  2.86e+01
sst          -3.59e+00 -1.79e-02  1.46e+00  1.21e+01
count         9.71e-03  1.45e-01  1.90e-01  2.71e-01
                   Max. Global
X.Intercept.  1.20e+02  -0.94
sst           2.88e+01   1.20
count         5.83e-01   0.23
```

The output repeats the function call, which includes the form of the model, the kernel function (here Gaussian), and the bandwidth (units of meters). The output also includes a summary of the regression parameter values across the hexagons. In general, intensity is positively related to SST, but the minimum parameter value indicates that in at least one hexagon the relationship is negative. The units on this regression parameter are meter per second per degree celsius. The summary also includes the parameter values from a standard regression under the column heading Global. These values are the same as those output previously using the lm function.

The object model$SDF inherits the spatial polygon data frame class from the model object along with the corresponding map projection. A summary method on this object provides additional information about the GWR.

```
> summary(model$SDF)
Object of class SpatialPolygonsDataFrame
Coordinates:
       min     max
x -4356736 4341671
y  1124111 6002653
Is projected: TRUE
proj4string :
[+proj=lcc +lat_1=60 +lat_2=30 +lon_0=-60
+ellps=WGS84]
Data attributes:
     sum.w          X.Intercept.          sst
 Min.   : 2.72   Min.   :-811.9   Min.   :-3.5920
 1st Qu.: 6.14   1st Qu.:-320.7   1st Qu.:-0.0179
 Median : 7.88   Median : -12.0   Median : 1.4619
 Mean   : 7.72   Mean   :-143.4   Mean   : 6.0174
 3rd Qu.: 9.55   3rd Qu.:  28.6   3rd Qu.:12.1371
 Max.   :11.11   Max.   : 119.5   Max.   :28.7864
     count              gwr.e                pred
 Min.   :0.00971   Min.   :-27.1580   Min.   :17.2
 1st Qu.:0.14454   1st Qu.: -3.5292   1st Qu.:29.9
 Median :0.19029   Median : -0.0526   Median :36.2
 Mean   :0.22379   Mean   : -0.1788   Mean   :39.2
 3rd Qu.:0.27065   3rd Qu.:  4.2381   3rd Qu.:46.3
 Max.   :0.58289   Max.   : 21.3694   Max.   :74.0
    localR2
 Min.   :0.237
 1st Qu.:0.464
 Median :0.543
 Mean   :0.566
 3rd Qu.:0.668
 Max.   :0.966
```

Here, the coordinate bounding box is given along with the projection information. Data attributes are stored in the data slot. The attributes include the model parameters (including the intercept term), the sum of the weights, predicted values, prediction errors, and R-squared values. A histogram of the R-squared values is made by typing

```
> hist(model$SDF$localR2, main="",
+   xlab="Local R-Squared Values")
```

The R-squared values are centered near a value 0.5, but there are three hexagons with a value above 0.9 (Fig. 9.6).

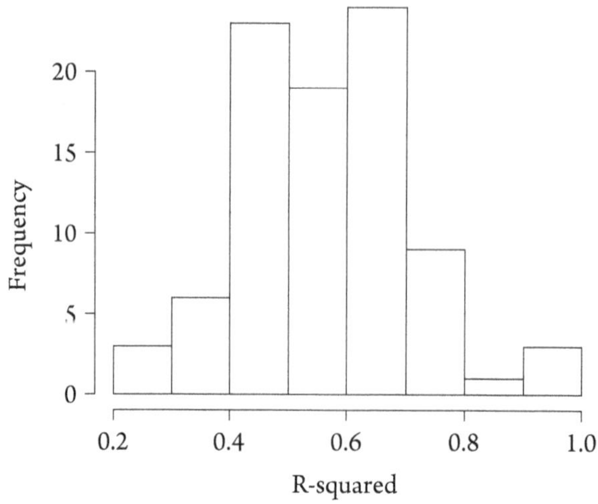

Figure 9.6 R-squared values from a GWR of intensity on SST and cyclone hours.

Additional insight is obtained by mapping the results. For instance, to make a choropleth map of the SST parameter values, type

```
> spplot(model$SDF, "sst", col="white",
+    col.regions=blue2red(10), at=seq(-25, 25, 5),
+    sp.layout=list(l2), colorkey=list(space="bottom"),
+     sub="SST Effect on Intensity (m/s per C)")
```

The results are shown in Figure 9.7. The marginal influence of SST on cyclone intensity is shown along with corresponding t values. The t value is calculated as the ratio of the parameter value to its standard error. Large values of $|t|$ indicate a statistically-significant relationship. Standard errors are available by specifying the argument `hatmatrix=TRUE` in the `gwr` function.

Hexagons are colored according to the SST parameter. The SST parameter (coefficient) represents a local "trend" of intensity as a function of SST holding cyclone hours constant. Hexagons with positive coefficients, indicating a direct relationship between cyclone strength and ocean warmth in meter per second per degree celsius are displayed using red hues and those with negative coefficients are shown with blue hues. The divergent color ramp `blue2red` creates the colors.

Hexagons with the largest positive parameters (greater than 5 m s^{-1}/°C) are found over the Caribbean Sea extending into the southeast Gulf of Mexico and east of the Lesser Antilles. Coefficients above zero extend over much of the Gulf of Mexico northeastward into the southwestern Atlantic. A region of coefficients less than zero is noted over the central North Atlantic extending from the middle Atlantic coast of the United States eastward across Bermuda. This map quantifies the grouped results shown in Figure 9.3.

Local statistical significance is estimated by dividing the parameter value by its standard error. The ratio, called the t value, is described by a t distribution under the null hypothesis of a zero-value (see Chapter 3). Regions of high t values (absolute value

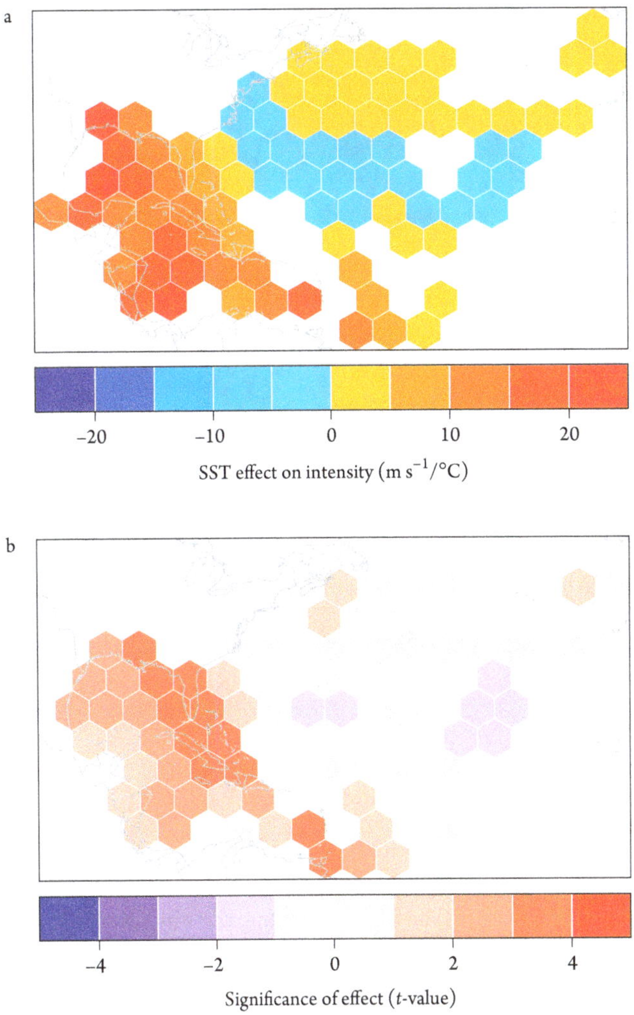

Figure 9.7 Effect of SST on cyclone intensity. (a) SST coefficient and (b) *t* value.

greater than 2) denote areas of statistical significance and generally correspond with regions of large upward trends including most of the Caribbean sea and the eastern Gulf of Mexico. Regions with negative trends over the central Atlantic extending to the coastline fail to show significance. To some degree, the results depend on hexagon size. One strategy is to rerun the GWR model with larger and smaller hexagons. In this way, you can check how much the results change and whether the changes influence your conclusions.

9.5.3 Model Fit

To assess model fit, you examine the residuals. Here, a residual is the difference between observed and predicted intensity in each hexagon. Residuals above zero

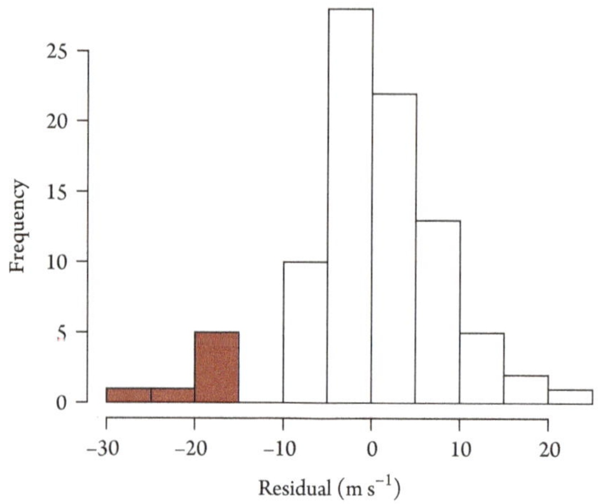

Figure 9.8 Histogram of GWR residuals. Residuals less than −10 m s^{-1} are brown.

indicate that the model underpredicts the intensity and residuals below zero indicate that the model overpredicts intensity. Residuals are saved in `model$SDF$gwr.e`. A histogram of the residuals (Fig. 9.8) shows the values are centered on zero. The left tail indicates that the model overpredicts intensity in some hexagons. Use the code below to check yourself.

```
> rsd = model$SDF$gwr.e
> hist(rsd, main="", xlab="Residual (m/s)")
> hist(rsd[rsd <= -10], add=TRUE, col="brown")
```

An interesting question concerns the location of these overpredictions. This is answered by coloring the hexagons having the largest negative residuals. First create a new vector for the spatial polygon data frame consisting of 1s indicating largest negative residuals and 0s indicating otherwise. You then use the `spplot` method and restrict the color region argument to `"brown"` and `"white"` and turn off the color key.

```
> model$SDF$op = as.integer(rsd <= (-10))
> spplot(model$SDF, "op", col="white",
+   col.regions=c("white", "brown"),
+   sp.layout=list(l2), colorkey=FALSE)
```

Results are shown in Figure 9.9. In general, the largest overpredictions occur at low latitudes in hexagons near land. This makes sense; in these regions, although SST values are high, cyclone intensities are limited by other environmental factors associated with land like drier air and greater friction. The map suggests a way the model can be improved. In this regard, a factor indicating the presence or absence of land or a covariate indicating the proportion of hexagon covered by land would be reasonable things to try.

Other types of spatial regression models fit this hexagon framework. For instance, if interest is cyclone counts, then a Poisson spatial model (Jagger et al., 2002) can be

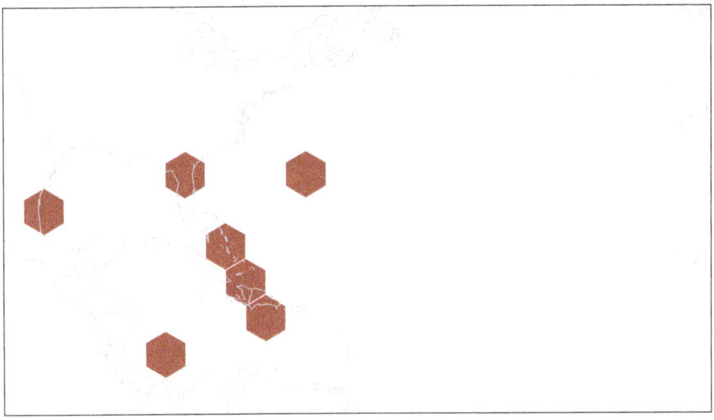

Figure 9.9 Regions of largest negative residuals.

used. And if interest is prediction rather than explanation a model that includes spatial autocorrelation through a correlated error term or a spatial, lag variable is possible. In Chapter 12, we show how to use this hexagon framework to construct a space-time model for hurricane occurrence.

9.6 SPATIAL INTERPOLATION

Spatial data often need to be interpolated. Cyclone rainfall is a good example. You know how much it rained at locations having rain gauges but not everywhere. Spatial interpolation uses the rainfall collected at the gauge sites to estimate rainfall at arbitrary locations. Estimates on a regular grid are used to create contour maps.

This can be done in various ways. Here we show how to do it statistically. Statistical interpolation is preferable because it includes uncertainty estimates. The procedure is called "kriging," after the name of a mining engineer (Matheron, 1963).

A kriged estimate (prediction) of some variable z at a given location is a weighted average of the z values over the entire domain where the weights are proportional to the spatial correlation. The estimates are optimal in the sense that they minimize the variance between the observed and interpolated values. In short, kriging involves estimating and modeling the spatial autocorrelation and then using the model together with the observations to interpolate valves at arbitrary locations.

Here you work through an example using rainfall from tropical cyclone Fay in 2008. Fay formed near the Dominican Republic as a tropical wave, passed over the island of Hispaniola and Cuba before making landfall on the Florida Keys. Fay then crossed the Florida peninsula and moved westward across portions of the Florida panhandle producing heavy rains along the way.

9.6.1 Preliminaries

Here you use a spatial interpolation model to generate a continuous isohyet surface of rainfall totals from Fay. The data are in *FayRain.txt* and are a compilation of reports from the U.S. NOAA/NWS official weather sites and coop sites. The coop sites are the Community Collaborative Rain, Hail and Snow Network (CoCoRaHS), a community-based, high-density precipitation cooperation made up of volunteers who take measurements of precipitation in their backyards. The data were obtained from NOAA/NCEP/HPC and from the Florida Climate Center (thanks to Josh Cossuth).

You make use of functions in the **gstat** package (Pebesma, 2004). Acquire the package and read in the data by typing

```
> require(gstat)
> FR = read.table("FayRain.txt", header=TRUE)
> names(FR)
[1] "lon" "lat" "tpi" "tpm"
```

The data frame contains 803 rain guage sites. Longitude and latitude coordinates of the sites are given in the first two columns and total rainfall in inches and millimeters are given in the second two columns.

Create a spatial points data frame by specifying columns that contain the spatial coordinates. Then assign a geographic coordinate system and convert the rainfall from millimeters to centimeters.

```
> coordinates(FR) = c("lon", "lat")
> ll2 = "+proj=longlat +datum=NAD83"
> proj4string(FR) = CRS(ll2)
> FR$tpm = FR$tpm/10
```

A summary of the observed rainfall totals in centimeters is given by typing

```
> summary(FR$tpm)
   Min. 1st Qu.  Median    Mean 3rd Qu.    Max.
    0.0     8.3    15.8    17.6    24.4    60.2
```

The median value is 15.8 cm and the maximum is 60.2 cm.

Next, import the Florida shapefile containing the county polygons as used in Chapter 5. You use the `readShapeSpatial` function from the **maptools** package to accomplish this task (see Chapter 5).

```
> require(maptools)
> FLpoly = readShapeSpatial("FL/FL",
+    proj4string=CRS(ll2))
```

Next, create a character string specifying the tags for a planar projection and transform the geographic coordinates of the site locations and map polygons to the projected coordinates. Here, you use Albers equal-area projection with true scale at latitudes 23 and 30°N and include a tag to specify meters as the unit of distance.

```
> require(rgdal)
> aea = "+proj=aea +lat_1=23 +lat_2=30 +lat_0=26,
+    +lon_0=-83 +units=m"
```

```
> FR = spTransform(FR, CRS(aea))
> FLpoly = spTransform(FLpoly, CRS(aea))
```

A map of the rain guage sites and storm totals with the state boundaries is made by typing

```
> l3 = list("sp.polygons", FLpoly, lwd=.3,
+    first=FALSE)
> spplot(FR, "tpm", sp.layout=l3)
```

Two areas of extreme rainfall are noted: one running north–south along the east coast and another one over the north. Rain gauges are clustered in urban areas.

9.6.2 Sample Variogram

Rainfall is an example of geostatistical data. In principle, it can be measured anywhere, but typically you have values at a sample of sites. The pattern of sites is not of much interest as it is a consequence of constraints (convenience, opportunity, economics, etc.) unrelated to the phenomenon. Instead, your interest centers on inference about how much rain fell across the region. This is done using kriging—the cornerstone of geostatistics (Cressie, 1993).

Kriging requires you to model the spatial autocorrelation with a variogram. The sample variogram $\hat{\gamma}(h)$ is given as

$$\hat{\gamma}(h) = \frac{1}{2N(h)} \sum_{i,j}^{N(h)} (z_i - z_j)^2 \tag{9.8}$$

where $N(h)$ is the number of distinct pairs of observation sites a lag distance h apart, and z_i and z_j are the rainfall totals at gauge sites i and j. Note that h is an approximate distance used with a lag tolerance δh. The assumption is that the rainfall field is stationary. This means that the relationship between rainfall at two locations depends only on the relative positions of the sites and not on where the sites are located. This relative position refers to distance and orientation.

You further assume that the variance of rainfall between sites depends only on their lag distance and not on their orientation relative to one another (isotropy assumption). Smaller variances are expected between nearby sites compared with variances between more distant sites. The variogram is the inverse of the spatial correlation function (correlogram). You expect larger correlation in rainfall amounts between sites that are nearby and smaller correlation between sites farther apart. This means that if a particular gauge has received a lot of rain, nearby gauges will also tend to be full resulting in small differences at short distances.

By definition, $\gamma(h)$ is the semivariogram and $2\gamma(h)$ is the variogram. However, for conciseness, $\gamma(h)$ is often referred to as the variogram. The variogram is a plot of the semivariance as a function of lag distance. Since your rainfall values have units of centimeters, the units of the semivariance are square centimeter.

The sample variogram is computed using the `variogram` function. The first argument is the model formula specifying the rainfall column from the data frame and the second argument is the data frame name. Here `~1` in the model formula indicates your assumption of no trend in the rainfall field. Trends can be included by specifying coordinate names (e.g., `~lon+lat`). Recall that although you specified a planar projection, the coordinate names are not changed from the original geographic latitude and longitude.

You compute the sample variogram for Fay's rainfall and save it by typing

```
> v = variogram(tpm ~ 1, data=FR)
```

To see the variogram values as a function of lag distance, use the plot method on the variogram object.

```
> plot(v)
```

The result is shown in Figure 9.10. Values start low (50 cm^2) at short lag distances, then increase to over 200 cm^2 at lag distances of about 200 km (Fig. 9.10). The zero-lag semivariance is called the "nugget" and the semivariance level where the variogram values no longer increase is called the "sill." The lag distance to the sill is called the "range." These three parameters (nugget, sill, and range) are used to model the variogram.

Semivariances are calculated using all pairs of rainfall observations within a lag distance (plus a lag tolerance). The number of pairs (indicated below each point on the plot) varies with lag distance. Generally there are more gauge pairs at medium distances.

You check on the assumption of isotropy by plotting sample variograms at specified azimuths. For example, variograms are estimated in four separate directions (0, 45, 90, and 135°) using the argument `alpha` in the `variogram` function, where 0° is

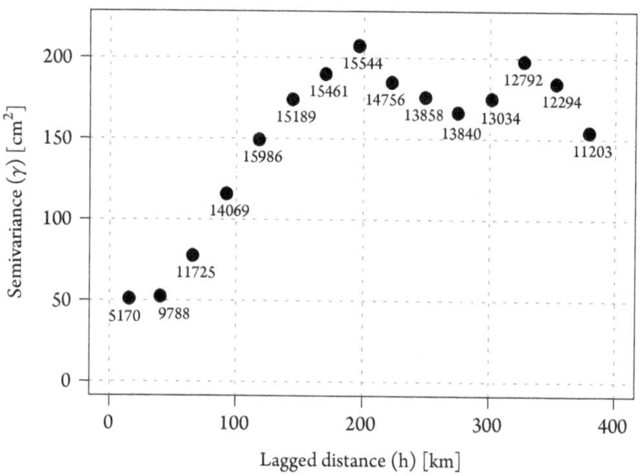

Figure 9.10 Empirical variogram of storm-total rainfall from tropical cyclone Fay.

north–south. These directional variograms restrict observational pairs to those having similar relative orientation within an angle tolerance (think of a pie slice). If the directional variograms look similar to each other, then the assumption of isotropy is valid.

9.6.3 Variogram Model

Next you fit a model to the sample variogram. The variogram model is a mathematical relationship defining the semivariance as a function of lag distance. There are several choices. To see the selection type `show.vgms()`. The nugget is shown as an open circle. The Gaussian function ("Gau") appears reasonable since it increases slowly at short lag distances then more rapidly at larger distances before leveling off at the sill.

You save the function and initial parameter values in a variogram model object by typing

```
> vmi = vgm(model="Gau", psill=150, range=200 * 1000,
+    nugget=50)
```

The `psill` argument is the partial sill as the difference between the sill and the nugget. You eyeball initial parameter values from your sample variogram plot.

You then use the `fit.variogram` function to fit the model to the sample variogram. Specifically, given the Gaussian function and initial parameter values, a weighted least-squares method improves these estimates. Note that ordinary least-squares is not appropriate in this case as the semivariances are correlated across the lag distances and the precision on the estimates varies depending on the number of site pairs for a given lag.

```
> v.fit = fit.variogram(v, vmi)
> v.fit
  model psill  range
1   Nug  46.8      0
2   Gau 157.1 128666
```

The result is a variogram model with a nugget of 46.8 cm^2, a partial sill of 157.1 cm^2, and a range of 128.7 km. You plot the variogram model on top of the sample variogram by typing

```
> plot(v, v.fit)
```

Let r be the range, c the partial sill, and c_o the nugget, then the equation defining the curve over the set of lag distances h in Figure 9.11 is

$$\gamma(h) = c\left(1 - \exp\left(-\frac{h^2}{r^2}\right)\right) + c_o \tag{9.9}$$

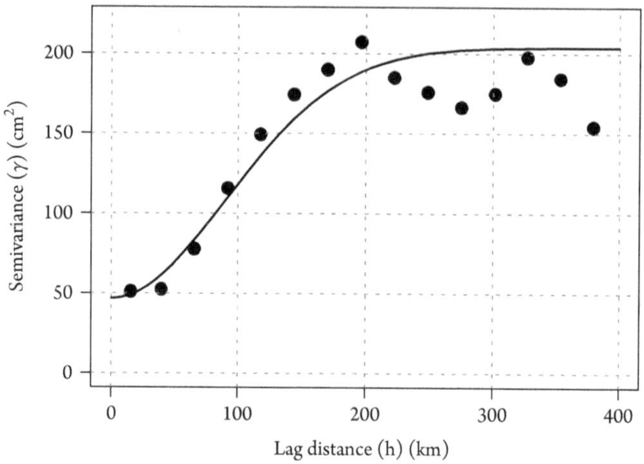

Figure 9.11 Gaussian variogram model for Fay's rainfall across Florida.

There are a variety of variogram functions. You can try the spherical function by replacing the `model="Gau"` with `model="Sph"` in the earlier `vgm` function.

```
> vmi2 = vgm(model="Sph", psill=150,
+     range=200 * 1000, nugget=50)
> v2.fit = fit.variogram(v, vmi2)
> plot(v, v2.fit)
```

The **geoR** package contains the `eyefit` function that can make choosing a function easier; however, the interpolated values are not overly sensitive, assuming a reasonable choice.

9.6.4 Kriging

The final step is to use the variogram model together with the rainfall values at the gauge sites to create an interpolated surface. The process is called kriging. As Edzer Pebesma notes, "krige" is to "kriging" as "predict" is to "prediction." Here, you use ordinary kriging as there are no spatial trends in the rainfall. Universal kriging is used when trends are present.

Interpolation is done using the `krige` function. The first argument is the model specification and the second is the data. Two other arguments are needed. One is the variogram model using the argument name `model` and the other is a set of locations, identifying where the interpolations are to be made. This is specified with the argument name `newdata`.

Here you interpolate first to locations (point kriging) on a regular grid and then to the county polygons (block kriging). To create a grid of locations within the boundary of Florida, type

```
> grd = spsample(FLpoly, n=5000, type="regular")
```

You specify the number of locations using the argument n. The actual number will be slightly different because of the irregular boundary. To view the locations simply, use the plot method on the grid object. More options for plotting are available if the spatial points object is converted to a spatial pixels object. You do this by typing:

```
> grd = as(grd, "SpatialPixels")
```

First use the krige function to interpolate (predict) the observed rainfall at the grid locations. For a given location, the interpolation is a weighted average of the rainfall across the entire region where the weights are determined by the variogram model.

```
> ipl = krige(tpm ~ 1, FR, newdata=grd, model=v.fit)
[using ordinary kriging]
```

The function recognizes the type of kriging being performed. Inverse distance-weighted interpolation is performed if the variogram model is not specified. The function will not work if there are multiple values at a given location.

The object (ipl) inherits the spatial pixels object specified in the newdata argument, but extends it to a spatial pixels data frame by adding a data slot. The data slot is a data frame with two variables. The first var1.pred is the interpolated rainfall and the second var1.var is the prediction variance.

You plot the interpolated field using the spplot method. You specify an inverted topographic color ramp to highlight in blue regions with the highest rain totals.

```
> spplot(ipl, "var1.pred", col.regions=
+    rev(topo.colors(20)), sp.layout=l3)
```

The map (see Fig. 9.12) makes it easy to see that parts of east central and north Florida were deluged by Fay.

You use block kriging to estimate rainfall amounts within each county. The county-wide rainfall average is relevant for water resource managers. Block kriging produces a smoothed estimate of this area average, which will differ from a simple arithmetic average over all sites within the county.

You use the same function to interpolate but specify the spatial polygons rather than the spatial grid as the new data.

```
> ipl2 = krige(tpm ~ 1, FR, newdata=FLpoly,
+    model=v.fit)
[using ordinary kriging]
```

You then map the results using the same plot arguments as before.

```
> spplot(ipl2, "var1.pred", col.regions=
+    rev(topo.colors(20)))
```

Rainfall maps using point and block kriging are shown in Figure 9.12. The overall pattern of rainfall from Fay featuring the largest amounts along the central east coast and over the eastern panhandle is similar on both maps.

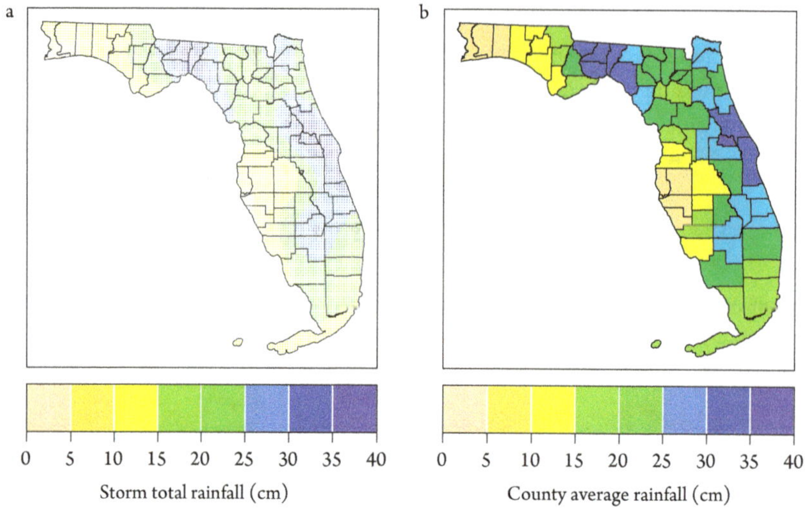

Figure 9.12 Rainfall from tropical cyclone Fay. (a) Point and (b) block kriging.

You compute the arithmetic average of county-wide rainfall, again using the `over` function by typing

```
> ipl3 = over(x=FLpoly, y=FR, fn=mean)
```

The function returns a data frame of the average rainfall in each county. The statewide mean of the kriged estimates is 20.8 cm, which compares with a statewide mean of the arithmetic averages of 20.9 cm. The correlation between the two estimates across the 67 counties is 0.87. The variogram model reduces the standard deviation of the kriged estimate (7.77 cm) relative to the standard deviation of the simple averages (9.93 cm) because of the smoothing.

9.6.5 Uncertainty

An advantage of kriging as a method of spatial interpolation is the accompanying uncertainty estimates. The prediction variances are listed in a column in the spatial data frame saved from your application of the `krige` function. Variances are smaller in regions with a greater number of rainfall gauges. Prediction variances are also smaller with block kriging as much of the variability within the county gets average out. To compare the distribution characteristics of the prediction variances for the point and block kriging of the rainfall observations, type

```
> round(summary(ipl$var1.var), 1)
   Min. 1st Qu.  Median    Mean 3rd Qu.    Max.
   47.4    48.8    49.6    50.6    50.8   179.0
> round(summary(ipl2$var1.var), 1)
   Min. 1st Qu.  Median    Mean 3rd Qu.    Max.
    0.5     1.4     2.0     2.4     2.8     8.5
```

The median prediction variance for your point kriging is 49.6 cm^2, which is close to the value of the nugget. By contrast, the median prediction variance for your block kriging is a much smaller 1.9 cm^2.

Simulations exploit this uncertainty and provide synthetic data for use in deterministic models. A rainfall–runoff model, for instance, can be run with simulated rainfall fields providing a realistic representation of the variation in the amount of runoff resulting from the uncertainty in the rainfall field (Bivand et al., 2008).

Conditional simulation, where the simulated field (realization) is generated given the data and the variogram model, is done using the same `krige` function by adding the argument `nsim` to specify the number of simulations. To speed computation when making many simulations you limit the number of site pairs used in the kriging.

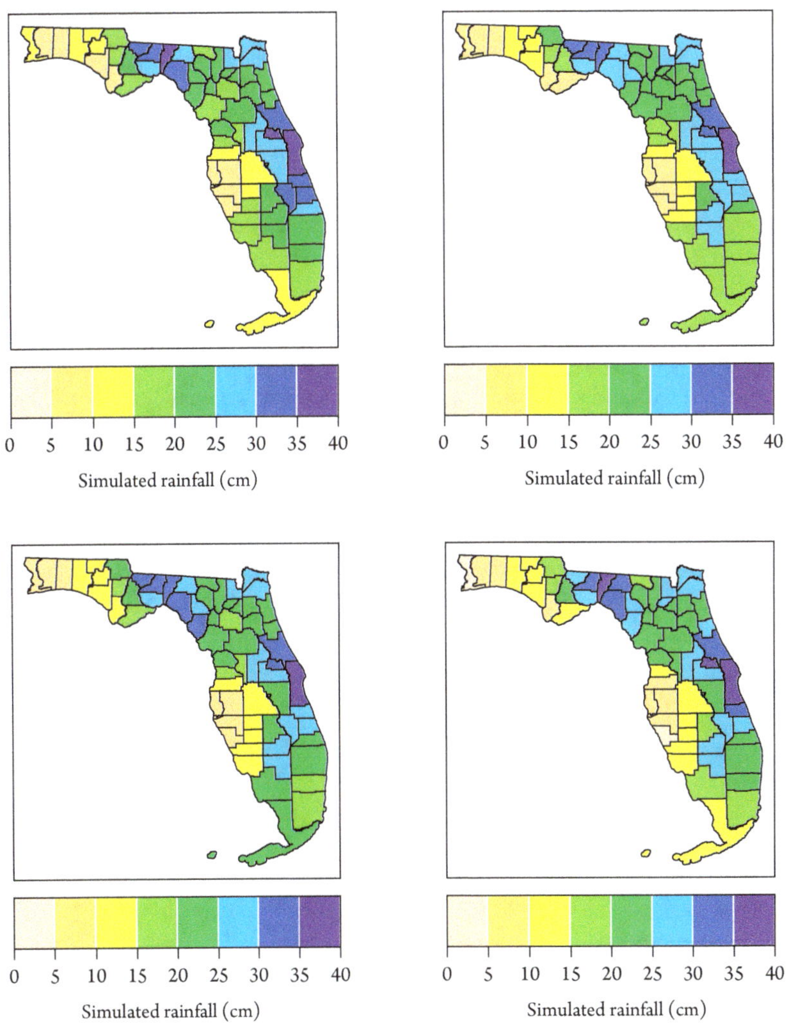

Figure 9.13 Rainfall simulations from tropical cyclone Fay.

This is done using the nmax argument. For a given location, the weights assigned to observations far away are very small, so it is efficient to limit how many are used in the simulation.

As an example, here you generate four realizations of the county-level storm total rainfall for Fay and limit your neighborhood to 50 of the closest observation sites. Note that it may take a few minutes to finish processing this function.

```
> ipl.sim = krige(tpm ~ 1, FR, newdata=FLpoly,
+    model=v.fit, nsim=4, nmax=50)
```

Maps of the four realizations are shown in Figure 9.13. Simulations are conditional on the observed rainfall and the variogram model using block kriging on the counties. Note that the overall pattern of rainfall remains the same, but there are differences especially in counties with relatively few observations and where the rainfall gradients are steep. The functionality is limited to Gaussian simulation, but the **RandomFields** package (Schlather, 2011) provides additional simulation algorithms.

In summary, kriging is a statistical method for spatial data interpolation involving three steps. First, you estimate a sample variogram that describes the spatial autocorrelation structure of your observations. This step includes checking for trends and isotropy. Second, you determine a variogram model using the method of weighted least squares. This step involves a somewhat subjective choice of the model, but the interpolation is not too sensitive to this choice assuming that it is reasonable. Third, you use the variogram model together with your observations to interpolate values at specified locations, on a grid, or over an area. Finally, simulation methods generate synthetic data for expressing and propagating uncertainty associated with the spatial variability in the data and with model specification. Kriging is a nuanced process, but with practice, it can be an important tool in your toolbox.

This chapter demonstrated ways to analyze and model hurricane data spatially. We began by showing how to combine hurricane track data with climate data. We then showed how to analyze the data including estimating the spatial autocorrelation. This was followed by a demonstration of how to use GWR to model hurricane intensity across space. Quantifying and mapping the changing relationship between the response variable and covariates can lead to new insights about how hurricanes operate across space. This research area is wide open. Finally, we showed how to perform a spatial interpolation of tropical cyclone rainfall using a prediction method from geostatistics. In the next chapter, we will look at some ways to analyze and model time series data.

10

TIME SERIES MODELS

"A big computer, a complex algorithm and a long time does not equal science."
—Robert Gentleman

In this chapter, we consider time series models. A time series is an ordered sequence of numbers with respect to time. In climatology, you encounter time-series data in a format given by

$$\{h\}_{t=1}^{T} = \{h_1, h_2, \ldots, h_T\} \tag{10.1}$$

where the time t is over a given season, month, week, or day and T is the time series length. The aim is to understand the underlying physical processes that produced the series. A trend is an example. Often by simply looking at a time series plot, you can pick out a trend that tells you that the process generating the data is changing.

A single time series gives you a sample from the process. Yet under the ergodic hypothesis, a single time series of infinite length contains the same information (loosely speaking) as the collection of all possible series of finite length. In this case, you can use your series to learn about the nature of the process. This is analogous to spatial interpolation encountered in Chapter 9, where the variogram was computed under the assumption that the rainfall field is stationary.

Here we consider a selection of techniques and models for time series data. We begin by showing you how to overlay plots as a tool for exploratory analysis. This is done to compare the variation between two series qualitatively. We demonstrate large variation in hurricane counts arising from a constant rate process. We then show techniques for smoothing. We continue with a change-point model and techniques for decomposing a continuous-valued series. We conclude with a unique way to create a network graph from a time series of counts and suggest a new definition of a climate anomaly.

10.1 TIME SERIES OVERLAYS

A plot showing your variables on a common time axis is an informative exploratory graph. Values from two different series are scaled to have the same relative range so the covariation in the variables can be compared visually. Here you do this with hurricane counts and sea-surface temperature (SST). Begin by loading *annual.RData*. These data were assembled in Chapter 6. Subset the data for years starting with 1900 and rename the year column.

```
> load("annual.RData")
> dat = subset(annual, Year >= 1900)
> colnames(dat)[1] = "Yr"
```

Plot the basin-wide hurricane count by year, then overlay a plot of the North Atlantic SST. You do this by keeping the current graphics device open with the `new=TRUE` argument in the `par` function.

```
> par(las=1, mar=c(5, 4, 2, 4) + .1)
> plot(dat$Yr, dat$B.1, xlab="Year",
+   ylab="Hurricane Count", lab=c(10, 7, 20),
+   type="h", lwd=2)
> par(new=TRUE)
> plot(dat$Yr, dat$sst, type="l", col="red", xaxt="n",
+   yaxt="n", xlab="", ylab="", lwd=2)
> axis(4)
> mtext(expression(paste("SST [",degree,"C]")),
+   side=4, line=2.5, las=0)
> legend("topleft", col=c("black", "red"), lty=1,
+   legend=c("Hurricanes","SST"))
```

You turn off the axis labels in the second plot call and then add them using the `axis` function where 4 references the vertical axis on the right side of the graph. Axes are numbered clockwise starting from the bottom of the plot. The axis is labeled using the `mtext` function.

The plot is shown in Figure 10.1. The correspondence between the two series is clear. There tends to be more hurricanes in periods of higher SST and fewer hurricanes in periods of lower SST. You retain the distinction between the two series by using bars for the discrete counts and lines for the continuous SST values.

10.2 DISCRETE TIME SERIES

Your hurricane counts arise from a rate process that is described as Poisson. More precisely, the number of occurrences over an interval is quantified using a Poisson distribution with a rate parameter proportional to the time interval. The counts in nonoverlapping intervals are independent. Since the rate of hurricanes can change

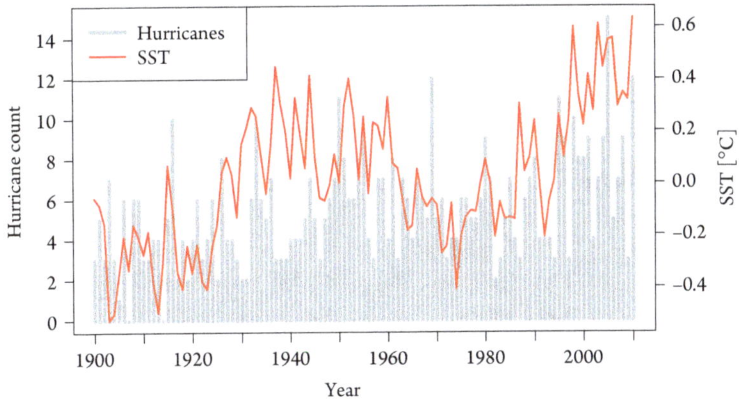

Figure 10.1 Hurricane counts and August–October SST anomalies.

from day to day and from year to year, you assume that the process has a rate that is a function of time ($\lambda(t)$).

Note that if you are interested in yearly counts, you focus on modeling the underlying yearly rate (more precisely, the integral of the underlying instantaneous rate over a year). You can integrate the rate over any time period and obtain the hurricane count over that period. For example, you can integrate to find the expected number of hurricanes after September 15th.

Here you examine methods for estimating the rate process. You first consider running averages to get a smoothed estimate of the annual rate. You then consider a change-point model where the period rate is constant but changes abruptly between periods. Running averages and change-point models provide a description of your series, but they are not very useful for predictions. You begin with a look at interannual count variability.

10.2.1 Count Variability

The time series of hurricane counts appears to have large interannual variable as seen in Figure 10.1. But this might simply be a consequence of the randomness in the counts given the rate. In fact, large variations in small-count processes are often misdiagnosed as physically significant. As an example, consider hurricane counts over a sequence of `N` years with a constant annual Poisson rate `lambda`. What is the probability that you will find at least `M` of these years with a count less than `X` (described as an inactive season) or a count greater than `Y` (described as an active season)?

Here we write it out in steps using R notation.

1. Assign the probability of the count `h` less than `X` or greater than `Y` as `PXY = 1 - ppois(Y) + ppois(X - 1))`. In other words, assign it as one minus the probability that `h` lies between `X` and `Y`, inclusive.
2. Assign an indicator `I = 1` for each year with `h < X` or `h > Y`.

3. Let the sum of `I` have a binomial distribution (Chapter 3) with probability `PXY` and `N`.
4. Then the probability of observing at least `M` of these years be given as
   `PM = 1 - pbinom(M - 1, N, PXY)`

You create the following function to perform these computations.

```
> PM = function(X, Y, lambda, N, M){
+   PXY = 1 - diff(ppois(c(X - 1, Y), lambda))
+   return(1 - pbinom(M - 1, N, PXY))
+   }
```

Arguments for `ppois` are `q` (quantile) and `lambda` (rate) and the arguments for `pbinom` are `q`, `size`, and `prob`.

You use your function to answer the following question. Given an annual rate of 6 hurricanes per year (`lambda`), what is the probability that in a random sequence of 10 years (`N`) you will find at least 2 years (`M`) with a hurricane count less than 3 (`X`) or greater than 9 (`Y`)?

```
> PM(X=3, Y=9, lambda=6, N=10, M=2)
[1] 0.441
```

With your function and a set of input values you find a 44 percent chance that 2 years out of the 10 will have large departures from the mean rate.

Your function is handy. It protects you against getting fooled by randomness. Indeed, the probability that at least 1 year in 10 falls outside the range of ± 2 standard deviations from the mean is 80 percent. This compares to 37 percent for a set of variables described by a normal distribution and underscores the limitation of using a concept that is relevant for continuous distributions on count data.

By contrast, if you consider the annual global counts over the period 1981–2006[1], you find a mean of 80.7 tropical cyclones per year with a range between 66 and 95. Assuming the global counts are Poisson, you use your function to determine the probability that no years have less than 66 or more than 95 tropical cyclones in the 26-year sample.

```
> 1 - PM(X=66, Y=95, lambda=80.7, N=26, M=1)
[1] 0.0757
```

This low probability provides suggestive evidence to support the notion that the physical processes governing global hurricane activity is more regular than Poisson. The regularity could be due to feedbacks in the climate system. For example, the cumulative effect of many hurricanes over a particular basin might make the atmosphere less conducive for activity in other basins. Or it might be related to a few governing mechanism like the North Atlantic Oscillation (Elsner and Kocher, 2000).

[1] From Elsner et al. (2008b).

10.2.2 Moving Average

A moving average removes year-to-year fluctuation in counts. The assumption is that of a smoothly varying rate process. You use the `filter` function to compute running averages. The first argument in the function is a univariate or multivariate time series and the second is the filter as a vector of coefficients in reverse time order.

For a moving average of length N, the coefficients all have the same value of $1/N$. For example, to compute the 5-year running average of the basin-wide hurricane counts, type

```
> ma = filter(dat$B.1, rep(1, 5)/5)
> str(ma, strict.width="cut")
 Time-Series [1:111] from 1 to 111: NA NA 4.2 3.8 4..
```

The output is an object of class `ts` (time series). Note, the filtering is not performed on values at the ends of the time series, so NAs are returned. If you use an odd number of years, then the number of values missing at the start of the filtered series matches the number of values missing at the end of the series.

Here you create a new function called `moveavg` and use it to compute the moving averages of basin counts over 5, 11, and 21 years.

```
> moveavg = function(X, N){filter(X, rep(1, N)/N)}
> h.5 = moveavg(dat$B.1, 5)
> h.11 = moveavg(dat$B.1, 11)
> h.21 = moveavg(dat$B.1, 21)
```

Then you plot the moving averages on top of the observed counts.

```
> plot(dat$Yr, dat$B.1, ylab="Hurricane Count/Rate",
+    xlab="Year", col="grey", type="h", lwd=1)
> cls = c("grey", "red", "blue", "green")
> lg = c("Count", "5-Yr Rate", "11-Yr Rate",
+   "21-Yr Rate")
> lines(dat$Yr, h.5, col="red", lwd=2)
> lines(dat$Yr, h.11, col="blue", lwd=2)
> lines(dat$Yr, h.21, col="green", lwd=2)
> legend("topleft", lty=1, lwd=2, col=cls, legend=lg)
```

Figure 10.2 shows the results. Note the reduction in the year-to-year variability as the length of the moving average increases. Note also that the low-frequency variation is not affected. Check this yourself by comparing the means (the mean is the zero frequency) of the moving averages. Thus a moving average is a low-pass "boxcar" filter.

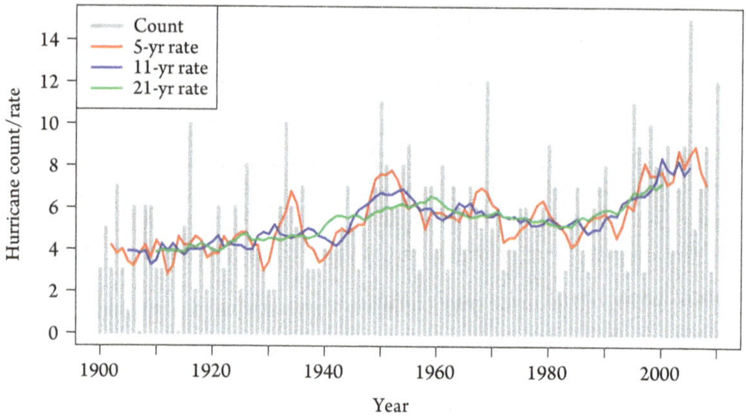

Figure 10.2 Hurricane counts and rates.

10.2.3 Seasonality

Hurricanes are seasonal. Very few occur before July, September is the most active month, and the season is typically over by November. In general, the ocean is too cool and the wind shear too strong during the months of January through May and after November. Ocean warmth peaks in early September. Seasonality is evident in plots showing the historical number of hurricanes that have occurred on each day of the year. Here we show you how to model this seasonality to produce a probability of hurricane occurrence as a function of the day of year.

You use the hourly interpolated best-track data described in Chapter 6 and saved in *best.use.RData*. The data span the years from 1851 to 2010. Import the data frame and subset on hurricane-force wind speeds.

```
> load("best.use.RData")
> H.df = subset(best.use, WmaxS >= 64)
> head(H.df)
     Sid Sn SYear       name   Yr Mo Da hr   lon lat
1      1  1  1851 NOT NAMED  1851  6 25  0 -94.8  28
1.1    1  1  1851 NOT NAMED  1851  6 25  1 -94.9  28
1.2    1  1  1851 NOT NAMED  1851  6 25  2 -95.0  28
1.3    1  1  1851 NOT NAMED  1851  6 25  3 -95.1  28
1.4    1  1  1851 NOT NAMED  1851  6 25  4 -95.2  28
1.5    1  1  1851 NOT NAMED  1851  6 25  5 -95.3  28
    Wmax WmaxS DWmaxDt Type Shour maguv diruv  jd
1   80.0  79.8  0.0860    *     0  5.24   271 175
1.1 80.0  79.9  0.0996    *     1  5.25   271 175
1.2 80.1  80.0  0.1114    *     2  5.26   271 175
1.3 80.1  80.2  0.1197    *     3  5.29   270 175
1.4 80.1  80.3  0.1227    *     4  5.32   270 175
1.5 80.0  80.4  0.1187    *     5  5.37   269 175
```

```
         M
1     FALSE
1.1 FALSE
1.2 FALSE
1.3 FALSE
1.4 FALSE
1.5 FALSE
```

Next, create a factor variable from the day-of-year column (`jd`). The day of year starts on the first of January. You use only the integer portion as the rows correspond to separate hours.

```
> jdf = factor(trunc(H.df$jd), levels=1:365)
```

The vector contains the day of year (1 through 365) for all 83,151 hurricane hours in the data set. You could use 366, but there are no hurricanes on December 31 during any leap year over the period of record.

Next, use the `table` function on the vector to obtain total hurricane hours by day of year and create a count of hurricane days by dividing the number of hours and rounding to the nearest integer.

```
> Hhrs = as.numeric(table(jdf))
> Hd = round(Hhrs/24, 0)
```

The vector `Hd` contains the number of hurricane days for each day of the year.

A plot of the raw counts shows that the variation from day to day is large. Here you create a model that smooths these variations. This is done with the `gamlss` function (see Chapter 8) in the **gamlss** package (Rigby and Stasinopoulos, 2005). You model your counts using a Poisson distribution with the logarithmic link as a function of day of year.

```
> require(gamlss)
> julian = 1:365
> sm = gamlss(Hd ~ pb(julian), family=PO, trace=FALSE)
```

Here you use a nonparametric smoothing function on the Julian day. The function is a penalized B-spline (Eilers and Marx, 1996) and is indicated as `pb()` in the model formula. The penalized B-spline is an extension of a Poisson regression that preserves the mean and variance of the daily hurricane counts and that has a polynomial curve as the limit. The Poisson distribution is specified in the `family` argument with `PO`.

Although there are days with hurricanes outside the main season, your interest centers on the months of June through November. Here you create a sequence of Julian days defining the hurricane season and convert them to dates.

```
> hs = 150:350
> doy = as.Date("1970-12-31") + hs
```

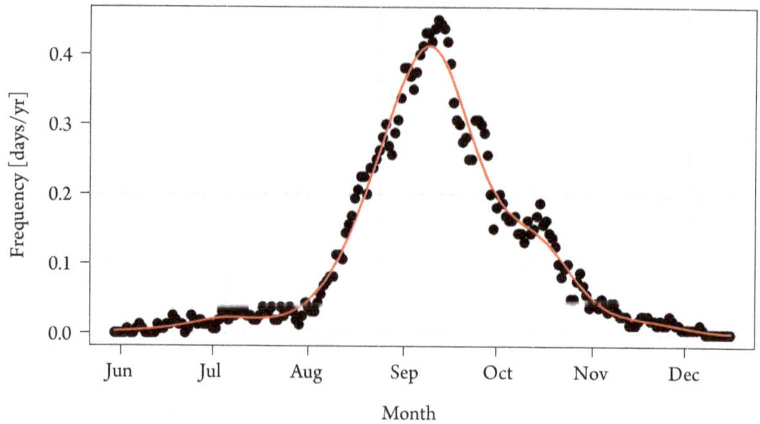

Figure 10.3 Seasonal occurrence of hurricanes.

You then convert the hurricane days to a relative frequency to allow for a probabilistic interpretation. This is done for the actual counts and the smoothed modeled counts.

```
> ny = (2010 - 1851) + 1
> Hdm = Hd[hs]/ny
> smf = fitted(sm)[hs]/ny
```

Finally, you plot the modeled and actual daily frequencies by typing

```
> plot(doy, Hdm, pch=16, xlab="",
+   ylab="Frequency (days/yr)")
> lines(doy, smf, lwd=2, col="red")
```

The results are shown in Figure 10.3. Points show the observed relative frequency of hurricanes by day of year. The red line is the fitted values of a model for the frequencies. Horizontal tic marks indicate the first day of the month.

On average, hurricane activity increases slowly until the beginning of August as the ocean warms and wind shear subsides. The increase is more pronounced starting in early August and peaks around the first or second week in September. The decline starting in mid-September is somewhat less pronounced than the increase and is associated with ocean cooling. There is a minor secondary peak during the middle of October related to hurricane genesis over the western Caribbean Sea. The climate processes that make this part of the basin relatively active are likely somewhat different than the processes occurring during the peak of the season.

10.3 CHANGE POINTS

Hurricane activity can change from inactive to active abruptly. In this case, a change-point model is appropriate for describing the time series. Here a change point refers to a jump in the rate of activity from one time to the next. The underlying assumption is a discontinuity in the rates. For example, suppose hurricanes suddenly become more frequent in the years 1934 and 1990, then the model would still be Poisson, but with different rates in the periods (epochs) 1900–1933, 1934–1989, and 1990–2010. Here you use the annual data loaded in §10.1 to build a change-point model.

10.3.1 Counts

The simplest model is one with a single change point. For instance, you check to see whether a model that has a rate change during a given year is better than a model that does not have a change during that year. Thus you have two models: one with a change point and the other without one. To make a choice, you check to see which model has the lower Schwarz Bayesian Criterion (SBC).

The SBC is proportional to $-2\log[p(\text{data}|\text{model})]$, where $p(\text{data}|\text{model})$ is the probability of the data given the model (see Chapter 4). This is done using the `gamlss` function in the **gamlss** package. Acquire the package and obtain the SBC value for each of three models by typing

```
> require(gamlss, quiet=TRUE)
> gamlss(B.1 ~ 1, family=PO, data=dat,
+    trace=FALSE)$sbc
[1] 529
> gamlss(B.1 ~ I(Yr >= 1910), family=PO, data=dat,
+    trace=FALSE)$sbc
[1] 529
> gamlss(B.1 ~ I(Yr >= 1940), family=PO, data=dat,
+    trace=FALSE)$sbc
[1] 515
```

Here the Poisson family is given as `PO` with the logarithm of the rate as the default link (Stasinopoulos and Rigby, 2007). The first model is one with no change point. The next two are change-point models with the first having a change point in the year 1910 and the second having a change point in 1940. The change-point models use the indictor function `I` to assign a `TRUE` or `FALSE` to each year based on logical expression involving the variable `Yr`.

The SBC value is 528.5 for the model with no change points. This compares with an SBC value of 528.7 for the change-point model where the change occurs in 1910 and a value of 514.8 for the change-point model where the change occurs in 1940. Since the SBC is lower in the latter case, 1940 is a candidate year for a change point.

You apply this procedure successively where each year gets considered in turn as a possible change point. You then plot the SBC as a function of year (Fig. 10.4). The horizontal line is the SBC for a model with no change points and tick marks are local minimum of SBC. Here the SBC for the model without a change point is adjusted by adding $2\log(20)$ to account for the prior possibility of five or six equally likely change points over the period of record. Here you find four candidate change points based on local minima of the SBC. The years are 1995, 1948, 1944, and 1932 (red tic marks).

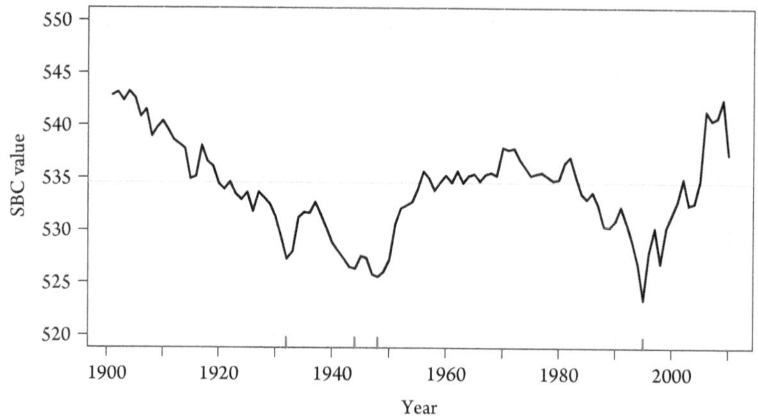

Figure 10.4 Schwarz Bayesian criterion (SBC) for change points.

You assume a priori that there is at most one change point per decade and that the posterior probability of the intercept model is 20 times that of the change-point model. This gives you 12 possible models (1995 only, 1995 & 1948, 1995 & 1948 & 1932, etc.) including the intercept-only model but excludes models with both 1944 and 1948 as the changes occur too close in time.

Next, you estimate the posterior probabilities for each of the 12 models using

$$\Pr(M_i|\text{data}) = \frac{\exp(-.5 \cdot \text{SBC}(M_i))}{\sum_{j=1}^{12} \exp(-.5 \cdot \text{SBC}(M_j))} \tag{10.2}$$

where the models are given by M_i, for $i = 1,\ldots,12$. The results are shown in Table 10.1. The top three models have a total posterior probability exceeding 80%.

Table 10.1 Model posterior probabilities from most (top) to least probable.

	Formula	*Probability*
X10	B.1~I(Yr>=1995)+I(Yr>=1932)	0.43
X6	B.1~I(Yr>=1995)+I(Yr>=1944)	0.20
X4	B.1~I(Yr>=1995)+I(Yr>=1948)	0.18
X12	B.1~I(Yr>=1995)+I(Yr>=1948)+I(Yr>=1932)	0.07
X14	B.1~I(Yr>=1995)+I(Yr>=1944)+I(Yr>=1932)	0.06
X3	B.1~I(Yr>=1948)	0.02
X5	B.1~I(Yr>=1944)	0.01
X2	B.1~I(Yr>=1995)	0.01
X9	B.1~I(Yr>=1932)	0.01
X11	B.1~I(Yr>=1948)+I(Yr>=1932)	0.01
X13	B.1~I(Yr>=1944)+I(Yr>=1932)	0.00
X1	B.1~1	0.00

Table 10.2 Best model coefficients and standard errors.

	Estimate	*Std. Error*	*t value*	*Pr(>|t|)*
(Intercept)	3.9063	0.4101	9.53	0.0000
I(Yr >= 1995)TRUE	2.5069	0.6494	3.86	0.0002
I(Yr >= 1932)TRUE	1.6493	0.5036	3.28	0.0014

These models all include 1995 with 1932, 1944, and 1948 competing as the second most important change-point year. You can select any of the models, but it makes sense to choose one with a high posterior probability. Note the weaker support for the single change-point models and even less support for the no change-point model.

The single best model has change points in 1932 and 1995. The coefficients of this model are shown in Table 10.2. The model predicts a rate of 3.9 hur/yr in the period 1900–1931. The rate jumps to 6.4 hur/yr in the period 1931–1994 and jumps again to 8.1 in the period 1995–2010.

10.3.2 Covariates

To better understand what might be causing the shifts in hurricane activity, here you include climate covariates in the model. The idea is that if the shift is no longer significant after adding a covariate, then you conclude that the likely causal mechanism is related to climate variation.

The two important covariates for annual basin-wide hurricane frequency are SST and the SOI as used throughout this book. You first fit and summarize a model using the two change points and the two covariates.

```
> model1 = gamlss(B.1 ~ I(Yr >= 1932) + I(Yr >= 1995) +
+    sst + soi, family=PO, data=dat, trace=FALSE)
> summary(model1)
```

You find that the change point at 1995 has the largest p-value among the variables. You also note that the model has an SBC of 498.5.

You consider whether the model can be improved by removing the change point at 1995, so you remove it and refit the model.

```
> model2 = gamlss(B.1 ~ I(Yr >= 1932)
+    + sst + soi, family=PO, data=dat, trace=FALSE)
> summary(model2)
```

With the reduced model, you find all variables statistically significant (p-value less than 0.1) and the model has an SBC of 496.3, which is lower than the SBC of your first model that includes 1995 as a change point.

Thus you conclude that the shift in the rate at 1995 is more likely the result of a synchronization (Tsonis et al., 2006) of the effects of SST and ENSO on hurricane activity than is the shift in 1932. The shift in 1932 is important after including SST and ENSO influences providing evidence that the increase in activity at this time is likely due, at least in part, to improvements in observing technologies.

A change-point model is useful for detecting rate shifts caused by climate and observational improvements. A model that includes terms for both can help you differentiate between those two possibilities. However, change-point models are not particularly useful for predicting when the next shift will occur.

10.4 CONTINUOUS TIME SERIES

The SST temperature, the SOI, and the NAO are continuous time series. Values fluctuate over a range of scales often without abrupt changes. In this case, it can be useful to split the series into a few components where each component has a smaller range of scales.

Here your goal is to decompose the SST time series as an initial step in creating a time-series model. The model can be used to make predictions of future SST values. Predicted SST values are subsequently used in your frequency model to forecast the probability of hurricanes (Elsner et al., 2008a).

You return to your monthly SST values over the period 1856–2010. As you did with the NAO values in Chapter 5, you input the data and create a continuous-valued time series object (`sst.ts`) containing monthly SST values beginning with January 1856.

```
> SST = read.table("SST.txt", header=TRUE)
> sst.m = as.matrix(SST[6:160, 2:13])
> sst.v = as.vector(t(sst.m))
> sst.ts = ts(sst.v, frequency=12, start=c(1856, 1))
```

First you plot your SST time series by typing

```
> plot(sst.ts, ylab="SST (C)")
```

The graph shows that the SST is dominated by interannual variability (Fig. 10.5). The ocean is coldest in February and March and warmest in August and September. The average temperature during March is 18.6°C and during August is 23.1°C. There also appears to be a trend toward greater warmth, although it is difficult to see because of the larger interannual variations.

The SST times series can be decomposed into components using the `stl` function. The function accepts a time series object as its first argument and the type of smoothing window is specified through the `s.window` argument.

```
> sdts = stl(sst.ts, s.window="periodic")
```

The seasonal component is found by a local regression smoothing of the monthly means. The seasonal values are then subtracted, and the remainder of the series smoothed to find the trend. The overall time-series mean value is removed from the seasonal component and added to the trend component. The process is iterated a few times. What remains is the difference between the actual monthly values and the sum

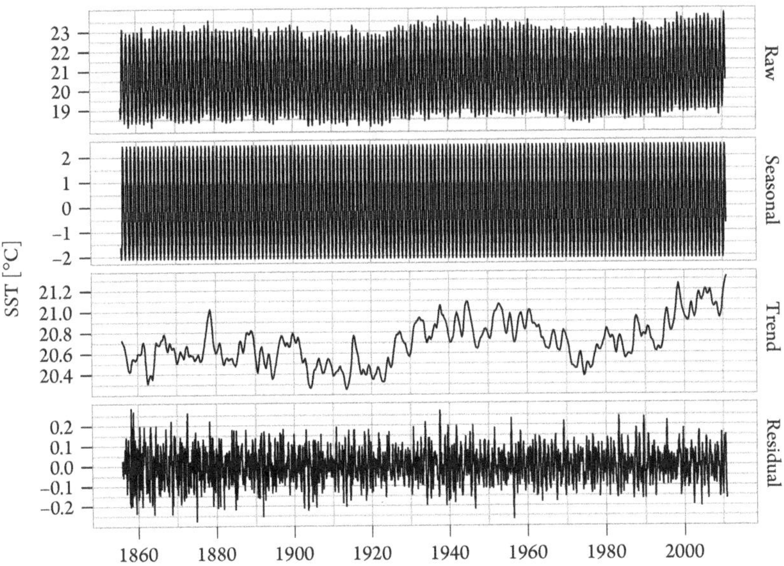

Figure 10.5 Monthly raw and component SST values.

of the seasonal and trend components. Note that if you have change points in your time series, you can use the `bfast` function in the **bfast** package to decompose your time series. In this case, the trend component has the change points.

The raw and component series are plotted in Figure 10.5. The data are prepared as follows. First a vector of dates is constructed using the `seq.dates` function from the **chron** package. This allows you to display the graphs using a scale that corresponds to real dates.

```
> require(chron)
> date = seq.dates(from="01/01/1856", to="12/31/2010",
+   by="months")
> #dates = chron(dates, origin=c(1, 1, 1856))
```

Next, a data frame is constructed that contains the vector of dates, the raw monthly SST time series, and the corresponding components from the seasonal decomposition.

```
> datw = data.frame(Date=as.Date(date),
+   Raw=as.numeric(sst.ts),
+   Seasonal=as.numeric(sdts$time.series[, 1]),
+   Trend=as.numeric(sdts$time.series[, 2]),
+   Residual=as.numeric(sdts$time.series[, 3]))
> head(datw)
        Date  Raw Seasonal Trend Residual
1 1856-01-01 19.1   -1.621  20.7  0.02587
2 1856-02-01 18.6   -2.060  20.7 -0.04036
```

```
3 1856-03-01 18.7    -2.068  20.7  0.02453
4 1856-04-01 19.0    -1.640  20.7 -0.05631
5 1856-05-01 19.9    -0.756  20.7 -0.01318
6 1856-06-01 21.2     0.478  20.7  0.00159
```

Here the data are in the "wide" form like a spreadsheet. To make them easier to plot as separate time-series graphs, you create a "long" form of the data frame with the `melt` function in the **reshape** package. The function melds your data frame into a form suitable for casting (Wickham, 2007). You specify the data frame and your `Date` column as your id variable. The function assumes that remaining variables are measure variables (non id variables) with the column names turned into a vector of factors.

```
> require(reshape)
> datl = melt(datw, id="Date")
> head(datl); tail(datl)
        Date variable value
1 1856-01-01      Raw  19.1
2 1856-02-01      Raw  18.6
3 1856-03-01      Raw  18.7
4 1856-04-01      Raw  19.0
5 1856-05-01      Raw  19.9
6 1856-06-01      Raw  21.2
           Date variable   value
7435 2010-07-01 Residual  0.0807
7436 2010-08-01 Residual  0.1489
7437 2010-09-01 Residual  0.0601
7438 2010-10-01 Residual -0.0501
7439 2010-11-01 Residual -0.1157
7440 2010-12-01 Residual -0.1666
```

Here you make use of the **ggplot2** package (see Chapter 5) to create a facet grid to display your time-series plots with the same time axis. The `qplot` function graphs the decomposed time-series values grouped by variable. The argument `scale="free_y"` allows the y axes to have different scales. This is important as the decomposition results in a large seasonal component centered on zero, while the trend component is much smaller.

```
> require(ggplot2)
> qplot(Date, value, data=datl, geom="line",
+   group=variable) + facet_grid(variable ~.,
+   scale="free_y")
```

The monthly time-series components are shown in Figure 10.5. The observed (raw) values are shown in the top panel. The seasonal component, trend component, and residuals are also shown in separate panels on the same time-series axis. Temperatures increase by more than 0.5°C over the past 100 years. But the trend is not monotonic. The residuals show year-to-year variation generally between −0.15 and +0.15°C with somewhat larger variation before about 1871.

You can build separate time series models for each component. For example, an autoregressive moving average (ARMA) model can be built for the residual component (R_t). An ARMA model with p autoregressive terms and q moving average terms (ARMA(p, q)) is given by

$$R_t = \sum_{i=1}^{p} \phi_i R_{t-i} + \sum_{i=1}^{q} \theta_i \varepsilon_{t-i} + \varepsilon_t \tag{10.3}$$

where the ϕ_i's and the θ_i's are the parameters of the autoregressive and moving average terms, respectively and ε_t is random white noise assumed to be described by a normal distribution with mean zero and variance σ^2. For the trend component, an ARIMA model is more appropriate. An ARIMA model generalizes the ARMA model by removing the nonstationarity through an initial differencing step (the "integrated" part of the model).

Here you use the `ar` function to determine the autoregressive portion of the series using the AIC.

```
> ar(datw$Trend)
Call:
ar(x = datw$Trend)

Coefficients:
      1        2        3        4        5        6
  1.279   -0.058   -0.101   -0.062   -0.045   -0.032
      7        8        9       10       11
 -0.008   -0.023    0.010   -0.034    0.067

Order selected 11  sigma^2 estimated as  0.000298
```

Result shows an autoregressive order of 11 months. Continuing, you assume that the integrated and moving average orders are both one.

```
> model = arima(datw$Trend, order=c(11, 1, 1))
```

You then use the model to make monthly forecasts out to 36 months using the `predict` method. Predictions are made at times specified by the `newxreg` argument.

```
> nfcs = 36
> fcst = predict(model, n.ahead=nfcs)
```

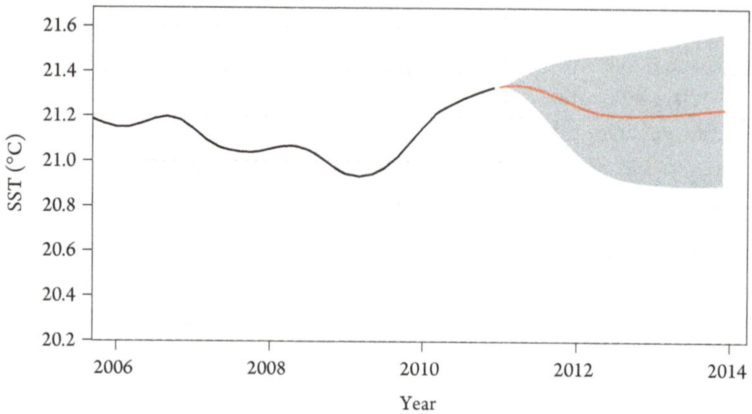

Figure 10.6 Observed and forecast SST trend component.

You plot the forecasts along the corresponding date axis by typing

```
> newdate = seq.dates(from="01/01/2011",
+   to="12/01/2013", by="months")
> plot(newdate, fcst$pred, type="l", ylim=c(21, 21.5))
```

The last 5 years of SST trend and the 36-month forecast are shown in Figure 10.6. The observed values are in black and the forecast values are in red. A 95 percent confidence band is shown in gray. Here you use the same scale. The band is quite large after a few months. A forecast of the actual SST must include forecasts for the seasonal and residual components as well.

10.5 TIME-SERIES NETWORK

Here, we show you an interesting way to characterize a time series. You first map the series to a network using geometry and then employ tools from graph theory to get a unique perspective on your data.

Network analysis is the application of graph theory. Graph theory is the study of mathematical structures used to model relations between objects. Objects and relations can be many things with the most familiar being people and friendships.

Network analysis was introduced into climatology by Tsonis and Roebber (2004). They used values of geopotential height on a spatial grid and the relations were based on correlation. Here you use network analysis to examine year-to-year relations in hurricane activity. The idea is new and requires mapping a time series to a network (Lacasa et al., 2008). The presentation here follows the work of Elsner et al. (2009).

10.5.1 Time Series Visibility

How can a time series of hurricane counts be represented as a network? Consider the plot in Figure 10.7. The time series of U.S. hurricane counts forms a discrete

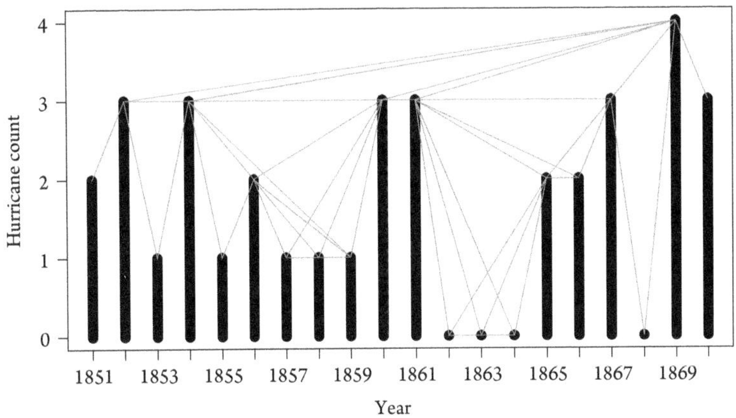

Figure 10.7 Visibility landscape for hurricane counts.

landscape. A bar is connected to another bar if there is a line of sight (visibility line) that allows the bars to "see" each other. Here visibility lines are drawn for all 10 bars. It is clear that 1869 by virtue of its high hurricane count (4) can see 1852, 1854, 1860, 1861, 1867, 1868, and 1870, while 1868 with its zero count can see only 1867 and 1869. Lines do not cut through bars. In this way, each year in the time series is linked in a network. The nodes are the years and the links (edges) are the visibility lines.

More formally, let h_a be the hurricane count for year t_a and h_b the count for year t_b, then 2 years are linked if for any other year t_i with count h_i

$$h_i \leq h_b + (h_a - h_b)\frac{t_b - t_i}{t_b - t_a} \tag{10.4}$$

By this definition, each year is visible to at least its nearest neighbors (the year before and the year after), but not itself. The network is invariant under rescaling the horizontal or vertical axes of the time series as well as under horizontal and vertical translations (Lacasa et al., 2008).

In network parlance, years are nodes and the visibility lines are the links (or edges). The network arises by releasing the years from chronological order and treating them as nodes linked by visibility lines. Here we see that 1869 is well connected, while 1853 is not. Years featuring many hurricanes generally result in more links, especially if neighboring years have relatively few hurricanes. This can be seen by comparing 1853 with 1858. Both years have only a single hurricane, but 1858 is adjacent to years that also have a single hurricane so it is linked to four other years. By contrast, 1853 is next to 2 years each with three hurricanes so it has the minimum number of two links. The degree of a node is the number of links connected to it.

The function `get.visibility` available in **get.visibility.R** computes the visibility lines. It takes a vector of counts as input and returns three lists: one containing the incidence matrix (`sm`), another a set of node edges (`node`), and the third a degree distribution (`pk`), which indicate the number of years with k number of edges. Source the code and compute the visibility lines by typing,

```
> source("get.visibility.R")
> vis = get.visibility(annual$US.1)
```

10.5.2 Network Plot

You use the `network` function from the **network** package (Butts et al., 2011) to create a network object from the incidence matrix by typing

```
> require(network)
> net = network(vis$sm, directed=FALSE)
```

Then use the plot method for network objects to graph the network.

```
> plot(net, label=1851:2010, label.cex=.6,
+   vertex.cex=1.5, label.pos=5, edge.col="grey")
```

The results are shown in Figure 10.8. Node color indicates the number of links (degree) going from light purple (few) to red. Here the placement of years on the network plot is based on simulated annealing (Kamada and Kawai, 1989), and the nodes are colored, based on the number of edges. Years with the largest number of edges are more likely to be found in dense sections of the network and are colored dark red. Years with fewer edges are found near the perimeter of the network and are colored light purple.

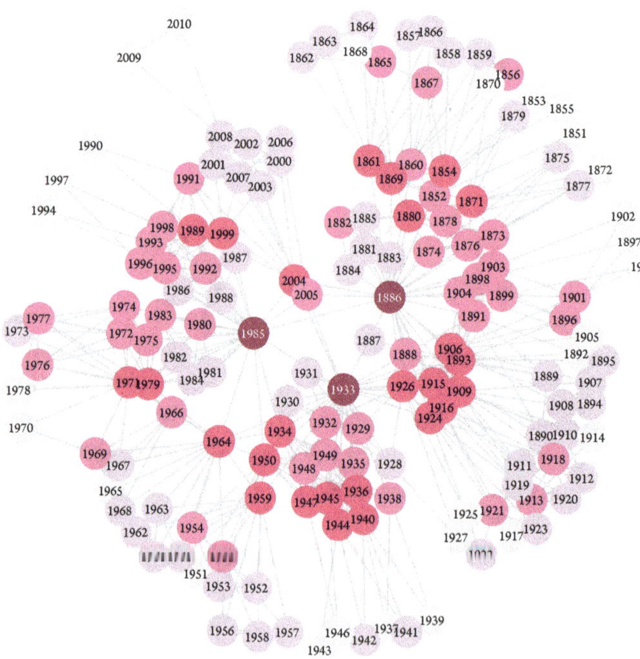

Figure 10.8 Visibility network of U.S hurricanes.

The **sna** package (Butts, 2010) contains functions for computing properties of your network. First create a square adjacency matrix where the number of rows is the number of years and each element is a zero or one depending on whether the years are linked and with zeros along the diagonal (a year is not linked with itself). Then compute the degree of each year indicating the number of years it can see and find which years can see the farthest.

The year with the highest degree is 1886 with 34 links. Two other years with high degree include 1933 with 31 links and 1985 with 28 links. Other relatively highly connected years are 1893, 1950 1964, and 1906, in that order. The average degree is 6.6, but the degree distribution is skewed so this number says little about a typical year.

10.5.3 Degree Distribution and Anomalous Years

The total number of links in the network (sum of the links over all nodes) is 1,054. There are 160 nodes, so 20 percent of the network consists of 32 of them. If you rank the nodes by the number of links, you find that the top 20 percent account for 40 percent of the links.

You plot the degree distribution of your network by typing

```
> plot(vis$pk$k, cumsum(vis$pk$P), pch=16, log="x",
+   ylab="Proportion of Years With k or Fewer Links",
+   xlab="Number of Links (k)")
```

The distribution (Fig. 10.9) is the cumulative percentage of years with *k* or fewer links as a function of the number of links. The horizontal axis is plotted using a log scale. Just over 80 percent of all years have 10 or fewer links, and over 50 percent have 5 or fewer. Although the degree distribution is skewed to the right, it does not appear to represent a small-world network (power-law distribution).

We perform a Monte Carlo (MC) simulation by randomly drawing counts from a Poisson distribution with the same number of years and the same hurricane rate

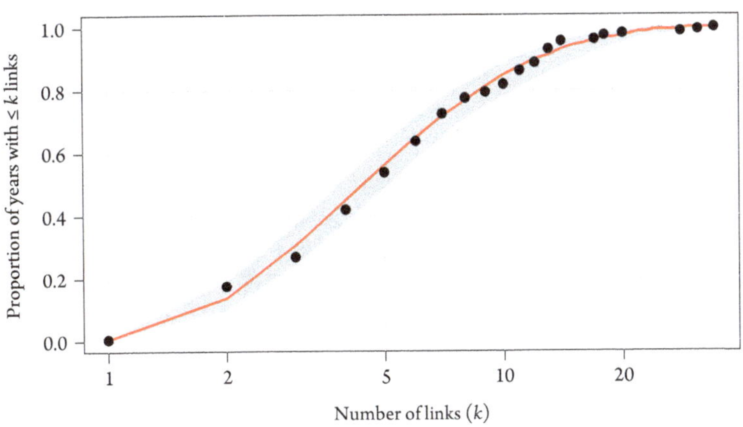

Figure 10.9 Degree distribution of the visibility network.

as the observations. A visibility network is constructed from the random counts and the degree distribution computed as before. The process is repeated 1,000 times after which the median and quantile values of the degree distributions are obtained. The median distribution is shown as a red line in Figure 10.9 and the 95 percent confidence interval shown as a gray band. Results indicate that the degree distribution of your hurricane count data does not deviate significantly from the degree distribution of a Poisson random time series.

However, it does suggest a new way to think about anomalous years. Years are anomalous not in a statistical sense of violating a Poisson assumption, but in the sense that the temporal ordering of the counts identifies a year that is unique in that it has a large count but is surrounded before and after by years with low counts. Thus we contend that node degree in a visibility network is a useful indicator of climatologically anomalous year. That is, a year that stands above most of the other years, but particularly above its "neighboring" years represents more of an anomaly in a physical sense than does a year that is simply well above the average. Node degree captures information about the frequency of hurricanes for a given year and information about the relationship of that frequency to the frequencies over the given year's recent history and near future.

The relationship between node degree and the annual hurricane count is tight but not exact. Years with a low number of hurricanes are ones that are not well connected to other years, while years with an above-normal number are ones that are more connected on average. The Spearman rank correlation between year degree and year count is 0.73. But this is largely a result of low count years. The correlation drops to 0.48 when considering only years with more than two hurricanes. Thus high count is necessary but not sufficient for characterizing the year as anomalous, as perhaps it should be.

10.5.4 Global Metrics

Global network metrics are used to compare different data sets. One example is the diameter of the network as the length of the longest geodesic path between any two years for which a path exists. A geodesic path (shortest path) is a path between two years such that no shorter path exists. For instance, in Figure 10.7, you see that 1861 is connected to 1865 directly and through a connection with 1862. The direct connection is a path of length one while the connection through 1862 is a path of length two.

The **igraph** package (Csardi and Nepusz, 2006) contains functions for computing network analytics. To find the diameter of your visibility network, load the package, create the network (graph) from the list of edges, then use the `diameter` function. Prefix the function name with the package name and two colons to avoid a conflict with the same name from another loaded package.

```
> require(igraph)
> vis = get.visibility(annual$US.1)
```

```
> g = graph.edgelist(vis$sm, directed=FALSE)
> igraph::diameter(g)
[1] 5
```

The result indicates that any 2 years are separated by at most 5 links, although there is more than one such geodesic.

Transitivity measures the probability that the adjacent nodes are themselves connected. Given that year *i* can see years *j* and *k*, what is the probability that year *j* can see year *k*? In a social network, transivity indicates the likelihood that any two of your friends are themselves friends. To compute the transitivity for your visibility network, type

```
> tran = transitivity(g)
> round(tran, 3)
[1] 0.468
```

The transitivity tells you that there is a 46.8 percent chance that two adjacent nodes of a given node are connected. The higher the probability, the greater the network density. The visibility network constructed from Gulf hurricane counts has a transitivity of 0.563, which compares with a transitivity of 0.479 for the network constructed from Florida counts. The network density is inversely related to interannual variance, but this rather large difference provides some evidence to support clustering of hurricanes in the vicinity of Florida relative to the Gulf coast region (see Chapter 11). A Monte Carlo simulation would help you interpret the difference against the backdrop of random variations.

Another global property is the minimum spanning tree. A tree is a connected network that contains no closed loops. By "connected," we mean that every year in the network is reachable from every other year via some path through the network (Newman, 2010). A tree is said to span if it connects all the years together. A network may have more than one spanning tree. The minimum spanning tree is the one with the fewest number of edges. A network may contain more than one minimum spanning tree. You compute the minimum spanning tree by typing

```
> mst = minimum.spanning.tree(g)
> net = network(get.edgelist(mst))
```

The result is an object of class `igraph`. This is converted to a network object by specifying the edge list in the `network` function. You plot the network tree by typing

```
> plot(net)
```

The graph is shown in Figure 10.10, where the nodes are labeled with their corresponding years and are colored according to the level of "betweenness." Arrows point toward later years. The node betweenness (or betweenness centrality) is the number

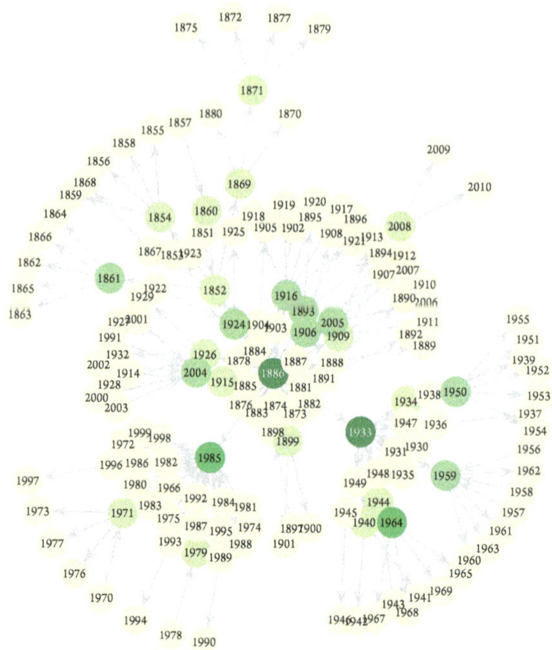

Figure 10.10
Minimum spanning tree of the hurricane visibility network.

of geodesics (shortest paths) going through it. By definition, the minimum spanning tree must have a transitivity of zero. You check this by typing

```
> transitivity(mst)
[1] 0
```

In summary, the visibility network is the set of years as nodes together with links defined by a sight line on the time series graph such that the line does not intersect a count bar. Analysis of the topological properties of the network, like betweenness, provide new insights into the relationship between hurricanes and climate.

This chapter showed you some methods and models for working with time-series data. We began by showing you how to overlay time-series plots. We then discussed the nature of discrete time series and showed how to compute moving averages and how to create a model for describing the day-to-day variation in hurricane activity. Next we showed how to analyze and model count data for change points and how to interpret the shifts in light of climate variability confounded by technological advances. We then looked at ways to analyze and model continuous time series. We showed how to decompose a series into its component parts and how to model the nonseasonal part with an ARMA model. We finished with a novel way to construct a network from time-series counts. We showed how metrics from the network provide insight into hurricane climatology. In the next chapter, we consider ways to analyze and model hurricane clusters.

11

CLUSTER MODELS

"There are in fact two things, science and opinion; the former begets knowledge, the latter, ignorance."
—Hippocrates

A cluster is a group of the same or similar events close together. Clusters arise in hurricane origin locations, tracks, and landfalls. In this chapter, we look at how to analyze and model clusters. We divide the chapter into methods for time, space, and feature clustering. Of the three feature clustering is best known to climatologists. We begin by showing you how to detect and model time clusters.

11.1 TIME CLUSTERS

Consecutive hurricanes originating in the same area often take similar paths. This grouping, or clustering, increases the potential for multiple landfalls above what you expect from random events. A statistical model for landfall probability captures clustering through covariates like the North Atlantic Oscillation (NAO), which relates a steering mechanism (position and strength of the subtropical high pressure) to coastal hurricane activity. But there could be additional time correlation not related to the covariates. A model that does not account for this extra variation will underestimate the potential for multiple hits in a season.

Following Jagger and Elsner (2006), you consider three coastal regions including the Gulf Coast, Florida, and the East Coast (Fig. 6.2). Regions are large enough to capture enough hurricanes, but not too large as to include many non-coastal strikes. Here you use hourly position and intensity data described in Chapter 6. For each hurricane, you note its wind speed maximum within each region. If the maximum wind exceeds 33 m s^{-1}, then you count it as a hurricane for the region. A tropical cyclone that affects more than one region at hurricane intensity is counted in each region. Because of this, the sum of the regional counts is larger than the total count.

Begin by loading **annual.RData**. These data were assembled in Chapter 6. Subset the data for years starting with 1866.

```
> load("annual.RData")
> dat = subset(annual, Year >= 1866)
```

The covariate Southern Oscillation Index (SOI) data begins in 1866. Next, extract all hurricane counts for the Gulf coast, Florida, and East coast regions.

```
> cts = dat[, c("G.1", "F.1", "E.1")]
```

11.1.1 Cluster Detection

You start by comparing the observed with the expected number of years for the two groups of hurricane counts. The groups include years with no hurricanes and years with three or more. The expected number is from a Poisson distribution with a constant rate. The idea is that for regions that show a cluster of hurricanes, the observed number of years with no hurricanes and years with three or more hurricanes should be greater than the corresponding expected number. Said another way, a Poisson model with a hurricane rate estimated from counts over all years in regions with clustering will underestimate the number of years with no hurricanes and years with many hurricanes.

For example, you find the observed number of years without a Florida hurricane and the number of years with more than two hurricanes by typing

```
> obs = table(cut(cts$F.1,
+    breaks=c(-.5, .5, 2.5, Inf)))
> obs
(-0.5,0.5]   (0.5,2.5]   (2.5,Inf]
        70          62          13
```

And the expected numbers for these three groups by typing

```
> n = length(cts$F.1)
> mu = mean(cts$F.1)
> exp = n * diff(ppois(c(-Inf, 0, 2, Inf), lambda=mu))
> exp
[1] 61.66 75.27  8.07
```

The observed and expected counts for the three regions are given in Table 11.1. In the Gulf and East coast regions, the observed numbers are relatively close to the expected numbers in each of the groups. By contrast, in the Florida region, you see the observed numbers exceeds the expected numbers for the no-hurricane and the three-or-more hurricanes groups.

Table 11.1 Observed (O) and expected (E) number of hurricane years by count groups.

Region	$O(=0)$	$E(=0)$	$O(\geq 3)$	$E(\geq 3)$
Gulf Coast	63	66.1	7	6.6
Florida	70	61.7	13	8.1
East Coast	74	72.3	7	4.9

Table 11.2 Observed versus expected statistics. The Pearson and χ^2 test statistics along with the corresponding p-values are given for each coastal region.

Region	*Pearson*	*p-value*	χ^2	*p-value*
Gulf Coast	135	0.6858	0.264	0.894
Florida	187	0.0092	6.475	0.039
East Coast	150	0.3440	1.172	0.557

The difference between the observed and expected numbers is used to assess the statistical significance of the clustering. This is done using Pearson residuals and the χ^2 statistic. The Pearson residual is the difference between the observed count and expected rate divided by the square root of the variance. The p-value is evidence in support of the null hypothesis of no clustering.

For example, to obtain the χ^2 statistic, type

```
> xis = sum((obs - exp)^2 / exp)
> xis
[1] 6.48
```

The p-value as evidence in support of the null hypothesis is given by

```
> pchisq(q=xis, df=2, lower.tail=FALSE)
[1] 0.0393
```

where `df` is the degrees of freedom equal to the number of groups minus one. The χ^2 and Pearson statistics for the three regions are shown in Table 11.2. The p-values for the Gulf and East coasts are greater than 0.05 indicating little support for the cluster hypothesis. By contrast, the p-value for the Florida region is 0.009 using the Pearson residuals and 0.037 using the χ^2 statistic. Those values provide evidence that hurricane occurrences in Florida are clustered in time.

11.1.2 Conditional Counts

What might be causing this clustering? The extra variation in annual hurricane counts might be due to variation in hurricane rates. You examine this possibility with a

Poisson regression model (see Chapter 7). The model includes an index for the NAO and the SOI after Elsner and Jagger (2006).

This is a generalized linear model (GLM) approach using the Poisson family that includes a logarithmic link function for the rate. The annual hurricane count model is

$$H_i \sim \text{dpois}(\lambda_i) \tag{11.1}$$
$$\log(\lambda_i) = \beta_0 + \beta_{\text{soi}}\,\text{SOI}_i + \beta_{\text{nao}}\,\text{NAO}_i$$

where H_i is the hurricane count in year i simulated ($\sim$) from a Poisson distribution with a rate λ_i that depends on the year i. The logarithm of the rate depends in a linear way on the SOI and NAO covariates. The code to fit the model, determine the expected count, and table the observed versus expected counts for each region is given below.

```
> pfts = list(G.1 = glm(G.1 ~ nao + soi,
+    family="poisson", data=dat),
+    F.1 = glm(F.1 ~ nao + soi, family="poisson",
+    data=dat),
+    E.1 = glm(E.1 ~ nao + soi, family="poisson",
+    data=dat))
> prsp = sapply(pfts, fitted, type="response")
> rt = regionTable(cts, prsp, df=3)
```

The count model provides an expected number of hurricanes in each year. The expected is compared to the observed as before. Results indicate that clustering is somewhat ameliorated by conditioning the rates on the covariates. In particular, the Pearson residual reduces to 172.4 with an increase in the corresponding p-value to 0.042. However, the p-value remains near 0.15 indicating the conditional model, while an improvement, fails to capture all the extra (beyond Poisson) variation in Florida hurricane counts.

11.1.3 Cluster Model

Having found evidence that Florida hurricanes arrive in clusters next you model this process. In the simplest case, you assume the following. Each cluster has either one or two hurricanes and the annual cluster count is described by a Poisson distribution with a rate r. Note the difference. Earlier, you assumed that each hurricane was independent and the annual hurricane count followed a Poisson distribution. Furthermore, now you let p be the probability that a cluster will have two hurricanes.

Formally, your model can be expressed as follows. Let N be the number of clusters in a given year and $X_i, i = 1,\ldots,N$ be the number of hurricanes in each cluster minus one. Then, the number of hurricanes in a given year is $H = N + \sum_{i=1}^{N} X_i$. Conditional on N, $M = \sum_{i=1}^{N} X_i$ is described by a binomial distribution since the X_is are independent Bernoulli variables and p is constant. That is, $H = N + M$, where the annual number of clusters N has a Poisson distribution with cluster rate r, and M

has a binomial distribution with proportion p and size N. Here the binomial distribution describes the number of occurrences of at least one hurricane in a sequence of N independent years, with each year having a probability p of observing at least one hurricane.

In summary, your cluster model has the following properties:

1. The expected number of hurricanes is given by $E(H) = r(1+p)$.
2. The variance of H is given by

$$\begin{aligned} \text{var}(H) &= E(\text{var}(H|N)) + \text{var}(E(H|N)) \\ &= E(N(p(1-p))) + \text{var}((1+p)N) \\ &= rp(1-p) + r(1+p)(1+p) = r(1+3p) \end{aligned}$$

3. The dispersion of H is given by $\text{var}(H)/E(H) = \phi = (1+3p)/(1+p)$, which is independent of cluster rate. Solving for p, you find $p = (\phi - 1)/(3 - \phi)$.
4. The probability mass function for the number of hurricanes, H, is given by

$$P(H = k|r,p) = \sum_{i=0}^{\lfloor i/2 \rfloor} \text{dpois}(k-i, r)\ \text{dbinom}(i, k-i, p); k = 0, 1, \ldots$$

$$\begin{aligned} P(H = 0|r,p) &= e^{-r}\text{dpois}(k-i, r) \\ &= e^{-r}\frac{r^{k-i}}{(k-i)!}\text{dbinom}(i, k-i, p) \\ &= \frac{k-i}{i}p^i(1-p)^{k-2i} \end{aligned}$$

5. The model has two parameters r and p. A better parameterization is to use $\lambda = r(1+p)$ with p to separate the hurricane frequency from the cluster probability. The parameters do not need to be fixed and can be functions of the covariates.
6. When $p = 0$, H is Poisson, and when $p = 1$, $H/2$ is Poisson, the dispersion is two, and the probability that H is even is 1.

You need a way to estimate r and p.

11.1.4 Parameter Estimation

Your goal is a hurricane count distribution for the Florida region. For that, you need to estimate the annual cluster rate (r) and the probability (p) that the cluster size is two. Continuing with the GLM approach, you separately estimate the annual hurricane frequency, λ, and the annual cluster rate r. The ratio of these two parameters minus one is an estimate of the probability p.

This is reasonable if p does not vary much, since the annual hurricane count variance is proportional to the expected hurricane count (i.e., $\text{var}(H) = r(1+3p) \propto r \propto E(H)$). You estimated the parameters of the annual count model using Poisson regression, which assumes that the variance of the count is proportional to the expected count. Thus under the assumption that p is constant, Poisson regression can be used to estimate λ in the cluster model.

You regress the logarithm of the link function for the cluster rate onto the predictors NAO and SOI. The model is given by

$$N_i \sim \text{dpois}(r_i) \tag{11.2}$$
$$\log(r_i) = \alpha_0 + \alpha_1 \text{ SOI}_i + \alpha_2 \text{ NAO}_i$$

The parameters of this annual cluster count model cannot be estimated directly, since the observed hurricane count does not furnish information about the number of clusters.

Consider the observed set of annual Florida hurricane counts. Since the annual frequency is quite small, the majority of years have either no hurricanes or a single hurricane. You can create a "reduced" data set by using an indictor of whether or not there was at least one hurricane. Formally, let $I_i = I(H_i > 0) = I(N_i > 0))$, then I is an indicator of the occurrence of a hurricane cluster for each year. You assume that I has a binomial distribution with size parameter of one and a proportion equal to π. This leads to a logistic regression model (see Chapter 7) for I.

Note that since $\exp(-r)$ is the probability of no clusters, the probability of a cluster π is $1 - \exp(-r)$. Thus the cluster rate is $r = -\log(1-\pi)$. If you use a logarithmic link function on r, then $\log(r) = \log(-\log(1-\pi)) = \text{cloglog}(\pi)$, where cloglog is the complementary log—log function. Thus, you model I using the cloglog function to obtain r.

Your cluster model is a combination of two models, one for the counts and another for the clusters. Formally, it is given by

$$I_i \sim \text{dbern}(\pi_i) \tag{11.3}$$
$$\text{cloglog}(\pi_i) = \alpha_0 + \alpha_1 \text{ SOI}_i + \alpha_2 \text{ NAO}_i$$

where `dbern` is the Bernoulli distribution with mean π. The covariates are the same as those used in the cluster count model. Given these equations, you have the following relationships for r and p.

$$\log(\hat{r}(1+\hat{p})) = \hat{\beta}_0 + \hat{\beta}_1 \text{ SOI} + \hat{\beta}_2 \text{ NAO} \tag{11.4}$$
$$\log(\hat{r}) = \hat{\alpha}_0 + \hat{\alpha}_1 \text{ SOI} + \hat{\alpha}_2 \text{ NAO} \tag{11.5}$$

By subtracting the coefficients in Eq. 11.5 for the annual cluster count model from the those in Eq. 11.4 for the annual hurricane count model, you have a regression model for the probabilities given by

$$\log(1+\hat{p}) = \hat{\beta}_0 - \hat{\alpha}_0 + (\hat{\beta}_1 - \hat{\alpha}_1) \text{ SOI} + (\hat{\beta}_2 - \hat{\alpha}_2) \text{ NAO} \tag{11.6}$$

11.1.5 Model Diagnostics

You start by comparing fitted values from the count model with fitted values from the cluster model. Let H_i be the hurricane count in year i and $\hat{\lambda}_i$ and $\hat{r}_i$ be the fitted annual count and cluster rates, respectively. Then let τ_0 be a test statistic given by

$$\tau_0 = \frac{1}{n}\sum_{i=1}^{n}(H_i - \hat{r}_i) = \frac{1}{n}\sum_{i=1}^{n}(\hat{\lambda}_i - \hat{r}_i). \tag{11.7}$$

The value of τ_0 is greater than one if there is clustering. You test the significance of τ_0 by generating random samples of length n from a Poisson distribution with rate λ_i and computing τ_j for $j = 1,\ldots,N$, where N is the number of samples. A test of the null hypothesis that $\tau_0 \leq 0$ is the proportion of simulated τ's that are at least as large as τ_0.

You do this with the `testfits` function in the **correlationfuns.R** package by specifying the model formula, data, and number of random samples.

```
> source("correlationfuns.R")
> tfF = testfits(F.1 ~ nao + soi, data=dat, N=1000)
> tfF$test; tfF$testpvalue
[1] 0.104
[1] 0.026
```

For Florida hurricanes, the test statistic τ_0 has a value of 0.104 indicating some difference in count and cluster rates. The proportion of 1,000 simulated τs that are at least as large as this is 0.026 providing sufficient evidence to reject the no-cluster hypothesis. Repeating the simulation using Gulf Coast hurricanes

```
> tfG = testfits(G.1 ~ nao + soi, data=dat, N=1000)
> tfG$test; tfG$testpvalue
[1] -0.0617
[1] 0.787
```

you find that, in contrast to Florida, there is little evidence against the no-cluster hypothesis. The results are consistent with what you found previously using the χ^2 statistic.

A linear regression through the origin of the fitted count rate on the cluster rate under the assumption that p is constant yields an estimate for $1 + p$. You plot the annual count and cluster rates and draw the regression line using the `plotfits` function.

```
> par(mfrow=c(1, 2), pty="s")
> ptfF = plotfits(tfF)
> mtext("a", side=3, line=1, adj=0, cex=1.1)
> ptfG = plotfits(tfG)
> mtext("b", side=3, line=1, adj=0, cex=1.1)
```

The regressions are shown in Figure 11.1 for Florida and the Gulf coast. The black line is the $y = x$ axis, and cluster and hurricane rates align along this axis if there is no clustering. The red line is the regression of the fitted hurricane rate onto the fitted cluster rate with the intercept set to zero. The slope is an estimate of $1 + p$.

Table 11.3 Coefficients of the count rate model.

	Estimate	*Std. Error*	*z Value*	*Pr(>|z|)*
(Intercept)	−0.27	0.11	−2.55	0.01
nao	−0.23	0.09	−2.50	0.01
soi	0.06	0.03	1.86	0.06

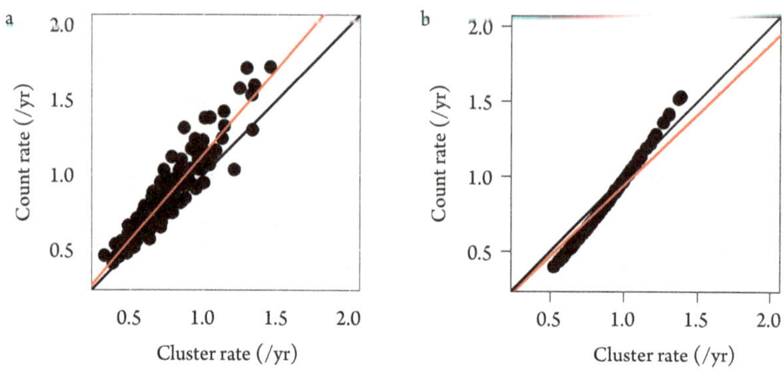

Figure 11.1 Count versus cluster rates for (a) Florida and (b) Gulf coast.

The regression slopes are printed by typing,

```
> coefficients(ptfF); coefficients(ptfG)
rate
1.14
 rate
0.942
```

The slope is 1.14 for the Florida region giving 0.14 as an estimate for p (probability that the cluster will have two hurricanes). The regression slope is 0.94 for the Gulf coast region, which you interpret as a lack of evidence for hurricane clusters.

Your focus is now on Florida hurricanes only. You continue by looking at the coefficients from both models. Type

```
> summary(tfF$fits$poisson)$coef
> summary(tfF$fits$binomial)$coef
```

The coefficient tables (Tables 11.3 and 11.4) show that the NAO and SOI covariates are significant in the hurricane count model, but only the NAO is significant in the cluster rate model.

The difference in coefficient values from the two models is an estimate of $\log(1+p)$, where again p is the probability that a cluster will have two hurricanes. The difference in the NAO coefficient is 0.043 and the difference in the SOI coefficient is 0.035 indicating that the NAO increases the probability of clustering more

Table 11.4 Coefficients of the cluster rate model.

	Estimate	*Std. Error*	*z Value*	*Pr(>|z|)*
(Intercept)	−0.42	0.13	−3.11	0.00
nao	−0.27	0.12	−2.23	0.03
soi	0.02	0.04	0.52	0.60

than ENSO. Lower values of the NAO lead to a larger rate increase for the Poisson model relative to the binomial model.

11.1.6 Forecasts

It is interesting to compare forecasts of the distribution of Florida hurricanes using your Poisson and cluster models. Here you set $p = .138$ for the cluster model. You can use the same two-component formulation for your Poisson model by setting $p = 0$.

You prepare the data using the `lambdapclust` function as follows:

```
> ctsF = cts[, "F.1", drop=FALSE]
> pars = lambdapclust(prsp[, "F.1", drop=FALSE],
+   p=.138)
> ny = nrow(ctsF)
> h = 0:5
```

Next, you compute the expected number of years with `h` hurricanes from the cluster and Poisson models and tabulate the observed number of years. You combine them in a data object.

```
> eCl = sapply(h, function(x)
+     sum(do.call('dclust',
+     c(x=list(rep(x, ny)), pars))))
> ePo = sapply(0:5, function(x)
+     sum(dpois(x=rep(x,ny),
+     lambda=prsp[, "F.1"])))
> o = as.numeric(table(ctsF))
> dat = rbind(o, eCl, ePo)
> names(dat) = 0:5
```

Finally, you plot the observed versus the expected from your cluster and Poisson models using a bar plot where the bars are plotted side by side.

```
> barplot(dat, ylab="Number of Years",
+     xlab="Number of Florida Hurricanes",
+     names.arg=c(0:5),
+     col=c("black", "red", "blue"),
```

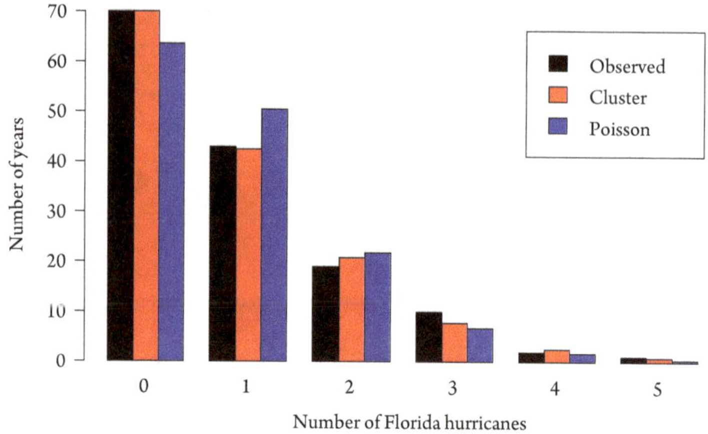

Figure 11.2 Observed versus expected number of Florida hurricane years.

```
+    legend=c("Observed", "Cluster", "Poisson"),
+    beside=TRUE)
```

Results are shown in Figure 11.2. The expected numbers are based on your cluster model (p = 0.137) and on your Poisson model (p = 0). The cluster model fits the observed counts better than does the Poisson model particularly at the low and high count years.

Florida had hurricanes in only 2 of the 11 years from 2000 through 2010. But these 2 years featured seven hurricanes. Seasonal forecast models that predict U.S. hurricane activity assume a Poisson distribution. You show here that this assumption applied to Florida hurricanes leads to a forecast that underpredicts both the number of years without hurricanes and the number of years with three or more hurricanes (Jagger and Elsner, 2012).

11.2 SPATIAL CLUSTERS

Is there a tendency for hurricanes to cluster in space? More specifically, given that a hurricane originates at a particular location, is it more (or less) likely that a later hurricane will form in the same vicinity? This question can be answered using spatial point pattern analysis. Here you consider models for analyzing and modeling events across space.

We begin with some definitions. An *event* is the occurrence of some phenomenon of interest. For example, an event can be a hurricane's origin or its lifetime maximum intensity. An *event location* is the spatial location of the event. For example, the latitude and longitude of the maximum intensity event. A *point* is any location where an event could occur. A *spatial point pattern* is a collection of events and event locations together with spatial domain. The *spatial domain* is defined as the region of interest over which events occur. For example, the North Atlantic basin is the spatial domain for hurricanes.

Complete spatial randomness (CSR) defines a situation where an event is equally likely to occur at any location within the study area regardless of the locations of other events. A spatial point pattern is said to be clustered if given an event at some location it is more likely than random that another event will occur nearby. Regularity is the opposite; given an event, if it is less likely than random that another event will occur nearby, then the spatial point pattern is regular.

A *realization* is a collection of spatial point patterns generated under a spatial point process model. To illustrate, consider four point patterns each consisting of events inside the unit plane. You generate the event locations and plot them by typing

```
> par(mfrow=c(2, 2), mex=.9, pty="s")
> for(i in 1:4){
+ u = runif(n=30, min=0, max=1)
+ v = runif(n=30, min=0, max=1)
+ plot(u, v, pch=19, xlim=c(0, 1), ylim=c(0, 1))
+ }
```

The pattern of events are shown in Figure 11.3. The plots which show events under CSR illustrate that some amount of clustering occurs by chance.

The **spatstat** package (Baddeley and Turner, 2005) contains a large number of functions for analyzing and modeling point pattern data. To make the functions available and obtain a citation, type

```
> require(spatstat)
> citation(package="spatstat")
```

CSR lies on a continuum between clustered and regular spatial point patterns. You use simulation functions in **spatstat** to compare the CSR plots in Figure 11.3 with regular and clustered patterns. Examples are shown in Figure 11.4. Here you see two

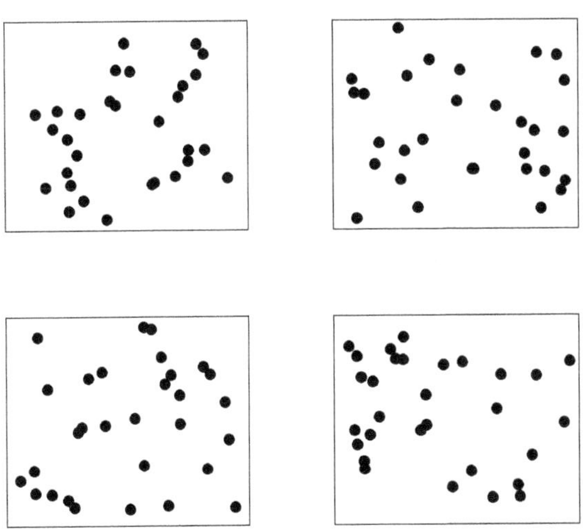

Figure 11.3 Point patterns exhibiting complete spatial randomness.

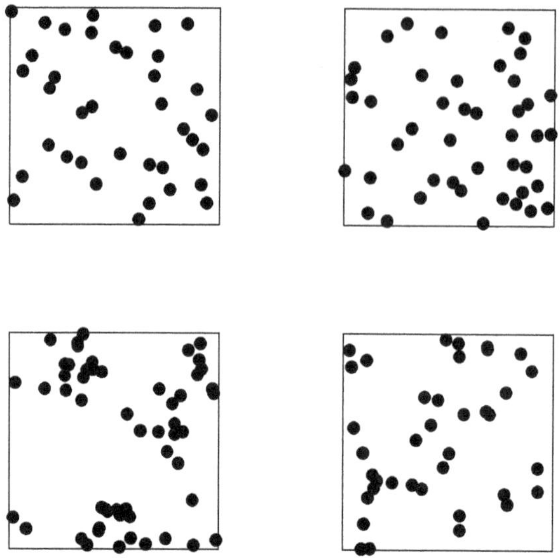

Figure 11.4 Regular (top) and clustered (bottom) point patterns.

realizations of patterns more regular than CSR (top row), and two realizations of patterns more clustered than CSR (bottom row). The simulations are made using a spatial point pattern model. Spatial scale also plays a role. A set of events might be clustered at one scale and regular at another.

11.2.1 Point Processes

Given a set of spatial events (spatial point pattern), your goal is to assess evidence for clustering (or regularity). Spatial cluster detection methods are based on statistical models for spatial point processes. The random variable in these models is the event locations. Here we provide some definitions useful for understanding point process models. Details on the underlying theory are available in Ripley (1981), Cressie (1993), and Diggle (2003).

A process generating events is *stationary* when it is invariant to translation across the spatial domain. This means that the relationship between two events depends only on their positions relative to one another and not on the event locations themselves. This relative position refers to (lag) distance and orientation between the events. A process is *isotropic* when orientation does not matter. Recall these same concepts applied to the variogram models you used in Chapter 9.

Given a single realization, the assumptions of stationarity and isotropy allow for replication. Two pairs of events in the same realization of a stationary process that are separated by the same distance should have the same relatedness. This allows you to use relatedness information from one part of the domain as a replicate for relatedness in another part of the domain. The assumptions of stationarity and isotropy are the starting point but they can be relaxed later on.

The Poisson distribution is used to define a model for CSR. A spatial point process is said to be *homogeneous* Poisson under the following two criteria:

- The number of events occurring within a finite region A is a random variable following a Poisson distribution with mean $\lambda \times A$ for some positive constant λ, with $|A|$ denoting the area of A.
- Given the total number of events N, their locations are an independent random sample of points where each point is equally likely to be picked as an event location.

The first criterion is about the density of the spatial process. For a given domain, it answers the question, "how many events?" It is the number of events divided by the domain area. The density is an estimate of the rate parameter of the Poisson distribution.[1] The second criterion is about homogeneity. Events are scattered throughout the domain and are not clustered or regular.

It is instructive to consider how you would create a homogeneous Poisson point pattern. The procedure follows from the definition. Step 1: Generate the number of events using a Poisson distribution with mean equal to λ. Step 2: Place the events inside the domain using uniform distributions for the spatial coordinates.

For example, let area of the domain be one and the density of events be 200, then type

```
> lambda = 200
> N = rpois(1, lambda)
> x = runif(N); y=runif(N)
> plot(x, y, pch=19)
```

Note that if your domain is not a rectangle, you can circumscribe it within a rectangle, then reject events sampled from inside the rectangle that fall outside the domain. This is called "rejection sampling."

By definition, these point patterns are CSR. However, as noted earlier, you are typically in the opposite position. You observe a set of events and want to know if they are regular or clustered. Your null hypothesis is CSR and you need a test statistic that will guide your inference. CSR models are easy to construct, so you can use Monte Carlo methods.

In some cases, the homogeneous assumption is too restrictive. Consider hurricane genesis as an event process. Event locations may be random, but the probability of an event is higher away from the equator. The constant risk hypothesis of the homogeneous Poisson point pattern model requires a generalization to include a spatially varying density function.

To do this, you define the density $\lambda(s)$, where s denotes spatial points. This is called an *inhomogeneous* Poisson model, and it is analogous to the universal kriging model used on field data (see Chapter 9). Inhomogeneity as defined by a spatially varying density implies nonstationarity as the number of events depends on location.

[1] In the spatial statistics, this is often called the *intensity*. Here we use the term *density* instead so as not to confuse it with hurricane strength.

11.2.2 Spatial Density

Density is a first-order property of a spatial point pattern. That is, the density function estimates the mean number of events at any point in your domain. Events are independent of one another, but event clusters might appear because of the varying density.

Given an observed set of events, how do you estimate $\lambda(s)$? One approach is to use kernel densities. Consider again the set of hurricanes over the period 1944–2000 that were designated tropical only and baroclinically enhanced (Chapter 7). Input these data and create a spatial points data frame of the genesis locations by typing

```
> bh = read.csv("bi.csv", header=TRUE)
> require(sp)
> coordinates(bh) = c("FirstLon", "FirstLat")
> ll = "+proj=longlat +ellps=WGS84"
> proj4string(bh) = CRS(ll)
```

Next convert the geographic coordinates using the Lambert conformal conic projection true at parallels 10 and 40°N and with a center longitude of 60°W. First save the reference system as a CRS object, then use the `spTransform` function from the **rgdal** package.

```
> lcc = "+proj=lcc +lat_1=40 +lat_2=10 +lon_0=-60"
> require(rgdal)
> bht = spTransform(bh, CRS(lcc))
```

The spatial distance unit is meters. Use the `map` (**maps**) and the `map2SpatialLines` (**maptools**) to obtain country borders by typing

```
> require(maps)
> require(maptools)
> brd = map("world", xlim=c(-100, -30), ylim=c(10, 48),
+   plot=FALSE)
> brd_ll = map2SpatialLines(brd, proj4string=CRS(ll))
> brd_lcc = spTransform(brd_ll, CRS(lcc))
```

You use the same coordinate transformation on the map borders as you do on the hurricane locations.

Next you convert your S4 class objects (`bh` and `clp`) into S3 class objects for use with the functions in the **spatstat** package.

```
> require(spatstat)
> bhp = as.ppp(bht)
> clpp = as.psp(brd_lcc)
```

The spatial point pattern object `bhp` contains marks. Mark are attributes attached to the events. By default the marks are the columns from the original data that are not used for location. You are interested only in the hurricane type (either tropical only or baroclinic), so you reset the marks accordingly by typing

```
> marks(bhp) = bht$Type == 0
```

You summarize the object with the `summary` method. The summary includes an average density over the spatial domain. The density is per unit area. Your length unit is meter from the map projection, so your density is per square meter. You retrieve the average density in units of per (1,000 km^2) by typing

```
> summary(bhp)$intensity * 1e+12
[1] 11.4
```

Thus, on average, each point in your spatial domain has slightly more than 10 hurricane origins per 1,000 km^2. This is the mean spatial density. Some areas have more or less than the mean so you would like an estimate of $\lambda(s)$.

You do this using the `density` method, which computes a kernel smoothed density function from your point pattern object. By default, the kernel is Gaussian with a bandwidth equal to the standard deviation of the isotropic kernel. The default output is a pixel image containing local density estimates. You again convert the density values to units of 1,000 km^2.

```
> den = density(bhp)
> den$v = den$v * 1e+12
```

Finally you use the plot method to plot the densities, the country borders, and the event locations.

```
> plot(den)
> plot(unmark(clpp), add=TRUE)
> plot(unmark(bhp), pch=19, cex=.3, add=TRUE)
```

Event density maps split by tropical-only and baroclinic hurricane genesis are shown in Figure 11.5. Here we converted the `im` object to a `SpatialGridDataFrame` object and used the `spplot` method. Regions of the central Atlantic extending westward through the Caribbean into the southern Gulf of Mexico have the greatest tropical-only density generally exceeding 10 hurricane origins per (1000 km^2). By contrast, regions along the eastern coast of the United States extending eastward to Bermuda have the greatest baroclinic density. The amount of smoothing is determined by the bandwidth and the type of kernel.

Densities at the genesis locations are made using the argument `at="points"` in the `density` function. Also the number of events in grid boxes across the spatial domain are obtained using the `quadratcount` or `pixellate` functions. For example, type

```
> plot(quadratcount(bhp))
> plot(pixellate(bhp, dimyx=5))
```

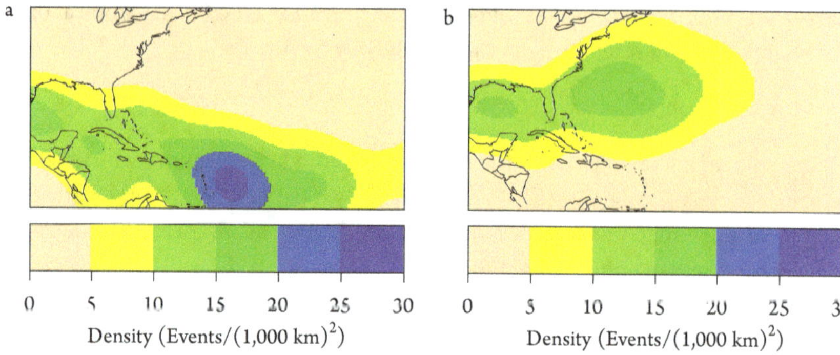

Figure 11.5 Genesis density for (a) tropical-only and (b) baroclinic hurricanes.

11.2.3 Second-Order Properties

The density function describes the rate (mean) of hurricane genesis events locally. Second-order functions describe the variability of events. Ripley's K function is an example. It is defined as

$$\hat{K}(s) = \lambda^{-1} n^{-1} \sum_{i \neq j} I(d_{ij} < s), \tag{11.8}$$

where d_{ij} is the Euclidean distance between the ith and jth events in a data set of n events, and λ is the average density of events, estimated as $n/|A|$, where $|A|$ is the area of the domain. I is an indicator that equals 1 if the distance is less than or equal to s.

The definition assumes stationarity and isotropy for the point pattern and implies that under CSR if the events are homogeneous, $\hat{K}(s)$ is approximately equal to πs^2. For point patterns more regular than CSR, you expect fewer events within distance s of a randomly chosen event than under CSR, so $\hat{K}(s) < \pi s^2$. For point patterns more clustered than CSR, you expect more events within a given distance than under CSR or $\hat{K}(s) > \pi s^2$.

The function `Kest` is used to compute $\hat{K}(s)$. Here you save the results by typing

```
> k = Kest(bhp)
```

The function takes an object of class `ppp` and computes $\hat{K}(s)$ using a formula that corrects for edge effects to reduce bias arising from counting hurricanes near the borders of your spatial domain (Ripley, 1991; Baddeley et al., 2000).

The K function is defined such that $\lambda\hat{K}(s)$ equals the expected number of additional hurricanes within a distance s of any other hurricane. You plot the expected number as a function of separation distance by typing

```
> lam = summary(bhp)$intensity
> m2km = .001
> plot(k$r * m2km, k$iso * lam, type="l", lwd=2,
```

```
+      xlab="Lag Distance (s) [km]",
+      ylab="Avg Number of Nearby Hurricanes")
```

You first save the value of λ and a conversion factor to go from meters to kilometers. The `iso` column refers to the K function computed using the isotropic correction for regular polygon domains.

The `Kest` function also returns theoretical values under the hypothesis of CSR. You add these to your plot by typing

```
> lines(k$r * m2km, k$theo * lam, col="red", lwd=2)
```

The empirical curve lies above the theoretical curve at all lag distances. For instance, at a lag distance of 471 km, there are on average about 16 additional hurricanes nearby. This compares with an expected number of eight additional hurricanes. This indicates that for a given hurricane, there are more nearby hurricanes than you would expect by chance under the assumption of CSR. This is indicative of clustering.

But as you have seen earlier, the clustering is related to the inhomogeneous distribution of hurricane events across the basin. So this result is not particularly useful. Instead you use the `Kinhom` function to compute a generalization of the K function for inhomogeneous point patterns (Baddeley et al, 2000). The results are shown in Figure 11.6. Black curves are the empirical estimates and red curves are the theoretical.

In the case of the inhomogeneous estimate the empirical and theoretical curves match better. The curves depart at large lag distance with the empirical curve falling below the theoretical curve indicating regularity. The regularity is likely due to land areas having no genesis points. The match can be improved by removing land areas from the analysis domain.

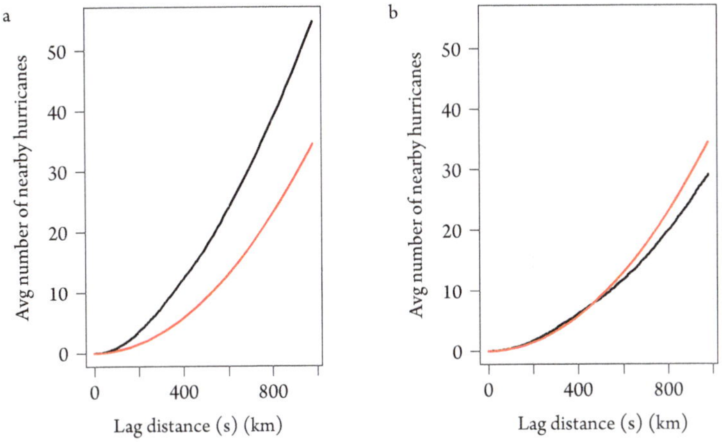

Figure 11.6 2nd-order genesis statistics. (a) Ripley's K and (b) generalization of K.

11.2.4 Models

You can also fit a parametric model to your point pattern. This allows you to predict the occurrence of hurricane events given covariates and the degree of clustering. The **spatstat** package contains options for fitting point pattern models with the `ppm` function.

Models can include spatial trend, dependence on covariates, and event interactions. Models are fit using the method of maximum pseudo-likelihood, or using an approximate maximum likelihood method. Model syntax is standard R. Typically, the null model is homogeneous Poisson without covariates.

11.3 FEATURE CLUSTERS

In the first two sections of this chapter, you examined clustering in time and geographic space. It is also possible to cluster in feature space. Indeed, this is the best known of the cluster methods. Feature clustering is called cluster analysis.

Cluster analysis searches for groups in feature (attribute) data. Objects belonging to the same cluster have similar features. Objects belonging to different clusters have dissimilar features. In two or three dimensions, clusters can be visualized. In more than three, analytic assistance is helpful.

Note the difference from the two earlier cluster methods. With those methods your initial goal is cluster detection with the ultimate goal a model to help you make inferences. Here, your goal is descriptive. Cluster analysis is a data-reduction technique. You start with the assumption that your data can be grouped based on feature similarity.

Cluster analysis methods fall into two categories: partition and hierarchical. Partition clustering divides your data into a prespecified number of groups. The number of groups is usually chosen by trial. An index of cluster quality can make it easier for you to choose an optimal number.

Hierarchical clustering creates ordered sets of clustered groups from your data. Agglomerative hierarchical starts with each object forming its own cluster. It then successively merges clusters until a single large cluster remains. Divisive hierarchical is the opposite. It starts with a single cluster that includes all objects and successively splits the clusters into smaller groups.

To cluster in feature space, you need to

1. create a dissimilarity matrix from a set of objects, and
2. group the objects based on the values of the dissimilarity matrix.

11.3.1 Dissimilarity and Distance

The dissimilarity between two objects measures how different they are. The larger the dissimilarity, the greater the difference. Objects are considered vectors in attribute (feature) space, where vector length equals the number of data set attributes.

Consider, for instance, a data set with three attributes and four objects (Table 11.5). Object 1 is a vector consisting of the triplet $(x_{1,1}, x_{1,2}, x_{1,3})$, object 2

Table 11.5 Attributes and objects. A data set ready for cluster analysis.

	Feature 1	*Feature 2*	*Feature 3*
Object 1	$x_{1,1}$	$x_{1,2}$	$x_{1,3}$
Object 2	$x_{2,1}$	$x_{2,2}$	$x_{2,3}$
Object 3	$x_{3,1}$	$x_{3,2}$	$x_{3,3}$
Object 4	$x_{4,1}$	$x_{4,2}$	$x_{4,3}$

is a vector consisting of the triplet $(x_{2,1}, x_{2,2}, x_{2,3})$, and so on. Then the elements of a dissimilarity matrix are pairwise distances between the objects in feature space. As an example, an object might be a hurricane track with features that include genesis location, lifetime maximum intensity, and lifetime maximum intensification.

Although distance is an actual metric, the dissimilarity function need not be. The minimum requirements for a dissimilarity measure d are:

- $d_{i,i} = 0$
- $d_{i,j} \geq 0$
- $d_{i,j} = d_{j,i}$.

The following axioms of a proper metric may be included but are not necessary:

- $d_{i,k} \leq d_{i,j} + d_{j,k}$ triangle inequality
- $d_{i,j} = 0$ implies $i = j$.

Before clustering, you arrange your data set as a $n \times p$ data matrix, where n is the number of rows (one for each object) and p is the number of columns (one for each attribute variable).

How you compute dissimilarity depends on your attribute variables. If the variables are numeric, you use Euclidean or Manhattan distance given as

$$d_{i,j}^E = \sqrt{\sum_{f=1}^{p} (x_{i,f} - x_{j,f})^2} \quad (11.9)$$

$$d_{i,j}^M = \sum_{f=1}^{p} |x_{i,f} - x_{j,f}| \quad (11.10)$$

The summation is over the number of attributes (features). With hierarchical clustering, the maximum distance norm, given by

$$d_{i,j}^{\max} = \max(|x_{i,f} - x_{j,f}|, f = 1, \ldots, p) \quad (11.11)$$

is sometimes used.

Measurement units on your features influence the distances in feature space which affects your clusters. The feature with the largest variance will have the greatest impact. If all features are equally important to your grouping, then the data need to be standardized. You can do this with the `scale` function, whose default method centers and scales the columns of your data matrix.

The `dist` function computes the distance between objects using a variety of methods including Euclidean (default), Manhattan, and maximum. To begin, create two feature vectors each having five values and plot them with labels that correspond to the object number.

```
> x1 = c(2, 1, -3, -2, -3)
> x2 = c(1, 2, -1, -2, -2)
> plot(x1, x2, xlab="Feature 1", ylab="Feature 2")
> text(x1, x2, labels=1:5, pos=c(1, rep(4, 4)))
```

From the plotted points, you can group the two features into two clusters by eye. There is an obvious distance separation between the clusters. To compute pairwise Euclidean distances between the five objects, type

```
> d = dist(cbind(x1, x2))
> d
     1    2    3    4
2 1.41
3 5.39 5.00
4 5.00 5.00 1.41
5 5.83 5.66 1.00 1.00
```

The result is a vector of distances, but printed as a table with the object numbers listed as row and column names. The values are the pairwise distances so, for instance, the distance between object 1 and object 2 is 1.41 units. You can see two distinct clusters of distances (dissimilarities) those less than or equal to 1.41 and those greater than or equal to 5. Objects 1 and 5 are the most dissimilar followed by objects 2 and 5. On the other hand, objects 3 and 5 and 4 and 5 are the most similar. You can change the default Euclidean distance to the Manhattan distance by including `method="man"` in the `dist` function.

If the feature vectors contain values that are not continuous numeric (e.g., factors, ordered, binary) or if there is a mixture of data types (one feature is numeric and the other is a factor, for instance), then dissimilarities need to be computed differently. The **cluster** package contains functions for cluster analysis including `daisy` for dissimilarity matrix calculations, which by default uses Euclidean distance as the dissimilarity metric.

To test, type

```
> require(cluster)
> d = daisy(cbind(x1, x2))
> d
Dissimilarities :
     1    2    3    4
2 1.41
3 5.39 5.00
```

```
4 5.00 5.00 1.41
5 5.83 5.66 1.00 1.00

Metric :   euclidean
Number of objects : 5
```

Note, the values that make up the dissimilarity matrix are the same as the earlier distance values, but additional information is saved including the metric used and the total number of objects in your data set. The function contains a logical flag called `stand` that when set to true standardizes the feature vectors before calculating dissimilarities. If some of the features are not numeric, the function computes a generalized coefficient of dissimilarity (Gower, 1971).

11.3.2 K-Means Clustering

Partition clustering requires you to specify the number of clusters beforehand. This number is denoted *k*. The algorithm allocates each object in your data frame to one and only one of the *k* clusters.

The *k*-means method is the most popular. Membership of an object is determined by its distance from the centroid of each cluster. The centroid is the multidimensional version of the mean. The method alternates between calculating the centroids based on the current cluster members and reassigning objects to clusters based on the new centroids.

For example, in deciding which of the two clusters an object belongs, the method computes the dissimilarity between the object and the centroid of cluster one and between the object and the centroid of cluster two. It then assigns the object to the cluster with the smallest dissimilarity and recomputes the centroid with this new object included. An object already in this cluster might now have greater dissimilarity due to the reposition of the centroid, in which case it might get reassigned. Assignments and reassignments continue in this way until all objects have a membership.

The `kmeans` function from the base **stat** package performs *k*-means clustering. By default, it uses the algorithm of Hartigan and Wong (1979). The first argument is the data frame (not the dissimilarity matrix) and the second is the number of clusters (`centers`). To perform a *k*-means cluster analysis on the example data above, type

```
> dat = cbind(x1, x2)
> ca = kmeans(dat, centers=2)
> ca
K-means clustering with 2 clusters of sizes 3, 2

Cluster means:
     x1    x2
1 -2.67 -1.67
2  1.50  1.50
```

```
Clustering vector:
[1] 2 2 1 1 1

Within cluster sum of squares by cluster:
[1] 1.33 1.00
 (between_SS / total_SS =  93.4 %)

Available components:

[1] "cluster"      "centers"      "totss"
[4] "withinss"     "tot.withinss" "betweenss"
[7] "size"
```

Initial centroids are chosen at random so it is recommended that you rerun the algorithm a few times to see if it arrives at the same groupings. While the algorithm minimizes within-cluster variance, it does not ensure that there is a global minimum across all clusters.

The first bit of output gives the number of clusters and the size of the clusters that results. Here cluster 1 has two members and cluster 2 has three. The next bit of output are the cluster centroids. There are two features labeled `x1` and `x2`. The rows list the centroids for clusters 1 and 2 as vectors in this two-dimensional feature space. The centroid of the first cluster is $(-2.67, -1.67)$ and the centroid of the second cluster is $(1.5, 1.5)$.

The centroid is the feature average using all objects in the cluster. You can see this by adding the centroids to your plot.

```
> points(ca$centers, pch=8, cex=2)
> text(ca$centers, pos=c(4, 1),
+   labels=c("Cluster 2", "Cluster 1"))
```

The next bit of output tags each object with cluster membership. Here you see that the first two objects belong to cluster 2 and the next three belong to cluster 1. The cluster number keeps track of distinct clusters, but the numerical order is irrelevant.

The within-cluster sum of squares is the sum of the distances between each object and the cluster centroid for which it is a member. From your earlier plot you can see that the centroids (stars) minimize the within-cluster distances while maximizing the between-cluster distances.

The function `pam` uses medoids rather than centroids. A medoid is a multidimensional center based on medians. The method accepts a dissimilarity matrix, tends to be more robust (converges to the same result), and provides for additional graphical displays.

11.3.3 Track Clusters

Here your interest is to group hurricanes according to common track features. These features include location of hurricane origin, location of lifetime maximum intensity, location of final hurricane intensity, and lifetime maximum intensity. The three location features are further subdivided into latitude and longitude making a total of seven attribute (feature) variables. Note that location is treated here as an attribute rather than as a spatial coordinate.

You first create a data frame containing the attributes you wish to cluster. Here you use the cyclones since 1950:

```
> load("best.use.RData")
> best = subset(best.use, Yr >= 1950)
```

You then split the data frame into separate cyclone tracks using `Sid` with each element in the list as a separate track. You are interested in identifying features associated with each track.

```
> best.split = split(best, best$Sid)
```

You then assemble a data frame that contains your attributes.

```
> best.c = t(sapply(best.split, function(x){
+    x1 = unlist(x[1, c("lon", "lat")])
+    x2 = unlist(x[nrow(x), c("lon", "lat")])
+    x3 = max(x$WmaxS)
+    x4 = unlist(x[rev(which(x3 == x$WmaxS))[1],
+      c("lon", "lat")])
+    return(c(x1, x2, x3, x4))
+ }))
> best.c = data.frame(best.c)
> colnames(best.c) = c("FirstLon", "FirstLat",
+    "LastLon", "LastLat", "WmaxS", "MaxLon", "MaxLat")
```

The data frame contains seven features from 667 cyclones.

Before clustering, you check the feature variances by applying the `var` function on the columns of the data frame using the `sapply` function.

```
> sapply(best.c, var)
FirstLon FirstLat  LastLon  LastLat    WmaxS
   499.2     57.0    710.4    164.8    870.7
  MaxLon   MaxLat
   356.7     75.9
```

The variances range from a minimum of 57 degrees squared for the latitude of origin feature to a maximum of 870.7 knots squared for the maximum intensity feature. Because of this large range, you scale the features so that each has the same influence on the clustering.

```
> best.cs = scale(best.c)
> m = attr(best.cs, "scaled:center")
> s = attr(best.cs, "scaled:scale")
```

The function `scale` centers and scales the columns of your numeric data frame. The center and scale values are saved as attributes in the data frame. Here you save them to rescale the centroids after the cluster analysis.

You perform a *k*-means cluster analysis setting the number of clusters to three by typing

```
> k = 3
> ct = kmeans(best.cs, centers=k)
> summary(ct)
             Length Class  Mode
cluster      667    -none- numeric
centers       21    -none- numeric
totss          1    -none- numeric
withinss       3    -none- numeric
tot.withinss   1    -none- numeric
betweenss      1    -none- numeric
size           3    -none- numeric
```

The output is a list of length 7 containing the cluster vector, cluster means (centroids), the total sum of squares, the within-cluster sum of squares by cluster, the total within sum of squares, the between sum of squares, and the size of each cluster.

Your cluster means are scaled so they are not directly interpretable. The cluster vector gives the membership of each hurricane in order as they appear in your data set. The ratio of the between sum of squares to the total sum of squares is 44.7 percent. This ratio will increase with the number of clusters, but at the expense of having clusters that may not be interpretable. With four clusters, the increase in this ratio is smaller than the increase going from two to three clusters. So you are content with your three-cluster result.

11.3.4 Track Plots

Since six of the seven features are spatial coordinates it is tempting to plot the centroids on a map and connect them with a track line. This would be misleading since the combination of attributes is not a spatial feature. Instead, you plot examples of cyclones that resemble each cluster.

First, you add cluster membership and distance to your original data frame, then split the data by cluster member.

```
> ctrs = ct$center[ct$cluster, ]
> cln = ct$cluster
> dist = rowMeans((best.cs - ctrs)^2)
> id = 1:length(dist)
```

```
> best.c = data.frame(best.c, id=id, dist=dist,
+   cln=cln)
> best.c.split = split(best.c, best.c$cln)
```

Next you subset your cluster data based on the tracks that come closest to the cluster centroids. That closeness occurs in feature space that includes latitude, longitude, and intensity features. Here you choose nine tracks for each centroid.

```
> te = 9
> bestid = unlist(lapply(best.c.split, function(x)
+   x$id[order(x$dist)[1:te]]))
> cinfo = subset(best.c, id %in% bestid)
```

Finally, you plot the tracks on a map. This requires a few steps to make the plot easier to read. Begin by setting a bounding box using latitude and longitude of your cluster data and renaming your objects.

```
> cyclones = best.split
> uselat = range(unlist(lapply(cyclones[cinfo$id],
+   function(x) x$lat)))
> uselon = range(unlist(lapply(cyclones[cinfo$id],
+   function(x) x$lon)))
```

Next, order the clusters by number and distance and set the colors for plotting. Use the `brewer.pal` function in the **RColorBrewer** package (Neuwirth, 2011). You use a single hue sequential color ramp that allows you to highlight the tracks that are closest to the centroids of the feature clusters.

```
> cinfo = cinfo[order(cinfo$cln, cinfo$dist), ]
> require(RColorBrewer)
> blues = brewer.pal(te, "Blues")
> greens = brewer.pal(te, "Greens")
> reds = brewer.pal(te, "Reds")
> cinfo$colors = c(rev(blues), rev(greens), rev(reds))
```

Next, reorder the tracks so that those farthest from the centroids get plotted first.

```
> cinfo = cinfo[order(-cinfo$dist, cinfo$cln), ]
```

Finally, plot the tracks and add world and country borders. Results are shown in Figure 11.7. Track color is based on attribute proximity to the cluster centroid using a color saturation that decreases with distance.

```
> plot(uselon, uselat, type="n", xaxt="n", yaxt="n",
+   ylab="", xlab="")
> for(i in 1:nrow(cinfo)){
+   cid = cinfo$id[i]
+   cyclone = cyclones[[cid]]
```

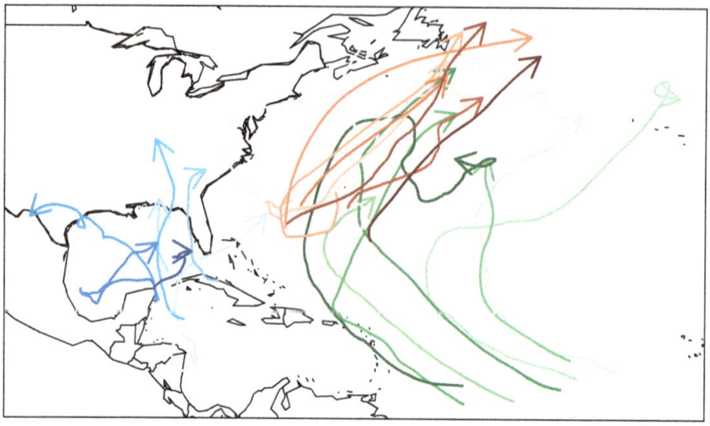

Figure 11.7 Tracks by cluster membership.

```
+    lines(cyclone$lon, cyclone$lat, lwd=2,
+      col=cinfo$colors[i])
+ }
> require(maps)
> map("world", add=TRUE)
> map("usa", add=TRUE)
```

The analysis splits the cyclones into a group over the Gulf of Mexico, a group at high latitudes, and a group that begins at low latitudes but ends at high latitudes generally east of the United States. Some of the cyclones that begin at high latitude are baroclinic (see Chapter 7).

Track membership can change depending on the initial random centroids particularly for tracks that are farthest from a centroid. The two- and three-cluster tracks are the most stable. The approach can be extended to include other track features including velocity, acceleration, and curvature representing additional dimensions of the space and time behavior of hurricanes.

Model-based clustering, where the observations are assumed to be quantified by a finite mixture of probability distributions, is an attractive alternative to cluster analysis. That is especially true if you want to make predictions of future activity. It is good to keep in mind the advice from Bill Venables and Brian Ripley, who note that you should not assume cluster analysis is the best way to discover interesting groups. You should not assume cluster analysis is the best way to discover interesting groups. Indeed, visualization methods are often more effective in this task.

This chapter showed you how to detect, model, and analyze clusters in hurricane data. We began by showing you how to detect and model the arrival of hurricanes along the coast. We then showed you how to detect and analyze the presence of spatial clusters in hurricane genesis locations. We also looked at the first- and second-order statistical properties of hurricane genesis. Finally, we showed you how to apply cluster analysis to hurricane track features and map representative tracks.

In the next chapter, we look at several applications of Bayesian models.

12

BAYESIAN MODELS

"Errors using inadequate data are much less than those using no data at all."
—Charles Babbage

In this chapter, we focus on Bayesian modeling. Information about past hurricanes is available from instruments and written accounts. Written accounts are generally less precise than instrumental observations, which tend to become even more precise as technology advances. Here we show you how to build Bayesian models that make use of the available information while accounting for differences in levels of precision. We begin with a model for U.S. hurricane frequency and finish with a space–time model for basin-wide occurrences.

12.1 LONG-RANGE OUTLOOK

We start with a model for predicting U.S. hurricane activity over the next three decades. The model is useful as a benchmark for climate change studies. The methodology was originally presented in Elsner and Bossak (2001) based on the formalism given in Epstein (1985).

12.1.1 Poisson-Gamma Conjugate

As you have seen throughout this book, the arrival of hurricanes on the coast is usefully considered a stochastic process, where the annual counts are described reasonably well by a Poisson distribution. The Poisson distribution is a limiting form of the binomial distribution with no upper bound on the number of occurrences and where the parameter λ characterizes the rate process. Knowledge of λ allows you to make statements about future hurricane frequency. Since the process is stochastic, your statements will be given in terms of probabilities (see Chapter 7).

For example, the probability of $\hat{h}$ hurricanes occurring over the next T years (e.g., 1, 5, 20, etc.) is

$$f(\hat{h}|\lambda, T) = \exp(-\lambda T)\frac{(\lambda T)^h}{h!} \quad \text{for} \quad h = 0, 1, \ldots, \quad \lambda > 0, \quad \text{and} \quad T > 0 \tag{12.1}$$

The hat notation is used to indicate future values.

The parameter λ and statistic T appear in the formula as a product, which is the mean and variance of the distribution. Knowledge about λ can come from historical written archives and instrumental records. It is logical for you to want to use as much of this information as possible before inferring something about future activity.

This requires you to treat λ as a parameter that can be any positive real number, rather than as a fixed constant. One form for expressing your judgment about the values λ can take is through the gamma distribution. The numbers that are used to estimate λ from a set of data are the time interval T' and the number of hurricanes h' that occurred during this interval.[1] For instance, observations from the hurricane record since 1851 indicate 15 hurricanes over the first 10 years, so $T' = 10$ and $h' = 15$. To verify this, type

```
> H = read.table("US.txt", header=TRUE)
> sum(H$All[1:10])
[1] 15
```

The gamma distribution of possible future values for λ is given by

$$f(\hat{\lambda}|h', T') = \frac{T'^{h'}\lambda^{h'-1}}{\Gamma(h')}\exp(-\lambda T') \tag{12.2}$$

with the expected value $\mathrm{E}(\hat{\lambda}) = h'/T'$, and the gamma function $\Gamma(x)$ given by

$$\Gamma(x) = \int_0^\infty t^{x-1}\mathrm{e}^{-t}dt \tag{12.3}$$

Of importance here is the fact that if the probability density on $\hat{\lambda}$ is a gamma distribution, with initial numbers (prior parameters) h' and T', and the numbers h and T are later observed, then the posterior density of $\hat{\lambda}$ is also gamma with parameters $h + h'$ and $T + T'$. In other words, the gamma density is the conjugate prior for the Poisson rate λ (Chapter 4).

12.1.2 Prior Parameters

The above formulation gives you a convenient way to combine earlier less reliable information with later information. You simply add the prior parameters h' and T' to the sample numbers h and T to get the posterior parameters.

But how do you estimate the prior parameters? You have data but the values could be too low (or too high) due to missing or misclassified hurricanes. One way is to

[1] The prime notation indicates prior (here, earlier) information.

use bootstrapping. Bootstrapping is sampling from your sample (resampling) to provide an estimate of the variation about your statistic of interest (see Chapter 3). Here you use the `bootstrap` function in the **bootstrap** package (Tibshirani and Leisch, 2007) to obtain a confidence interval about λ from data before 1899.

First, load the package and save the counts over the earlier period of record as a vector. Then to get a bootstrap sample of the mean, use the `bootstrap` function on this vector of counts.

```
> require(bootstrap)
> early = H$All[H$Year < 1899]
> bs = bootstrap(early, theta=mean, nboot=1000)
```

To obtain a 90 percent bootstrapped confidence interval about the mean, type

```
> qbs = quantile(bs$thetastar, prob=c(.05, .95))
> qbs
  5%  95%
1.42 2.10
```

Although you cannot say with certainty what the true hurricane rate was over this early period, you can make a sound judgment that you are 90 percent confident that the interval contains it. In other words you are willing to admit a 5 percent chance that the true rate is less than 1.42 and a 5 percent chance that it is greater than 2.1.

Given this appraisal of your belief about the early hurricane, rate you need to obtain an estimate of the parameters of the gamma distribution. Said another way, given your 90 percent confidence interval for the rate, what is your best estimate for the number of hurricanes and the length of time over which those hurricanes occurred?

You do this with the optimization function `optim`. You start by creating an objective function defined as the absolute value of the difference between gamma quantiles and your target quantiles.

```
> obj = function(x){
+    sum(abs(pgamma(q=qbs, shape=x[1], rate=x[2]) -
+    c(.05, .95)))
+ }
```

You then apply the optimization function to your objective function starting with reasonable initial values for the gamma parameters given in the `par` argument and save the solution in the vector `theta`.

```
> theta = optim(par = c(2, 1), obj)$par
```

Store these parameters as separate objects by typing

```
> hp = theta[1]
> Tp = theta[2]
```

This procedure quantifies your judgment about hurricanes before the reliable set of counts. It does so in terms of the shape and rate parameter of the gamma distribution.

12.1.3 Posterior Density

You now have two distinct pieces of information from which to obtain a posterior distribution for λ (landfall rate). Your prior parameters $h' = 69.7$ and $T' = 39.9$ from above and your likelihood statistics based on the data over the reliable period of record (1899–2010). The total number of hurricanes over this reliable period and the record length are

```
> late = H$All[H$Year >= 1899]
> h = sum(late)
> T = length(late)
> h; T
[1] 187
[1] 112
```

The posterior parameters are therefore $h'' = h + h' = 256.7$ and $T'' = T + T' = 151.9$. Note that although the likelihood parameters h and T must be integers, the prior parameters can take on any real value depending on your degree of belief.

Since the prior, likelihood, and posterior are in the same gamma family, you can use dgamma to compute the densities.

```
> curve(dgamma(x, shape=h + hp, rate=T + Tp), from=1,
+   to=3, xlab="Landfall Rate [hur/yr]",
+   ylab="Density", col=1, lwd=4, las=1)
> curve(dgamma(x, shape=h, rate=T), add=TRUE,
+   col=2, lwd=4)
> curve(dgamma(x, shape=hp, rate=Tp), add=TRUE,
+   col=3, lwd=4)
> legend("topright", c("Prior", "Likelihood",
+   "Posterior"), col=c(3, 2, 1), lwd=c(3, 3, 3))
```

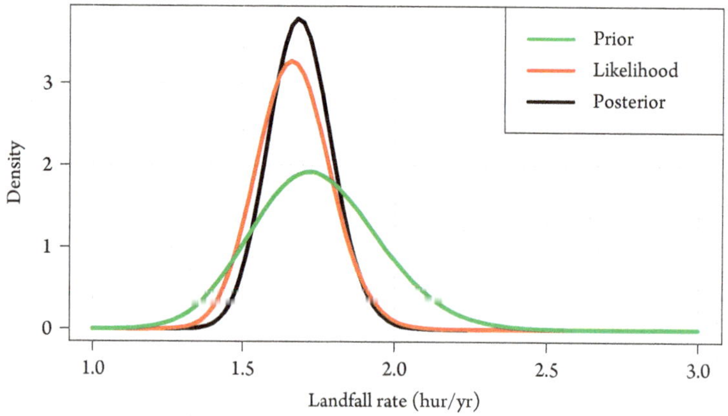

Figure 12.1 Gamma densities for the landfall rate.

The densities are shown in Figure 12.1. Note that the posterior density resembles the likelihood but is shifted in the direction of the prior. It is also narrower. The posterior is a weighted average of the prior and the likelihood where the weights are proportional to the precision. The greater the precision, the more weight it carries in determining the posterior. The relatively broad density on the prior estimate indicates low precision. Combining the prior and likelihood results in a posterior distribution that represents your best information about λ.

12.1.4 Predictive Distribution

The information you have about λ is contained in the two parameters h'' and T'' of the gamma density. Of practical interest is how to use this information to predict future hurricane activity. The answer lies in the fact that the *predictive density* for observing $\hat{h}$ hurricanes over the next $\hat{T}$ years is a negative binomial distribution, with parameters h'' and $\frac{T''}{\hat{T}+T''}$ given by

$$f(\hat{h}|h'', \frac{T''}{\hat{T}+T''}) = \frac{\Gamma(\hat{h}+h'')}{\Gamma(h'')\hat{h}!}\left[\frac{T''}{\hat{T}+T''}\right]^{h''}\left[\frac{\hat{T}}{\hat{T}+T''}\right]^{\hat{h}} \tag{12.4}$$

The mean and variance of the negative binomial are $\hat{T}\frac{h''}{T''}$ and $\hat{T}\frac{h''}{T''}(\frac{\hat{T}+T''}{T''})$, respectively. Note that the variance of the predictive distribution is larger than it would be if λ were known precisely. If you are interested in the probability of a hurricane *next* year, then $\hat{T}$ is one and small compared with T'' so it makes little difference, but if you are interested in the distribution of hurricane activity over the next 20 years, then it is important.

You plot the posterior probabilities and cumulative probabilities for the number of U.S. hurricanes over the next 10 years by typing,

```
> Th = 10
> m = Th * (h + hp)/(T + Tp)
> v = Th * m * (Th + T + Tp)/(T + Tp)
> nl = 0:32
> hh = dnbinom(nl, mu=m, size=v)
> par(las=1, mar=c(5, 4, 2, 4))
> plot(nl, hh, type="h",
+   xlab="Number of U.S. Hurricanes",
+   ylab="Probability", col="gray", lwd=4)
> par(new=TRUE)
> p = pnbinom(nl, mu=m, size=v)
> plot(nl, p, type="l", col="red", xaxt="n", yaxt="n",
+    xlab="", ylab="", lwd=2)
> axis(4)
> mtext("Probability of h less than or equal to H",
+    side=4, line=2.5, las=0)
```

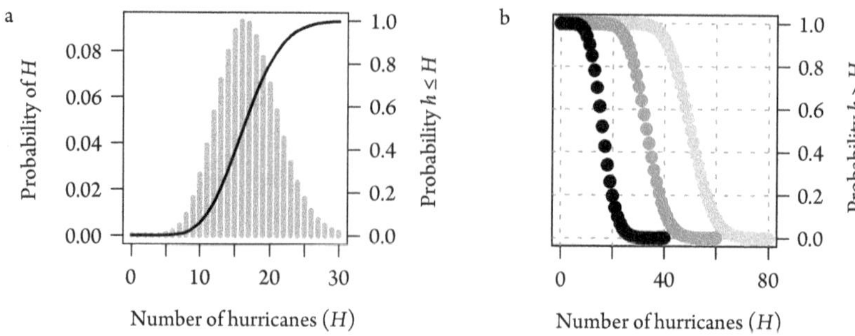

Figure 12.2 Predictive probabilities. (a) 10 years and (b) 10, 20, and 30 years.

Results are shown in Figure 12.2, where the probabilities of H hurricanes in 10 years are displayed with vertical bars and with a scale plotted along the left vertical axis, and the probability that the number of hurricanes will be less than or equal to H is displayed as a solid curve and with a scale along the right axis. Probabilities that the number of hurricanes will exceed H in 10-, 20-, and 30-year periods are shown in the right panel. The expected number of U.S. hurricanes over the next 30 years is 51, of which 18 (not shown) are anticipated to be major hurricanes. Those probabilities are the best estimate of future hurricane climatology.

The above approach is a rational and coherent foundation for incorporating all available information about hurricane occurrences, while accounting for differences in data precision over the years. It could be used to account for the influence of climate change by discounting older information. That is, records influenced by recent global warming can be given more weight than records from the pre-industrial era.

12.2 SEASONAL MODEL

Here you create a Bayesian model for predicting annual basin-wide hurricane counts. The counts are reliable starting in the middle twentieth century. But data records on past hurricanes extend farther back, and these earlier records are useful to understand and predict seasonal activity. The logarithm of the annual rate is linearly related to sea-surface temperature (SST) and the Southern Oscillation Index (SOI) as discussed in Chapter 7.

Samples from the posterior distribution are generated using the Markov chain Monte Carlo approach discussed in Chapter 4. The JAGS code is given below. You copy and paste the code into a text file in your working directory with the name *JAGSmodel2.txt*.

```
___JAGS Code___
model {
for(i in 1:N) {
   h[i] ~ dpois(lambda[i])
   mu[i] <- b0 + b1*SOI[i] + b2*SST[i] + b3*RI[i] +
```

```
            eta[i]
   lambda[i] <- exp(mu[i])
   tt[i] <- tau[RI[i] + 1]
   eta[i] ~ dnorm(0, tt[i])
   }
  b0 ~ dnorm(0, .0001)
  b1 ~ dnorm(0, .0001)
  b2 ~ dnorm(0, .0001)
  b3 <- log(pm)
  pm ~ dunif(lo, hi)
  tau[1] ~ dgamma(.001, .001)
  tau[2] ~ dgamma(.001, .001)
}
```

The code specifies the regression model and associated priors. In particular, the annual hurricane count `h[i]` is stochastically generated from a Poisson distribution with a rate `lambda[i]`. The parameter `lambda[i]` is deterministically linked to `mu[i]` through the exponential function, where `mu[i]` is linearly related to the SST and SOI covariates and a "reliability" index (`RI[i]`). The reliability index is coded as 1 for years prior to aircraft and 0 for years after. Here you use 1943 as the cutoff year.

Your model includes a random effects term (`eta[i]`) that quantifies the extra variation in annual hurricane rates not modeled by the covariates. It accommodates two variance distributions to account for the different levels of uncertainty before and after the cutoff year. JAGS is a declarative language so the order of the assignment statements is irrelevant. Your model is a directed acyclic graph (DAG), where the nodes are the parameters and data and the arrows indicate the conditional dependency, either stochastic or deterministic (see Chapter 4).

Load the annual data and create a data frame including the basin-wide hurricane count `H`, the August–October value of SOI, the August through October value of SST, and a reliability index `RI` that is coded as 1 for years prior to aircraft and 0 for years after.

```
> load("annual.RData")
> dat = data.frame(Yr=annual$Year, H=annual$B.1,
+   SOI=annual$soi, SST=annual$sst,
+   RI=as.numeric(annual$Year < 1943))
> dat = dat[16:160, ]
```

You execute the model using the `jags.model` function in **rjags** (Plummer, 2011). The function specifies the file name, the data as a list, initial values as a list, as well as the number of chains and the number of initial updates (`n.adapt`).[2]

[2] You can monitor the progress by specifying `options(jags.pb="text")` [or ="gui"] before your call to `jags.model`.

```
> require(rjags)
> model = jags.model('JAGSmodel2.txt',
+     data=list(N=length(dat$H), h=dat$H,
+     SST=dat$SST, SOI=dat$SOI, RI=dat$RI,
+     lo=.2, hi=.95),
+     inits = list(b0=0, b1=0, b2=0,
+     tau=c(.1, .1), pm=.9, .RNG.seed=3042,
+     .RNG.name="base::Super-Duper"),
+     n.chains = 2,
+     n.adapt = 1000)
```

You continue sampling by applying the `update` function on the model object and specifying the number of additional samples. You then save the last 1,000 samples of the SST (`b1`) and SOI (`b2`) coefficients.

```
> update(model, 2000)
> out = coda.samples(model, c('b1', 'b2'), 1000)
```

The `coda.samples` is a wrapper function[3] for `jags.samples` that monitors your requested nodes, updates the model, and outputs the samples to a single `mcmc.list` object. The first argument is the model object, the second is a vector of variable names, and the third is the number of iterations. To examine the samples, you plot histograms (Fig. 12.3).

```
> par(mfrow=c(1, 2))
> hist(out[[1]][, 1], xlab=expression(beta[1]))
> hist(out[[1]][, 2], xlab=expression(beta[2]))
```

The chain number is given inside the double brackets and the sample values are given as a matrix where the rows are the consecutive samples and the columns are the

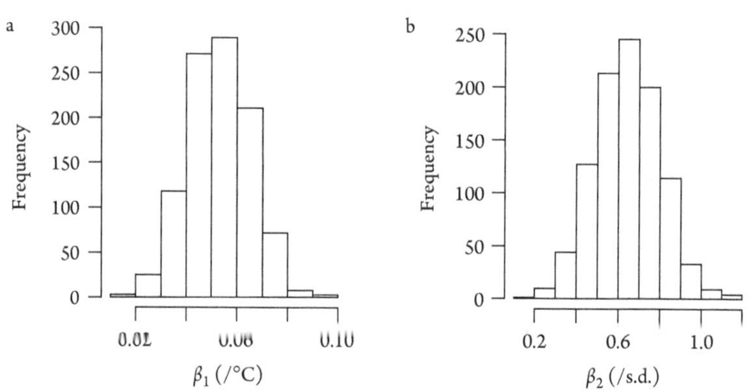

Figure 12.3 Posterior samples for the (a) SST (β_1) and (b) SOI (β_2) parameters.

[3] A function whose main purpose is to call another function.

parameters. The distributions are shifted to the right of zero verifying that both the SST and the SOI are important in modulating annual hurricane rates across the North Atlantic.

A time series of box plots shows the distribution of the annual rate parameter (λ) as a function of year. First, generate additional samples from the posterior, this time monitoring `lambda`.

```
> out = coda.samples(model, c('lambda'), 1000)
```

Then use the `fivenum` function with the `points` and `lines` functions within a loop over all years.

```
> plot(c(1866, 2010), c(0, 20), type="n", bty="n",
+   xlab="Year", ylab="Annual Rate (hur/yr)")
> for(i in 1:dim(dat)[1]){
+ points(dat$Yr[i], fivenum(out[[1]][, i])[3], pch=16)
+ lines(c(dat$Yr[i], dat$Yr[i]),
+   c(fivenum(out[[1]][, i])[1],
+   fivenum(out[[1]][, i])[5]))
+ }
```

The resulting plot is shown in Figure 12.4. The model is useful for describing the annual rate variation and the associated uncertainty levels. The median rate is given as a point and the range is given as a vertical line. Hurricane rates appear to fluctuate about the value of 5 hur/yr until about 1945 when they appear to increase slightly before falling back again during the 1970s and 1980s. There is a more substantial increase beginning in the middle 1990s.

The extra variation in the rate specific to each year that is not modeled by the two covariates is quantified with the term `eta`. The variation includes the two levels of data precision before and after the start of the aircraft reconnaissance era. A time series plot of posterior samples of `eta` are shown in Figure 12.5. The posterior median is plotted as a point and the vertical lines extend from the 25th to the 75th percentiles. The red line is a local regression through the median values. As expected, the graph

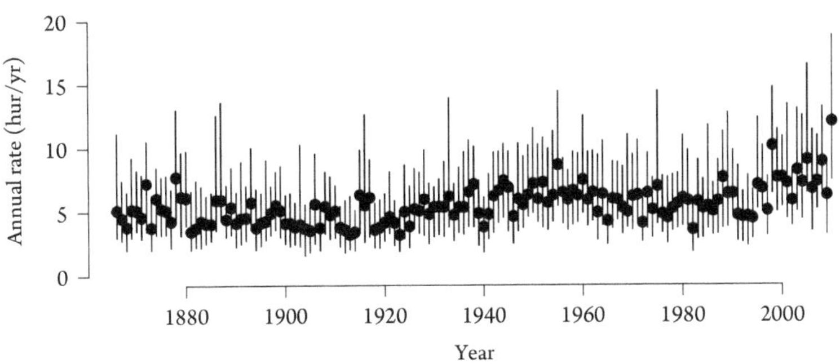

Figure 12.4 Modeled annual hurricane rates.

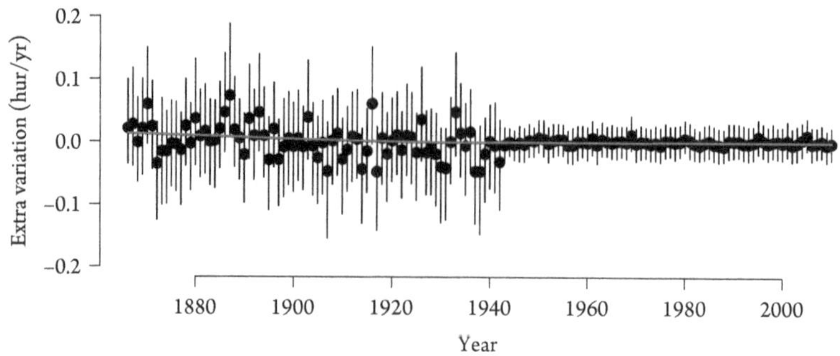

Figure 12.5 Extra variation in hurricane rates.

shows larger unexplained variations and a tendency for underprediction of the rates (positive values) during the earlier years, but much less so afterward.

12.3 CONSENSUS MODEL

In choosing one model over another you typically apply a selection procedure on a set of covariates. You then make predictions using the best model as if it had generated the data. Unfortunately, this approach ignores the uncertainty in your model selection procedure resulting in over confidence in your predictions.

For example, given a pool of covariates for hurricane activity, a stepwise regression procedure is typically employed to search through hundreds of candidate models (combinations of covariates). The model that provides the best level of skill is subsequently chosen. The best model is then subjected to a leave-one-out cross-validation (LOOCV) to obtain an estimate of how well it will predict future data. This procedure unfortunately does not result in a true cross-validation as the process of selecting the covariates is itself not cross-validated. A true cross-validation assesses how well an algorithm for choosing a particular model (including the predictor selection phase) will do in forecasting the unknown future (see Chapter 7).

An alternative to choosing a single best model is to use Bayesian model averaging (BMA). BMA works by assigning a probability to each model (combination of covariates), then averaging over all models weighted by their probability (Raftery et al., 2005). Here you produce a consensus forecast of seasonal hurricane activity using BMA. In doing so you we show how the approach can facilitate a physical interpretation of your modeled relationships. The presentation follows closely the work of Jagger and Elsner (2010).

12.3.1 Bayesian Model Averaging

Let H_i, $i = 1,\ldots,N$ denote your set of observed hurricane counts by year. Assume that your model has k covariates, then let $\mathbf{X}$ be the covariate matrix with components

$X[i,j+1], i=1,\ldots,N, j=1,\ldots k$ associated with the ith observation of the jth covariate and with the intercept term $X[i,1]=1$ for all i. Associated with the intercept and k covariates are $k+1$ parameters $\beta_j, j=1,\ldots,k+1$.

You assume that the counts are adequately described by a Poisson distribution. The logarithm of the rate is regressed onto the covariates as

$$H_i \sim \text{pois}(\lambda_i)$$

$$\log(\lambda_i) = \sum_{j=1}^{k+1} X[i,j]\beta_j$$

This is a generalized linear model (GLM) and the method of maximum likelihoods is used to estimate the parameters (Chapter 7). From these parameter estimates and the values of the corresponding covariates, you infer λ from the regression equation. The future hurricane count conditional on these covariates is described by a Poisson distribution with a mean of λ. Thus your model is probabilistic, and the average count represents a single forecast.

A full model is defined as one that uses all k covariates. However, it is usual that some of the covariates do not contribute much to the model. In frequentist statistics, these are the ones that are not statistically significant. You can choose a reduced model by setting some of the k parameters to zero. Thus, with k covariates, there are a total of $m=2^k$ possible models. The idea behind BMA is that all the m models are used with a probability assigned to each. Predictions are made by a weighted average over the predictions made with each model and where the weights are the model probabilities. Models with greater probability carry proportionally more weight in the average.

Consider a simple case. You have observations of Y arising from either one of two possible regression models. Let $Y_1=\alpha_1+\epsilon_1$ be a constant mean model and $Y_2=\alpha_2+\beta x+\epsilon_2$ be a simple regression model where x is a single covariate. The residual terms ϵ_1, ϵ_2 are independent and normally distributed with means of zero and variances of σ_1^2 and σ_2^2, respectively.

Suppose that you assign a probability p that the constant mean model generated the observed data. Then there is a probability $1-p$ that the simple regression model generated the data instead. With BMA, the posterior predictive expectation (mean) of Y is $\mu=p\mu_1+(1-p)\mu_2=p\alpha_1+(1-p)(\alpha_2+\beta x)$. This represents a consensus opinion that combines information from both models as opposed to choosing one over the other.

The posterior predictive distribution of Y given the data is not necessarily normal. Instead it is a mixture of normal distributions with a posterior predicted variance of $p\sigma_1^2+(1-p)\sigma_2^2+p(1-p)(\alpha_2+\beta x-\alpha_1)^2$. This variance under BMA is larger than a simple weighted sum of the individual model variances by an amount $p(1-p)(\alpha_2+\beta x-\alpha_1)^2$ that represents the uncertainty associated with model choice. Thus, the predictive distribution under BMA has a larger variance than the predictive distribution from a single model.

Over a set of competing models, you need a way to assign a probability to each. You start with a collection of models $M_i, i=1,\ldots,m$, where each is a unique description of your data. In the above example, you assign a probability to the constant mean model and a probability to the simple regression model under the constraint that the total probability over both is one.

Now with your data D and a set of proposed models M_i, you determine the probability of your data given each model $[P(D|M_i)]$. You also assign a prior probability to each $[P(M_i)]$ representing your belief that the model generated your data. Under the situation in which you are maximally noncommittal on a model before hand you assigned $1/m$ to each model's prior probability. For example, in the above case, if you believe both models are equally likely, then you assign $P(M_1) = P(M_2) = 0.5$. Using Bayes rule (Chapter 4), you find the probability of the model given, the data as $P(M_i|D) = P(D|M_i) \times P(M_i)/P(D)$ since $P(D)$ is fixed for all models you can let $W_i = P(D|M_i) \times P(M_i)$ be the model weights with probabilities $P(M_i|D) = W_i/\sum_{i=1}^{m} W_i$.

Let the random variable H represent the prediction of a future hurricane count. The posterior distribution of H at h under each model is given by $f(h|D,M_i)$. The marginal posterior probability over all models is given by

$$f(h|D) = \sum_{i=1}^{m} f(h|D,M_i)P(M_i|D) \tag{12.5}$$

A point estimate for the future count (i.e., the posterior mean of H over the models) is obtained by taking the expectation of H given the data as

$$\mathbb{E}(H|D) = \sum_{h=0}^{\infty} hf(h|D) \tag{12.6}$$

Expanding $f(h|D)$, you get

$$\mathbb{E}(H|D) = \sum_{i=1}^{m} P(M_i|D) \sum_{h=0}^{\infty} hf(h|D,M_i) \tag{12.7}$$

which is

$$\sum_{i=1}^{m} P(M_i|D)\mathbb{E}(H|D,M_i) \tag{12.8}$$

where $\mathbb{E}(H|D,M_i) = \mu_i$. For a given model, $P(D|M_i)$ is the marginal likelihood over the parameter space. In other words, $P(D|M_i) = \int P(D|M_i,\theta)f(\theta|M_i)d\theta$, where $f(\theta|M_i)$ is the prior distribution of the parameters for model M_i and $P(D|M_i,\theta)$ is the likelihood of the data given the model $[L(\theta;M_i,D)]$.

The above integral cannot always be evaluated analytically or it may be infinite as when an improper prior is put on the parameter vector θ. In these cases, approximation methods can be used (Hoeting et al., 1999). Here you use the Bayesian Information Criterion (BIC) approximation, which is based on a Laplace expansion of the integral about the maximum likelihood estimates (Madigan and Raftery, 1994).

In summary BMA keeps all candidate models assigning a probability based on how likely it would be for your data to have come from each model. A consensus model, representing a weighted average of all models, is then used to make predictions. If values for the prior parameters come from reasonably well-behaved distributions, then a consensus model from a BMA procedure yields the lowest mean square error of any single best model (Raftery and Zheng, 2003).

BMA provides better coverage probabilities on the predictions than any single model (Raftery and Zheng, 2003). Consider a data record split into a training and testing set. Using the training set, you can create $1-\alpha$ credible intervals on the predictions. Then, using the testing set, you can calculate the proportion of observations that lie within the credible intervals. This is called the coverage probability. In standard practice with a single best model, the credible intervals are too small resulting in coverage probabilities less than $1-\alpha$. Since BMA provides a larger variance than any model individually, the coverage probabilities on the predictions are greater or equal to $1-\alpha$.

12.3.2 Data Plots

You use the data saved in file *annual.RData* and described in Chapter 6. Load the data and create a new data frame that is a subset of the data for years since 1866.

```
> load("annual.RData")
> dat = annual[annual$Year >= 1866, ]
```

The counts are the number of near-coastal hurricanes passing through the regions shown in Figure 6.2. You consider monthly values of SST, SOI, NAO, and sunspots as covariates.

The monthly covariate values are shown in Figure 12.6 as image plots. The monthly values for May through October displayed on the vertical axis are plotted as a function of year displayed on the horizontal axis. The values are shown using a color ramp from blue (low) to yellow (high). The SST and sunspot number (SSN) covariates are characterized by high month-to-month correlation as can be seen by the vertical striations.

12.3.3 Model Selection

You assume that the logarithm of the annual hurricane rate is a linear combination of a fixed subset of your covariates (Poisson generalized linear model). With six months and four environmental variables per month, you have 2^{24} or more than 16.7 million possible models.

Model selection is done with functions in the **BMA** package (Raftery et al., 2009). Obtain the package and source additional functions from **bic.glm.R** that allows you to make posterior predictions.

```
> require(BMA)
> source("bic.glm.R")
```

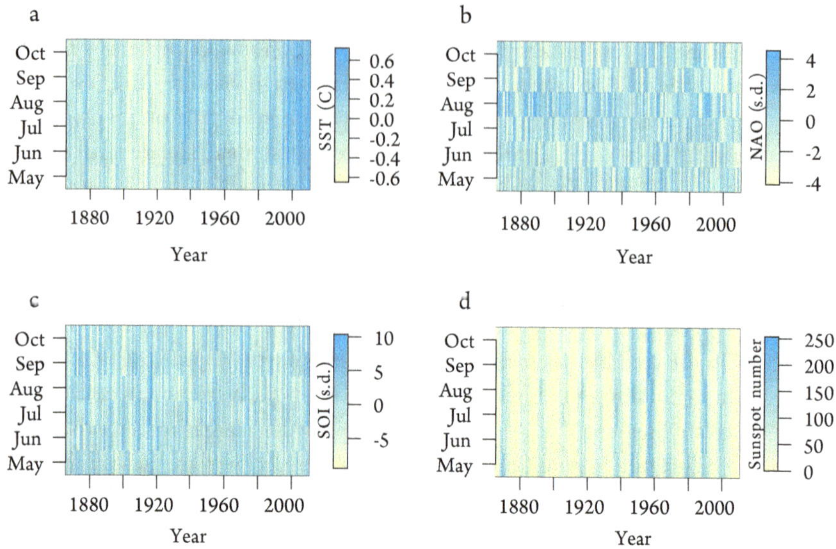

Figure 12.6 Covariates by month. (a) SST, (b) NAO, (c) SOI, and (d) sunspot number.

Specifically, you use the `bic.glm` function for determining the probability on each model and the `imageplot.bma` function for displaying the results. First save the model formula involving the response variable (here `US.1`) and all covariates.

```
> fml = US.1 ~ sst.Oct + sst.Sep + sst.Aug + sst.Jul +
+   sst.Jun + sst.May + nao.Oct + nao.Sep + nao.Aug +
+   nao.Jul + nao.Jun + nao.May + soi.Oct + soi.Sep +
+   soi.Aug + soi.Jul + soi.Jun + soi.May + ssn.Oct +
+   ssn.Sep + ssn.Aug + ssn.Jul + ssn.Jun + ssn.May
```

Then save the output from the `bic.glm` function by typing

```
> mdls = bic.glm(f=fml, data=dat, glm.family="poisson")
```

The function returns an object of class `bic.glm`. By default, only models with a BIC within a factor of 20 of the model with the lowest BIC are kept, and only the top 150 models of each size (number of covariates) are considered. The summary method is used to display the results. Information about the top three models with the lowest BIC values (highest posterior probabilities) is shown in Table 12.1.

```
> summary(mdls, n.models=3, digits=2)
```

The first column lists the covariates (and intercept). The second column gives the posterior probability that a given coefficient is not zero over all the 209 models. One could view this as the inclusion probability. That is, what is the probability that the associated covariate had a nonzero coefficient in the data generating model? For example, the posterior probability that the June NAO covariate is in the data

Table 12.1 Selected models from a BMA procedure.

	$p! = 0$	*EV*	*SD*	*Model 1*	*Model 2*	*Model 3*
Intercept	100	7.5e-01	0.10695	0.692	0.695	0.698
sst.Oct	3.2	7.5e-03	0.07167	.	.	.
sst.Sep	5.8	−3.3e-02	0.21834	.	.	.
sst.Aug	5.0	−3.4e-03	0.15889	.	.	.
sst.Jul	23.9	1.7e-01	0.39788	.	.	.
sst.Jun	5.5	1.5e-02	0.11962	.	.	.
sst.May	3.4	−5.9e-03	0.11293	.	.	.
nao.Oct	2.2	6.1e-04	0.00729	.	.	.
nao.Sep	3.2	1.2e-03	0.00973	.	.	.
nao.Aug	1.9	3.9e-04	0.00564	.	.	.
nao.Jul	3.4	1.6e-03	0.01198	.	.	.
nao.Jun	43.3	−4.3e-02	0.05662	−0.105	−0.096	−0.103
nao.May	5.4	−3.2e-03	0.01713	.	.	.
soi.Oct	37.7	2.0e-02	0.02841	0.055	.	.
soi.Sep	4.6	1.5e-03	0.00869	.	.	.
soi.Aug	25.0	1.3e-02	0.02519	.	.	0.054
soi.Jul	39.2	2.4e-02	0.03381	.	0.054	.
soi.Jun	9.1	−4.6e-03	0.01740	.	.	.
soi.May	1.4	2.0e-04	0.00402	.	.	.
ssn.Oct	7.0	−3.8e-04	0.00178	.	.	.
ssn.Sep	94.4	−1.1e-02	0.00442	−0.012	−0.012	−0.012
ssn.Aug	1.1	1.1e-05	0.00041	.	.	.
ssn.Jul	1.7	−3.7e-05	0.00055	.	.	.
ssn.Jun	88.2	8.5e-03	0.00440	0.010	0.011	0.011
ssn.May	5.7	3.0e-04	0.00149	.	.	.
nVar				4	4	4
BIC				171.072	171.511	171.528
post prob				0.040	0.032	0.031

generating model is 43.3 percent. The third and fourth columns are the posterior expected value (EV) and standard deviation (SD) across all models. Subsequent columns include the most probable models as indicated by values in rows corresponding to a covariate. The number of variables in the model, the model BIC, and the posterior probability are also given in the table.

Models are ordered by BIC values with the first model having the lowest BIC the second model having the second lowest BIC, and so on. The BIC value for a given model is

$$-2 \cdot \ln L + k \ln(n), \qquad (12.9)$$

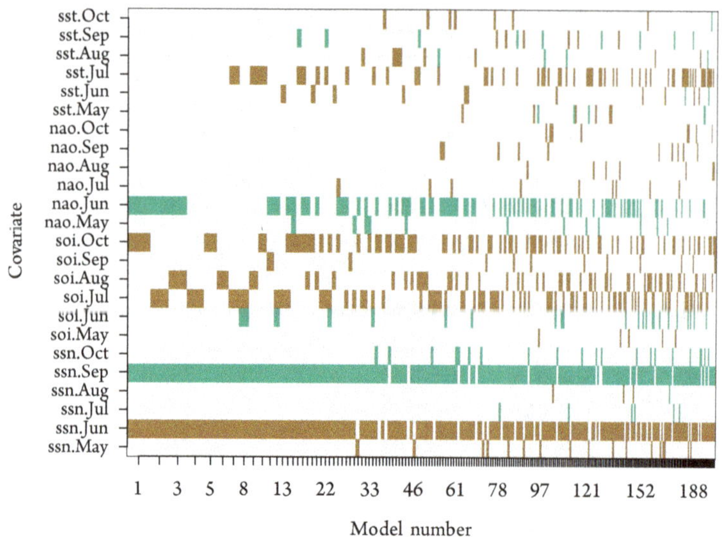

Figure 12.7 Covariates by model number.

where L is the likelihood evaluated at the parameter estimates, k is the number of parameters, and n is the number of years. BIC includes a penalty term $(k \ln(n))$, which makes it useful for comparing models with different sizes. If the penalty term is removed, BIC can be reduced simply by increasing the number of covariates. The BIC as a selection criterion results in choosing models that are parsimonious and asymptotically consistent, meaning that the model with the lowest BIC converges to the true model as the number of years increases.

You use a plot method to display the model coefficients (by sign) ordered by decreasing posterior probabilities as shown in Figure 12.7. Models are listed along the horizontal axis by decreasing posterior probability. Covariates in a model are shown with colored bars. A brown bar indicates a positive relationship with hurricane rate and a green bar indicates a negative relationship. Bar width is proportional to the model's posterior probability.

```
> imageplot.bma(mdls)
```

The plot makes it easy to see the covariates picked by the most probable models. They are the ones with the most consistent coloring from left to right across the image. A covariate with only a few gaps indicates that it is included in most of the higher probable models. These include September and June SSN, June NAO, July SST, and any of the months of July through September for the SOI.

You might ask why July SST is selected as a model covariate more often than August and September? The answer lies in the fact that when the hurricanes arrive in August and September, they draw heat from the ocean surface so the correlation between hurricane activity and SST weakens. The thermodynamics of hurricane intensification works against the statistical correlation. Said another way, July SST better relates to an active hurricane season not because a warm ocean in July *causes* tropical cyclones in August and September, but because hurricanes in August and September cool the ocean slightly.

The SOI covariates get chosen frequently by the most probable models but with a mixture across the months of July through October. The posterior probability is somewhat higher for the months of June and October and smallest for August and September. Thunderstorms over the eastern equatorial Pacific during El Niño produce increased shear and subsidence across the Atlantic especially over the western Caribbean where during the months of July and October a relatively larger percentage of the North Atlantic hurricane activity occurs. Moreover, the inhibiting influence of El Niño might be less effective during the core months of August and September when, on average, other conditions tend to be favorable.

The sign on the September SSN parameter is negative indicating that the probability of a U.S. hurricane decreases with increasing number of sunspots. This result accords with the hypothesis that increases in UV radiation from an active sun (greater number of sunspots) warms the upper troposphere resulting in greater thermodynamic stability and a lower probability of a hurricane over the western Caribbean and Gulf of Mexico (Elsner and Jagger, 2008; Elsner et al., 2010). The positive relationship between hurricane probability and June SSN is explained by the direct influence the sun has on ocean temperature. Alternative explanations are possible especially in light of the role the solar cycle likely plays in modulating the NAO (Kodera, 2002; Ogi et al., 2003).

You can find the probability that a covariate irrespective of month is chosen by calculating the total posterior probability over all models that include this covariate. First, use the `substring` function on the covariate names given in the `bic.glm` object under `namesx` to remove the last four characters in each name. Also create a character string containing only the unique names.

```
> cn = substring(mdls$namesx, 1, 3)
> cnu = unique(cn)
```

Next create a matrix of logical values using the `outer` function, which performs an outer product of matrices and arrays. Also assign column names to the matrix.

```
> mn = outer(cn, cnu, "==")
> colnames(mn) = cnu
```

Next perform a matrix multiplication of the matching names with the matrix of logical entries indicating which covariate was chosen. This returns a matrix with dimensions of number of models by number of covariate types. The matrix entries are the number of covariates in each model of that type. Finally multiply the posterior probabilities given under `postprob` by a logical version of this matrix using the condition that the covariate type is included.

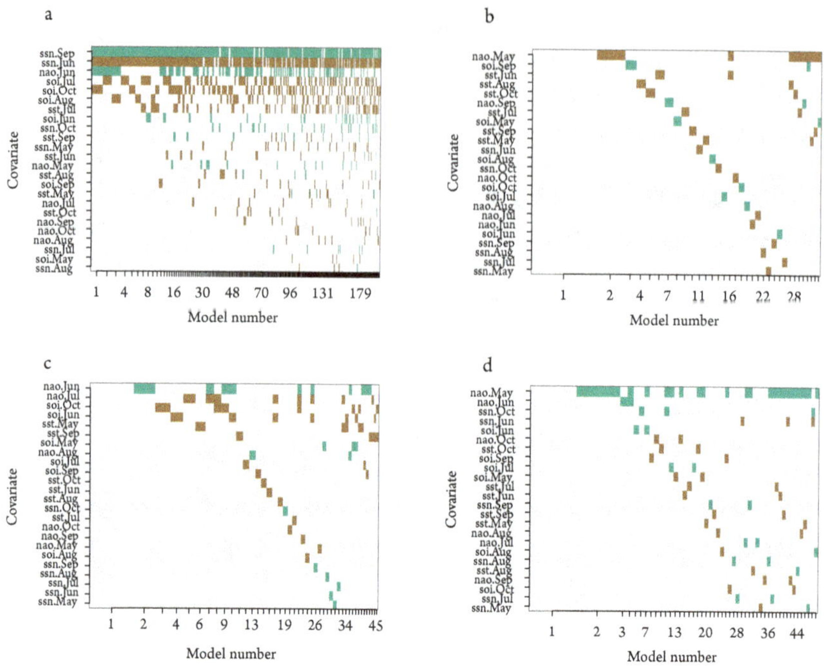

Figure 12.8 Covariates by model number. (a) Actual and (b–d) permuted series.

```
> xx = mdls$which %*% mn
> pc = 100 * mdls$postprob %*% (xx > 0)
> round(pc, 1)
      sst  nao  soi  ssn
[1,] 39.7 51.9 98.5 98.5
```

You see that the SOI and sunspot number have the largest probabilities at 98.5 percent while the NAO and SST have posterior probabilities of 51.9 percent and 39.7 percent, respectively. The lower probability of choosing the NAO reflects the rather large intraseasonal variability in this covariate as seen in Figure 12.6.

It is informative to compare the results of your BMA with a BMA performed on a random series of counts. Here you do this by resampling the actual hurricane counts. The randomization results in the same set of counts, but the counts are placed randomly across the years. The random series together with the covariates are used as before and the results are mapped in Figure 12.8.

The comparison shows that your set of covariates has a meaningful relationship with U.S. hurricane activity. There are fewer models chosen with the randomized data sets and the number of variables included in the set of most probable models is lower. In fact the averaged number of variables in the 20 most probable models is 4, which compares with an average of only one for the three randomized series. Moreover, there is little consistency in the variable selected from one model to the next as it should be when modeling randomized data.

12.3.4 Consensus Hindcasts

The BMA procedure assigns a posterior probability to a set of the most probable models. Each model can be used to make a prediction. But which forecast should you believe? Fortunately, no choice is necessary. Each model makes a prediction and then the forecasts are averaged. The average is weighted where the weights are proportional to the model's posterior probability. Here you assume perfect knowledge of the covariates and hindcasts are made in-sample. For an actual forecast situation, this is not available, but the method would be the same.

You use the prediction functions in **prediction.R** to hindcast annual rates, rate distributions, and posterior count distributions. First, source the code file, then compute the mean and standard deviation of the annual rate for each year. From the output, compute the in-sample average square error.

```
> source("prediction.R")
> ar = bic.poisson(mdls, newdata=mdls$x, simple=TRUE)
> sqrt(mean((ar[1, ] - mdls$y)^2))
[1] 1.36
```

Thus, on average, the consensus model results in a mean square error of 1.4 hurricanes per year.

Here you examine hindcast probabilities for the consecutive years of 2007 and 2008. You determine the posterior probabilities for the number of hurricanes for each year for hurricane numbers between zero and eight and display them using a side-by-side bar plot.

```
> yr1 = 2007; yr2 = 2008
> r1 = which(dat$Year==yr1)
> r2 = which(dat$Year==yr2)
> Pr = bic.poisson(mdls, newdata=mdls$x[c(r1, r2), ],
+   N=9)
> barplot(t(Pr), beside=TRUE, las=1,
+   xlab="Number of Hurricanes",
+   ylab="Probability", legend.text=
+   c(as.character(yr1), as.character(yr2)))
```

Results are shown in Figure 12.9. The vertical axis is the probability of observing h number of hurricanes. The model predicts a higher probability of at least one U.S. hurricane for 2008 compared with 2007. There is a 54 percent chance of three or more hurricanes for 2007 and a 57 percent chance of three or more hurricanes for 2008. There was one hurricane in 2007 and three hurricanes in 2008.

The consensus model hindcasts larger probabilities of an extreme year given the rate than would be expected from a Poisson process. That is, the consensus model is overdispersed with respect to a Poisson distribution. This is because model uncertainty is incorporated into the consensus hindcasts.

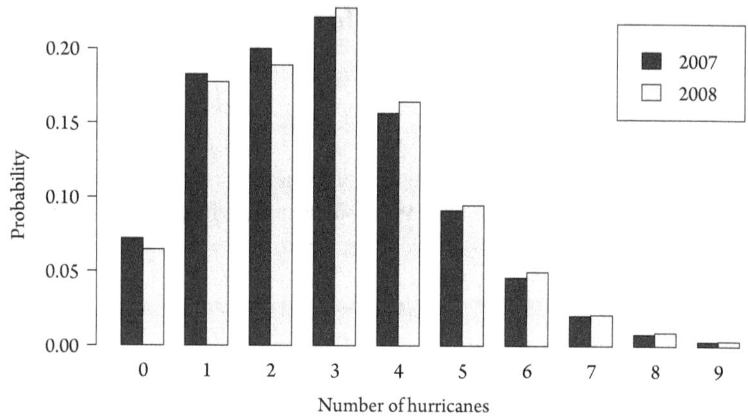

Figure 12.9 Forecasts from the consensus model.

A cross-validation of the BMA procedure is needed to get an estimate of how well the consensus model will do in predicting future counts. This is done in Jagger and Elsner (2010) using various scoring rules including the mean square error, the ranked probability score, the quadratic (Brier) score, and the logarithmic score. They find that the BMA procedure provides more accurate predictions compared with a procedure that selects a single best model using BIC or AIC irrespective of the scoring rule. The consensus forecast will not necessarily give you the smallest forecast error every year, but it will always provide a better assessment of forecast uncertainty compared to a forecast made from a single model. The BMA procedure provides a rational way for you to incorporate competing models into the forecast process.

12.4 SPACE–TIME MODEL

You save your most ambitious model for last. It draws on your knowledge of frequency models (Chapter 7), spatial models (Chapter 9), and Bayesian methods (Chapter 4). Substantial progress has been made in understanding and predicting hurricane activity on the seasonal and longer time scales over the basin as a whole. Much of this progress comes from statistical models described in this book.

However, significant gaps remain in our knowledge of what regulates hurricane activity *regionally*. Here your goal is a multilevel (hierarchical) statistical model to better understand and predict regional hurricane activity. Multilevel models are not new but have gained popularity with more computing power and better software. However, they have yet to be employed to study hurricane climate. For a comprehensive modern treatment of statistics for spatiotemporal data, see Cressie and Wikle (2011).

12.4.1 Lattice Data

Here you use the spatial hexagon framework described in Chapter 9 to create space–time lattice data consisting of hurricane counts at each hexagon for each year and accompanying covariate information. Some of the covariate information is at the local (hexagon) level and some of it is at the regional level (e.g., climate indices). Single-level models, are commonly used to describe basin-wide hurricane activity, but multilevel models allow you to model variations in relationships at the hexagon level and for individual years. Multilevel models also allow you to describe and explain hurricane variations across the basin.

Some data organization is needed. First input the hourly best-track data, the netCDF SST grids, and the annual aggregated counts and climate covariates that you arranged in Chapter 6. Specify the range of years over which you want to model the data and make a copy of the best-track data frame.

```
> load("best.use.RData")
> load("ncdataframe.RData")
> load("annual.RData")
> years = 1886:2009
> Wind.df = best.use
```

Next define the hexagon tiling. Here you follow closely the work flow outlined in Chapter 9. Acquire the **sp** package. Then designate the location columns in your data frame as spatial coordinates and add a map projection.

```
> require(sp)
> coordinates(Wind.df) = c("lon", "lat")
> coordinates(ncdataframe) = c("lon", "lat")
> ll = "+proj=longlat +ellps=WGS84"
> proj4string(Wind.df) = CRS(ll)
> proj4string(ncdataframe) = CRS(ll)
> slot(Wind.df, "coords")[1:3, ]
       lon lat
[1,] -94.8  28
[2,] -94.9  28
[3,] -95.0  28
```

With the `spTransform` function (**rgdal**), you change the geographic CRS to a Lambert conformal conic (LCC) planar projection using the parallels 15 and 45°N and a center longitude of 60°W.

```
> lcc = "+proj=lcc +lat_1=45 +lat_2=15 +lon_0=-60"
> require(rgdal)
> Wind.sdf = spTransform(Wind.df, CRS(lcc))
> SST.sdf = spTransform(ncdataframe, CRS(lcc))
> slot(Wind.sdf, "coords")[1:3, ]
         lon     lat
[1,] -3257623 3648556
[2,] -3266544 3651532
[3,] -3275488 3654496
```

The transformation does not rename the coordinates, but the values are the new LCC projected coordinates. You compare bounding boxes to ensure the SST data region extends beyond the cyclone data region.

```
> bbox(Wind.sdf)
        min     max
lon -4713727 4988922
lat  1020170 8945682
> bbox(SST.sdf)
        min     max
lon -4813071 8063178
lat    78055 9185849
```

Next, generate the hexagons. First, sample the hexagon centers using the bounding box from the cyclone data. Specify the number of centers to be 250 and fix the offset so that the sampler chooses the same set of centers given the number of centers and the bounding box. Then create a spatial polygons object:

```
> hpt = spsample(Wind.sdf, type="hexagonal",
+   n=250, bb=bbox(Wind.sdf) * 1.2, offset=c(1, -1))
> hpg = HexPoints2SpatialPolygons(hpt)
```

This results in 225 hexagons each with an area of approximately 511,457 km^2.

Next, overlay the hexagons on the cyclone and SST locations separately.

```
> Wind.hexid = over(x=Wind.sdf, y=hpg)
> SST.hexid = over(x=SST.sdf, y=hpg)
```

This creates a vector containing the hexagon identification number for each hourly observation. The length of the vector is the number of observations. Similarly, for the SST data, the vector has elements indicating in what hexagon the SST value belongs.

Then use the `split` function to divide the data frame into groups defined by the hexagon number. The groups are saved as lists. Each list is a data frame containing information specific to the cyclones occurring in each hexagon. You do this for the cyclone and SST data.

```
> Wind.split = split(Wind.sdf@data, Wind.hexid)
> SST.split = split(SST.sdf@data, SST.hexid)
```

You find that the first hexagon contains five cyclone observations. To view a selected set of the columns from this data frame, type

```
> Wind.split[[1]][c(1:2, 5:7, 9, 11)]
         Sid Sn   Yr Mo Da Wmax DWmaxDt
3398     185  5 1878  9  1 50.0   0.467
3398.1   185  5 1878  9  1 49.4   0.484
33539.5 1168  6 1990  8 13 34.8   1.222
```

```
33540    1168  6 1990  8 13 35.0    1.005
42583    1442 19 2010 10 29 30.0    0.690
```

A given hexagon tends to capture more than one cyclone hour. To view a selected set of SST grid values from the corresponding hexagon, type

```
> SST.split[[1]][, 1:5]
   Y1854M01 Y1854M02 Y1854M03 Y1854M04 Y1854M05
1      24.7     26.2     26.6     26.3     25.5
2      24.7     26.3     26.7     26.2     25.3
3      24.7     26.4     26.7     26.1     25.0
58     25.6     26.8     26.9     26.7     26.2
```

The hexagon contains four SST grid values and there are 1,871 months as separate columns starting in January 1854 (`Y1854M01`). Next, reassign names to match those corresponding to the hexagon identifications.

```
> names(Wind.split) = sapply(hpg@polygons, function(x)
+   x@ID)[as.numeric(names(Wind.split))]
> names(SST.split) = sapply(hpg@polygons, function(x)
+   x@ID)[as.numeric(names(SST.split))]
```

There are hexagon grids with cyclone data over land areas (no SST data), and there are areas over the ocean where no cyclones occurr. Thus, you subset each to match hexagons having cyclones and SST data.

```
> Wind.subset = Wind.split[names(Wind.split) %in%
+   names(SST.split)]
> SST.subset = SST.split[names(SST.split) %in%
+   names(Wind.split)]
```

The function `%in%` returns a logical vector indicating whether there is a match for the names in `Wind.split` from the set of names in `SST.split`. The data sets are now in synch. There are 109 hexagons with both cyclone and SST data. Note that for the cyclone data, you could subset `best.use` on `M==FALSE` to remove cyclone observations over land.

Next, compute the average SST within each hexagon by month and save them as a data frame.

```
> SST.mean = data.frame(t(sapply(SST.subset,
+   function(x) colMeans(x))))
> head(SST.mean)[1:5]
     Y1854M01 Y1854M02 Y1854M03 Y1854M04 Y1854M05
ID8      26.9     27.0     27.0     27.3     27.6
ID9      27.0     27.2     27.2     27.4     27.4
ID18     27.9     27.9     28.0     28.2     28.3
ID19     26.9     26.5     26.5     27.1     28.2
```

```
ID20      26.8      26.3      26.1      26.9      27.9
ID21      27.0      26.5      26.3      27.0      27.9
```

The data frame is organized with the hexagon identifier as the row (observation) and consecutive months as the columns (variables). Thus, there are 109 rows and 1,871 columns. Your interest in SST is more narrowly focused on the months of August through October when hurricanes occur. To generate these values by year, type

```
> SSTYearMonth = strsplit(substring(colnames(SST.mean),
+   2), "M")
> SSTYear = as.numeric(sapply(SSTYearMonth,
+   function(x) x[1]))
> SSTMonth = as.numeric(sapply(SSTYearMonth,
+   function(x) x[2]))
> SSTKeep = which(SSTMonth %in% c(8, 9, 10))
> SSTYear = SSTYear[SSTKeep]
```

The vector `SSTKeep` lists the column numbers corresponding to August, September, and October for each year and the vector `SSTYear` is the set of years for those months.

Next, subset `SST.mean` by `SSTKeep` and then compute the August–October average for each year.

```
> SST.mean.keep = SST.mean[, SSTKeep]
> SST.mean.year = sapply(unique(SSTYear), function(x)
+   as.vector(rowMeans(SST.mean.keep[,
+   which(x==SSTYear)])))
> dimnames(SST.mean.year) =
+   list(id=rownames(SST.mean.keep),
+   Year=paste("Y",unique(SSTYear), sep=""))
> SST.mean.year.subset = SST.mean.year[,
+   paste("Y", years, sep="")]
> SST.pdf = SpatialPolygonsDataFrame(
+   hpg[rownames(SST.mean.year.subset)],
+   data.frame(SST.mean.year.subset))
```

The data slot in the spatial polygon data frame `SST.pdf` contains the average SST during the hurricane season for each year. To list the first few rows and columns, type

```
> slot(SST.pdf, "data")[1:5, 1:6]
     Y1886 Y1887 Y1888 Y1889 Y1890 Y1891
ID8   28.4  28.2  28.5  28.4  27.7  28.4
ID9   28.0  27.8  28.1  28.0  27.3  28.1
ID18  27.7  28.0  28.5  27.5  27.6  27.7
ID19  28.5  28.4  28.6  28.5  27.9  28.5
ID20  28.4  28.4  28.1  28.2  27.6  28.6
```

You plot the SST data as you did in Chapter 9 using the spplot method. First, obtain the map borders in geographic coordinates and project them using the same CRS as your data.

```
> require(maps)
> require(maptools)
> require(colorRamps)
> cl = map("world", xlim=c(-120, 20),
+   ylim=c(-10, 70), plot=FALSE)
> clp = map2SpatialLines(cl, proj4string=CRS(ll))
> clp = spTransform(clp, CRS(lcc))
> l2 = list("sp.lines", clp, col="darkgray")
```

Then, obtain the color ramps and plot the values from the year 1959, for example.

```
> spplot(SST.pdf, "Y1959", col="white",
+     col.regions=blue2red(20), pretty=TRUE,
+     colorkey=list(space="bottom"),
+     sp.layout=list(l2),
+     sub="Sea Surface Temperature (C)")
```

Your plot shows how the data are organized.

Next, you need to generate a data set of that correspond to these hexagons in space and time. First, generate a list of the cyclones maximum intensity by hexagon using your get.max function and count the number of cyclones per hexagon per year.

```
> source("getmax.R")
> Wind.max = lapply(Wind.subset, function(x)
+   get.max(x, maxfield="WmaxS"))
> Wind.count = t(sapply(Wind.max, function(x)
+   table(factor(subset(x, WmaxS >= 33 &
+   Yr %in% years)$Yr, level=years))))
> colnames(Wind.count) = paste("Y",
+   colnames(Wind.count), sep="")
> Wind.count = data.frame(Wind.count)
> Wind.pdf = SpatialPolygonsDataFrame(
+   hpg[rownames(Wind.count)], Wind.count)
```

As an example, to list the hurricane counts by hexagon for 1959, type

```
> slot(Wind.pdf, "data")["Y1959"]
```

12.4.2 Local Independent Regressions

With your annual hurricane counts and seasonal SST collocated spatially, you build local (hexagon-level) independent Poisson regressions (LIPR). Here "independent" refers to separate regressions one for each hexagon. Along with the hexagon-level SST variable, you include the SOI as a covariate. The SOI varies by year but not by hexagon. The SOI covariate was organized in Chapter 6 and saved in *annual.RData*.

Load the annual climate covariates and subset the SOI and SST columns for the years specified in the previous section. These are your regional covariates.

```
> load("annual.RData")
> Cov = subset(annual, Year %in% years)[,
+   c("soi", "sst", "Year")]
```

Then create a data frame of your hexagon-level SST covariate and the hurricane counts from the data slots in the corresponding spatial polygon data frames.

```
> LSST = slot(SST.pdf, "data")
> Count = slot(Wind.pdf, "data")
```

Here you build 109 separate regressions, one for each hexagon. Your annual count, indicating the number of hurricanes, varies by hexagon and by year and you have local SST and regional SOI as covariates. Both are averages over the months of August–October.

The model is a Poisson regression with a logarithmic link function.

```
> lipr = lapply(1:nrow(Count), function(i)
+   glm(unlist(Count[i, ]) ~ unlist(LSST[i, ]) +
+   Cov$soi + Cov$Year, family="poisson"))
```

The standardized coefficients indicate the strength of the relationship between the covariate and the annual count. The coefficients are saved in the matrix `zvals` that you turn into a spatial polygon data frame.

```
> zvals = t(sapply(lipr, function(x)
+   summary(x)$coef[, 3]))
> rownames(zvals) = rownames(Count)
> colnames(zvals) = c("Intercept", "Local.SST",
+   "SOI", "Year")
> zvals.pdf = SpatialPolygonsDataFrame(
+   hpg[rownames(zvals)], data.frame(zvals))
```

To map the results, first generate a color ramp function, then use the `spplot` method.

```
> al = colorRampPalette(c("blue","white","red"),
+   space="Lab")
> spplot(zvals.pdf, c("Local.SST","SOI"),
+    col.regions=al(20), col="white",
+    names.attr=c("SST", "SOI"),
+    at=seq(-5, 5),
+    colorkey=list(space="bottom",
```

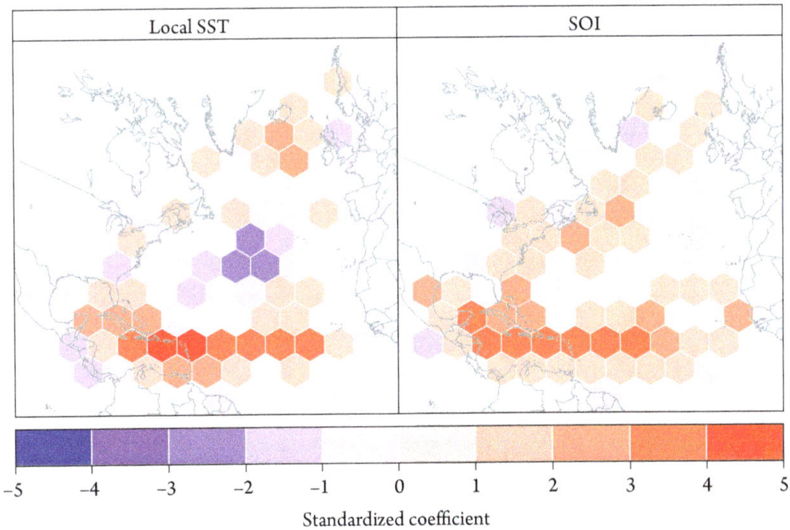

Figure 12.10 Poisson regression coefficients of counts on local SST and SOI.

```
+      labels=paste(seq(-5, 5))),
+      sp.layout=list(12),
+      sub="Standardized Coefficient")
```

The maps are shown in Figure 12.10. The standardized coefficient can be interpreted as a hypothesis test, one for each hexagon. Values more than 2 in absolute value greatly indicate a significant relationship between the covariate and hurricane probability. The probability of a hurricane is higher where the ocean is warm and when SOI is positive (La Niña conditions), a result you would anticipate from your basin-wide models (Chapter 7). The SST and SOI effects are strongest over the central and western North Atlantic at low latitudes. Curiously, the SST effect is muted along much of the eastern coast of the United States and along portions of the Mexican coast northward through Texas. This might explain why, despite warming seas, the U.S. hurricane counts are not increasing.

Your model contains the year as a covariate to address changes in occurrence rates over time. The exponent of the coefficient on the year term is the factor by which the occurrence rates are changing per year. The factor is shown for each grid in Figure 12.11. The factors range between 0.98 and 1.02 annually depending on location. A value of one indicates no change, less than one a decreasing trend, and greater than one an increasing trend. Since the data cover the period beginning in 1886, the increasing trends over the central and eastern North Atlantic are because of improving surveillance. However, the downward trend over a large part of the Caribbean Sea into the Gulf of Mexico is intriguing. It might be related to increasing wind shear or continental aerosols. In fact, you can see the downward trend along parts of the U.S. coastline from Louisiana to South Carolina and over New England.

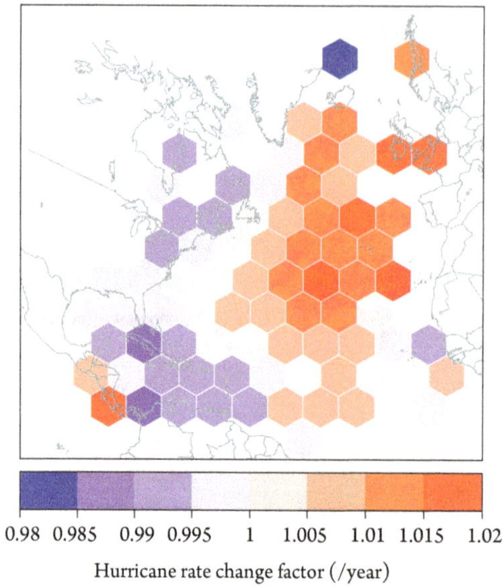

Figure 12.11 Factor by which hurricane rates have changed per year.

Model residuals, defined as the observed count minus the predicted rate for a given year and hexagon, can also be mapped. Here you map the residuals for the 2005 hurricane season. First, compute the residuals by typing

```
> preds = t(sapply(lipr, function(x)
+   predict(x, type="response")))
> rownames(preds) = rownames(Count)
> err2005 = Count[, "Y2005"] - preds[, "Y2005"]
> err.pdf = SpatialPolygonsDataFrame(
+   hpg[names(err2005)], data.frame(err2005))
```

Then select a color ramp function and create a choropleth map with the `spplot` method.

```
> al = colorRampPalette(c("blue", "white", "red"),
+   space="Lab")
> spplot(err.pdf, c("err2005"), col="white",
+   col.regions=al(20), at=seq(-5, 5, 1),
+   colorkey=list(space="bottom",
+   labels=paste(seq(-5, 5, 1))),
+   sp.layout=list(12),
+   par.settings=list(fontsize=list(text=10)),
+   sub="Observed [count] - Predicted [rate]")
```

Figure 12.12 shows where the model over-(blues) and under-(reds) predicts for the 2005 hurricane season. In particular the model underpredicted the large amount of hurricane activity over the western Caribbean and Gulf of Mexico.

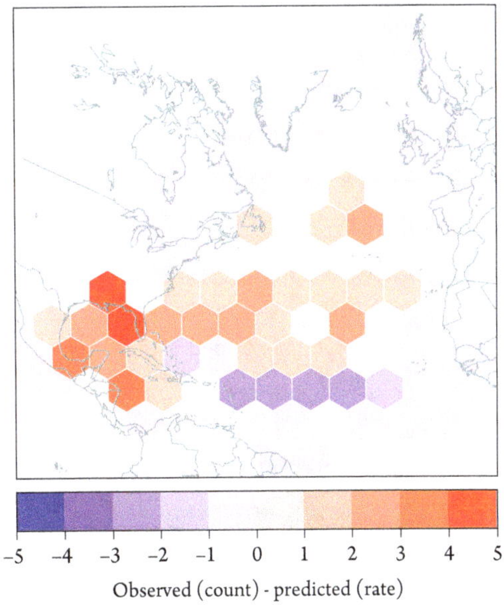

Figure 12.12 Model residuals for the 2005 hurricane season.

12.4.3 Spatial Autocorrelation

Residuals in neighboring hexagons tend to be more similar than residuals in hexagons farther away. This is spatial autocorrelation. To quantify the degree of spatial autocorrelation, you create a weights matrix indicating the spatial neighbors for each hexagon (see Chapter 9). The **spdep** package (Bivand et al., 2011a) has functions for creating weights based on contiguity neighbors. First, you use the `poly2nb` function (**spdep**) on the spatial polygons data frame to create a contiguity-based neighborhood list object.

```
> require(spdep, quietly=TRUE)
> hexnb = poly2nb(err.pdf)
```

The list is ordered by hexagon number. The first hexagon has three neighbors; hexagon numbers 2, 7, and 8. Hexagon numbers increase to the west and north. A hexagon has at most six contiguous neighbors. Hexagons at the borders have fewer neighbors. A graph of the hexagon connectivity, here defined by the first-order contiguity, is made by typing

```
> plot(hexnb, coordinates(err.pdf))
> plot(err.pdf, add=TRUE)
```

A summary method applied to the neighborhood list (`summary(hexnb)`) reveals the average number of neighbors and the distribution of connectivity among the hexagons.

You turn the neighborhood list object into a `listw` object using the `nb2listw` function that adds weights to the neighborhood list. The `style` argument determines the weighting scheme. With the argument value set to `W`, the weights are the inverse of the number of neighbors.

```
> wts = nb2listw(hexnb, style="W")
```

Next, you quantify the amount of spatial correlation through the value of Moran's I. This is done using the function `moran` (**spdep**). The first argument is the variable of interest followed by the name of the `listw` object. Also you need the number of hexagons and the global sum of the weights, which is obtained using the `Szero` function.

```
> m = length(err.pdf$err2005)
> s = Szero(wts)
> mI = moran(err.pdf$err2005, wts, n=m, S0=s)$I
```

The function returns a value for Moran's I of 0.39, which indicates spatial autocorrelation in model residuals for 2005. The expected value of Moran's I under the hypothesis of no spatial autocorrelation is $-1/(m-1)$, where m is the number of hexagons. This indicates that the model can be improved by including a term for the spatial correlation. Other years have differing degrees of spatial autocorrelation in the residuals, which might be interesting to examine in detail.

12.4.4 BUGS Data

Next you build a spatial regression model. The model allows you to borrow information from neighboring hexagons to improve the precision on the parameter estimates. The model is motivated by the fact that the residuals from your nonspatial regression above are spatially correlated and hurricanes are infrequent in most hexagons. The model is written in BUGS (see Chapter 4). Here you run BUGS (WinBUGS or OpenBUGS) outside of R.

As preparation you first convert your neighborhood list object to BUGS format. You then gather the hurricane counts and covariates as lists and put them together in the object you call `BUGSdata`.

```
> hexadj = nb2WB(hexnb)
> BUGSData = c(list(S=nrow(Wind.count),
+   T=ncol(Wind.count), h=as.matrix(Wind.count),
+   SST=as.matrix(SST.mean.year.subset),
+   SOI=Cov$soi), hexadj)
```

For each hexagon (`S` of them) and year (`T` of them), there is a count (`N`) that indicates the number of hurricanes. The associated covariates are the SOI and local SST. The local SST is constructed by averaging the August–October monthly gridded SST over each hexagon for each year. For each hexagon, there is a list indicating the neighboring hexagons by ID (adjacency `adj`), the weights for the neighbors (weights `wts`), and the number of neighbors (number `num`).

Finally, you use `writeDatafileR` from the file **writedatafileR.R** to write an ASCII text representation of your model to your working directory.

```
> source("writedatafileR.R")
> writeDatafileR(BUGSData,"BugsData.txt")
```

12.4.5 MCMC Output

Counts in each hexagon for each year are described by a Poisson distribution (`dpois`) with a rate that depends on hexagon and year (`lambda`). The logarithm of the rate is conditional on local SST and SOI. The error term (`error`) and the local effect terms include structured and unstructured components following Besag et al. (1991), where the structured component is an intrinsic conditional autoregressive (ICAR) specification. The adjacency matrix (`adj`), which gives your contiguity neighborhood is part of the ICAR specification.

The BUGS code for the model is

```
___BUGS code___
model {
for(hx in 1:S) {
  for(yr in 1:T) {
# Poisson likelihood for observed counts
   h[hx, yr] ~ dpois(lambda[hx, yr])
   log(lambda[hx, yr]) <- sst[hx] * SST[hx, yr] +
                           soi[hx] * SOI[yr] +
                           error[hx]
      }
# Error terms
   error[hx] <- u.error[hx] + s.error[hx]
   sst[hx] <- u.sst[hx] + s.sst[hx]
   soi[hx] <- u.soi[hx] + s.soi[hx]
# Unstructured errors
   u.error[hx] ~ dnorm(int, tau)
   u.sst[hx] ~ dnorm(int.sst, tau.sst)
   u.soi[hx] ~ dnorm(int.soi, tau.soi)
   }
# Structured errors
s.error[1:S] ~ car.normal(adj[], weights[],
                           num[], tau.s)
s.sst[1:S] ~ car.normal(adj[], weights[],
                           num[], tau.s.sst)
s.soi[1:S] ~ car.normal(adj[], weights[],
                           num[], tau.s.soi)
```

```
# Priors
int ~ dflat()
int.sst ~ dflat()
int.soi ~ dflat()
tau ~ dgamma(.5, .005)
tau.sst ~ dgamma(.5, .005)
tau.soi ~ dgamma(.5, .005)
tau.s ~ dgamma(.5, .005)
tau.s.sst ~ dgamma(.5, .005)
tau.s.soi ~ dgamma(.5, .005)
}
```

Think of the model as a directed acyclic graph (DAG) as shown in Chapter 4. Nodes are the parameters and data and the arrows indicate conditional dependency, either stochastic (~) or deterministic (<-).

You use uninformative (flat) priors for the intercept terms and a gamma distribution for priors on the precisions. A shape parameter of 0.5 and a scale of 0.005 translates to a 1 percent probability that the standard deviation is less than 0.04 and a 1 percent probability that the standard deviation is greater than .8, respectively. It would also be reasonable to combine the models into a multivariate CAR model, with a Wishart prior, having 4 degrees of freedom and a diagonal of 0.05.

Open BUGS (outside R) and copy the code to a new BUGS document saving it as (`BUGSmodel.odc`). Open the `BUGSdata.txt` file and copy the entire file to your BUGS document. Finally, copy the initial values below into the same document.

```
___Initial Values___
list(int=0, int.sst=0, int.soi=0, tau=100, tau.sst=100,
  tau.soi=100, tau.s=100, tau.s.sst=100, tau.s.soi=100)
list(int=.5, int.sst=.5, int.soi=.5, tau=100, tau.sst=100,
  tau.soi=100, tau.s=100, tau.s.sst=100, tau.s.soi=100)
```

It might be necessary to specify high values for the precisions since you need to have BUGS generate realistic initial samples for the uninitialized parameters. You use two sets of initial values (two separate `list` objects) to help monitor and diagnose convergence toward the posterior density. Each set will result in a separate chain of samples. Here you specify separate chains using a different value for the intercept terms.

In BUGS you first select Model > Specification to open the specification tool. Highlight the word `model` in your document then select `check` model. It will tell you that your model is syntactically correct in the lower left corner of the window. Next, highlight the word `list` in front of your data list and select `load data`. Then change the number of chains to two and select `compile`. It will tell you that the data are loaded and the model is compiled. Next, highlight the word `list` in front of your first set of initial values and select `load inits`. It will tell you that this chain contains uninitialized values. Repeat for your second set of initial values. It will again tell you about uninitialized values. Finally, select `gen inits` to initialize the remaining values (for both chains).

Next you tell BUGS what model parameters (nodes) you want to monitor. Select Inference > Samples to open the sample monitor tool. In the `node` window, type `sst` and then select `set`. These parameters are the coefficients on the SST variable. Repeat for the coefficients on the SOI variable by typing `soi` and selecting `set`. Note that there are 109 SST and SOI coefficients, one of each for each hexagon.

Next, you select Model > Update to open the update tool. In the `update` window, type 5000, and in the `refresh` window, type 10. Although you are monitoring only the SST and SOI coefficients, each update (iteration) represents new values for all parameters in your model using an MCMC algorithm (see Chapter 4). Select `update`. This will take time. The refresh number indicates the progress in intervals of 10 updates.

After the MCMC has finished updating, you output the monitored parameter values from both chains. In the `node` window, scroll to `sst`, then select `coda`. CODA is a suite of functions for analyzing outputs from BUGS software. Save the samples and index files separately (use the extension `txt`). The index file provides the order of the samples.

12.4.6 Convergence and Mixing

Before using your samples for inference, you need to check a few things. Starting from your initial values, the MCMC algorithm produces a new set of values (the first update sample) for each parameter (node). In most cases (except in simple models), these new values will not represent the posterior density. As updating continues the MCMC samples are guaranteed to *converge* to a stationary distribution representing samples from your posterior. It is useful to know the minimum number of updates needed until convergence. The period before convergence is called "burn-in", where the samples are moving away from the initial set of values and toward the posterior distribution.

You can analyze your MCMC samples in BUGS directly, but there are more options in R. Input the BUGS-outputted CODA files corresponding to the SST parameter using the `read.coda` function (**coda**). The first argument is the name of the samples file and the second is the name of the index file. Your new objects (`chain1` and `chain2`) are of class `mcmc`.

```
> require(coda)
> chain1 = read.coda("Chain1.txt",
+    "Index.txt", quiet=TRUE)
> chain2 = read.coda("Chain2.txt",
+    "Index.txt", quiet=TRUE)
```

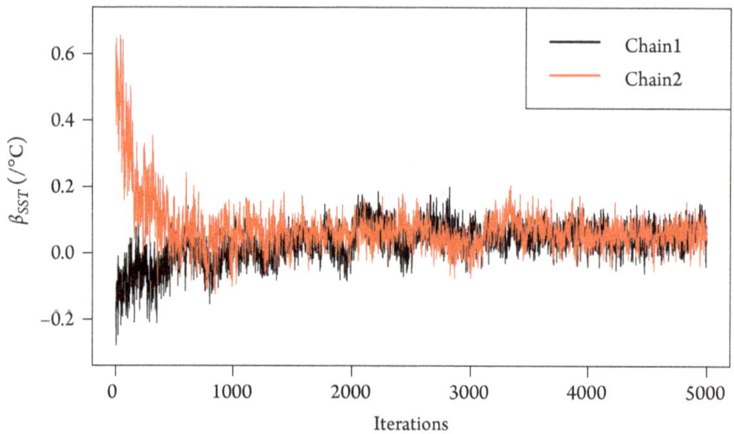

Figure 12.13 MCMC from a space–time model of cyclone counts in hexagon one.

To plot the MCMC samples of the SST coefficient in the first hexagon (hexagons are ordered from southwest to northeast) from both chains, type

```
> traceplot(chain1[, "sst[1]"], ylim=c(-.3, .7))
> traceplot(chain2[, "sst[1]"], col="red", add=TRUE)
```

This generates a graph showing the sequence of values (trace plot) from the first (black) and second (red) chains (Fig. 12.13). Values fluctuate from one iteration to the next but tend toward a stable distribution after about 3,000 updates. This tendency toward convergence is even more apparent by comparing the two trace plots. Initially, the values from chain one are quite different from those of chain two, but after about 4,000 iterations the distributions are visually indistinguishable. From this analysis, you estimate that convergence requires about 4,000 iterations.

You quantify the MCMC convergence with the potential-scale-reduction factor (PSRF) proposed by Gelman and Rubin (1992). After convergence, your two chains starting with different initial conditions should represent samples from the same distribution. You assess this by comparing the mean and variance of each chain to the mean and variance of the *combined* chain. Specifically, with two chains, the between-chain variance B/n and pooled within-chain variance W are defined by

$$\frac{B}{n} = \frac{1}{2-1}\sum_{j=1}^{2}(\bar{s}_{j.} - \bar{s}_{..})^2 \tag{12.10}$$

and

$$W = \frac{1}{2(n-1)}\sum_{j=1}^{2}\sum_{t=1}^{n}(s_{jt} - \bar{s}_{j.})^2 \tag{12.11}$$

where s_{jt} is the parameter value of the tth sample in the jth chain, $\bar{s}_{j.}$ is the mean of the samples in j, and $\bar{s}_{..}$ is the mean of the combined chains.

By taking the sampling variability of the combined mean into account, you get a pooled estimate for the variance:

$$\hat{V} = \frac{n-1}{n} W + \frac{B}{n} \tag{12.12}$$

Then an estimate $\hat{R}$ for PSRF is obtained by dividing the pooled variance with the pooled within-chain variance,

$$\hat{R} = \frac{\hat{V}}{W} = \frac{n-1}{n} + \frac{B}{nW} \tag{12.13}$$

If the chains have not converged, Bayesian credible intervals based on the t-distribution are too wide and have the potential to shrink by the PSRF. PSRF is sometimes called the shrink factor.

A value for $\hat{R}$ is available for each monitored node using the `gelman.diag` function (**coda**). Values substantially above 1 indicate lack of convergence. For example, to get the PSRF estimate on the SST parameter for the first hexagon, type

```
> gelman.diag(mcmc.list(chain1[, "sst[1]"],
+    chain2[, "sst[1]"]))
Potential scale reduction factors:

        Point est. Upper C.I.
[1,]          1.01       1.04
```

By default, only the second half of the chain is used. The point estimate indicates near convergence consistent with the evidence in the trace plots. To see the evolution of $\hat{R}$ as the number of iterations increase, type

```
> gelman.plot(mcmc.list(chain1[, "sst[1]"],
+    chain2[, "sst[1]"]))
```

The plot shows convergence after about 3,000 iterations.

Convergence does not guarantee that your samples have visited all (or even a large portion) of the posterior density. The speed with which your samples work their way through the posterior (mixing) depends on how far they advance from one update to the next. Mixing efficiency is inversely related to the between-sample correlation. Low values of correlation at small lag indicate that mixing is efficient.

The autocorrelation function shows the correlation as a function of consecutive sample lag. Here you use the `acf` function on the chain to examine the correlation by typing

```
> acf(chain1[, "sst[1]"], ci=0, lag.max=2000)
```

The results for the SST coefficient in hexagon 1 and hexagon 60 are shown in Figure 12.14. Depending on the hexagon, the correlation drops below zero after several hundred samples.

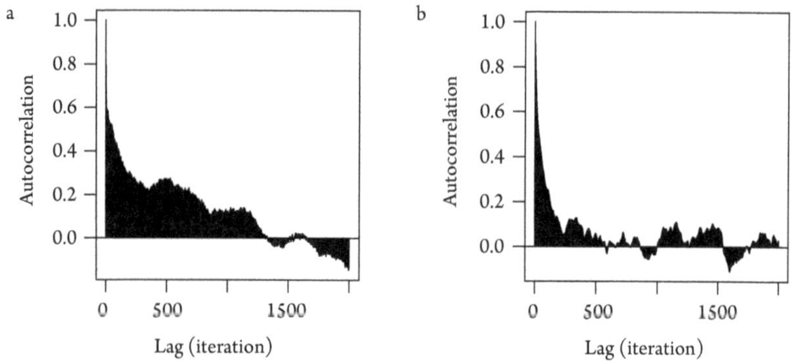

Figure 12.14 Autocorrelation of the SST coefficient in hexagon (a) 1 and (b) 60.

You use the effective chain size to quantify the decay. Because of the between-sample correlation, your effective chain size is less than the 5K updates. You use the `effectiveSize` function (**coda**) to estimate the effective chain size by typing

```
> effectiveSize(chain1[, "sst[1]"])
var1
  25
> effectiveSize(chain1[, "sst[60]"])
var1
59.7
```

The effective size is an estimate of how many independent samples you have given the amount of between-sample correlation. The slower the decay of the autocorrelation function, the lower the effective size. For chain one, the effective size on the SST coefficient is 25 for the first hexagon and 60 for hexagon 60.

The sample sizes numbers are too small to make reliable inferences, so you need more updates. Your goal is 1,000 independent samples. Since a conservative estimate of your effective sample size is 25 in 5K updates, you need 200K updates to reach your goal.

12.4.7 Updates

First return to BUGS. Next create a single chain and generate 205K updates. This time you monitor both the SST and SOI coefficients. After updating (this will take several hours),[4] you output every 200th sample over the last 200K updates. This is done with the sample monitor tool by setting `beg` equal to 5001 and `thin` equal to 200. Save the index files and chain values from the coefficients and read them into R.

[4] 200K updates took 10.5 hr on a 2 × 2.66 GHz dual-core Intel Xeon processor running on OS X version 10.6.8

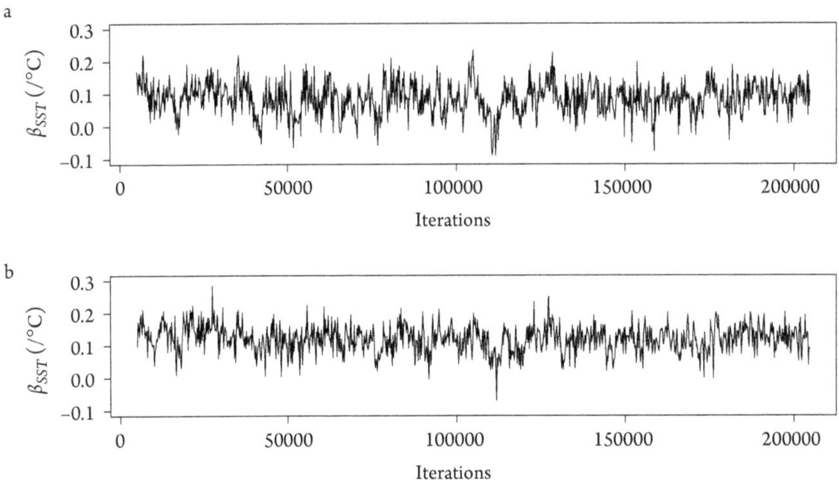

Figure 12.15 MCMC trace of the SST coefficients for hexagon (a) one and (b) 60.

```
> sstChain = read.coda("SSTChain.txt",
+    "SSTIndex.txt", quiet=TRUE)
> soiChain = read.coda("SOIChain.txt",
+    "SOIIndex.txt", quiet=TRUE)
```

You create a trace plot of your samples for hexagon one by typing

```
> traceplot(sstChain[, "sst[1]"], ylim=c(-.1, .3))
```

Note the apparent stationarity (see Fig. 12.15). Also note that since you saved only every 200th sample, the variation from one value to the next is much higher.

Now you use these samples to make inferences. For example, the probability that the SST coefficient for hexagon 1 is greater than zero is obtained by typing

```
> sum(sstChain[, "sst[1]"]>0)/
+   length(sstChain[, "sst[1]"]) * 100
[1] 95.2
```

You interpret the coefficient of the Poisson regression as a factor increase (or decrease) in the rate of occurrence per unit of the explanatory variable, given that the other covariates are held constant. This is a relative risk or the ratio of two probabilities. The relative risk of hurricanes per degree celsius in hexagon 1 is estimated using the posterior mean as

```
> exp(mean(sstChain[, "sst[1]"]))
[1] 1.09
```

A relative risk of one indicates no increase or decrease in occurrence probability relative to the long-term average. So you interpret the value of 1.09 to mean a 9 percent increase per degree celsius and only a 4.8 percent chance that the relative risk is one (no change).

12.4.8 Relative Risk Maps

Here you map the relative risk of hurricanes. You estimate the relative risk using the posterior median and you do this over all hexagons by typing

```
> RRsst = exp(apply(sstChain, 2, function(x)
+   median(x)))
```

Next, create a spatial data frame of the relative risk with row names equal to your hexagon identification numbers.

```
> RRsst.df = data.frame(RRsst)
> rownames(RRsst.df) = rownames(Wind.count)
> RRsst.pdf = SpatialPolygonsDataFrame(
+   hpg[rownames(RRsst.df)], RRsst.df)
```

Then choose a color ramp and create a choropleth map.

```
> al = colorRampPalette(c("#FEE8C8", "#FDBB84",
+   "#E34A33"), space="Lab")
> spplot(RRsst.pdf, col="white", col.regions=al(20),
+   at=seq(1, 1.25, .05), colorkey=list(space="bottom",
+   labels=paste(seq(1, 1.25, .05))),
+   sp.layout=list(l2),
+   sub="Relative Hurricane Risk [/C]")
```

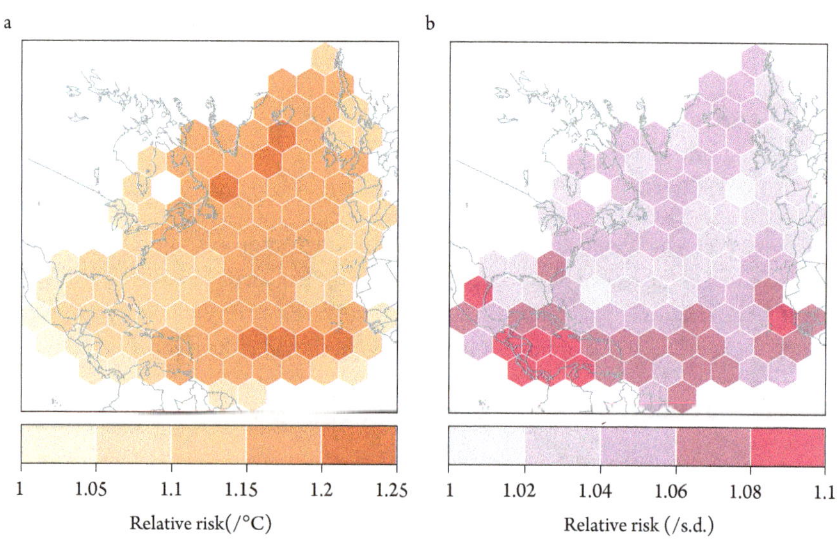

Figure 12.16 Hurricane risk per change in (a) local SST and (b) SOI.

Figure 12.16 maps the hurricane risk relative to a per degree celsius change in local SST and a per standard deviation change in SOI. Hurricane occurrence is most sensitive to rising ocean temperatures over the eastern tropical North Atlantic and less so across the western Caribbean Sea, Gulf of Mexico, and across much of the coast lines of Central America, Mexico, and the United States. In comparison, hurricane occurrence is most sensitive to ENSO over much of the Caribbean and less so over the central and northeastern North Atlantic.

This chapter demonstrated Bayesian models for hurricane climate research. We began by showing how to combine information about hurricane counts from the modern and historical cyclone archives to get a baseline estimate of future activity over the next several decades. We then showed how to create a Bayesian model for seasonal forecasts, where the model is based on an MCMC algorithm that allows you to exploit the older, less reliable cyclone information. We showed how to create a consensus model for seasonal prediction based on Bayesian model averaging. The approach circumvents the need to choose a single best model. Finally, we showed how to create a hierarchical model for exploiting the space–time nature of hurricane activity. Widespread application of Bayesian hierarchical models will undoubtedly lead to a better understanding of hurricane climate. In the final chapter, we consider a few practical applications of our statistical approach to hurricane climatology.

13

IMPACT MODELS

"The skill of writing is to create a context in which other people can think."
—Edwin Schlossberg

In this chapter, we show some broader applications of our models and methods. We focus on impact models. Hurricanes are capable of generating large financial losses. We begin with a model that estimates extreme losses conditional on climate covariates. We then describe a method for quantifying the relative change in potential losses over the decades.

13.1 EXTREME LOSSES

Financial losses from hurricanes are to some extent directly related to fluctuations in climate. Environmental factors influence the frequency and intensity of hurricanes at the coast as detailed throughout this book (see for example Chapters 7 and 8). So, it is not surprising that these same environmental signals appear in estimates of losses.

Here loss is the economic damage associated with a hurricane's direct impact.[1] A normalization procedure adjusts the loss estimate from a past hurricane to what it would be if the same cyclone struck in a recent year by accounting for inflation and changes in wealth and population over the intervening time, plus a factor to account for changes in the number of housing units exceeding population growth. The method produces loss estimates that can be compared over time (Pielke et al. 2000).

[1] Direct impact losses do not include losses from business interruption or other macroeconomic effects including demand surge and mitigation efforts.

13.1.1 Exploratory Analysis

Here you focus on losses exceeding one billion ($ U.S.) that have been adjusted to 2005. The loss data are available in *Losses.txt* in JAGS format (see Chapter 9). Input the data by typing

```
> source("Losses.txt")
```

The log-transformed loss amounts are in the column labeled 'y'. The annual number of loss events are in the column labeled 'L'. The data cover the period 1900–2007. More details about these data are given in Jagger et al. (2011). You begin by plotting a time series of the number of losses and a histogram of total loss per event.

```
> layout(matrix(c(1, 2), 1, 2, byrow=TRUE),
+     widths=c(3/5, 2/5))
> plot(1900:2007, L, type="h", xlab="Year",
+     ylab="Number of Loss Events")
> grid()
> mtext("a", side=3, line=1, adj=0, cex=1.1)
> hist(y, xlab="Loss Amount ($ log)",
+     ylab="Frequency", main="")
> mtext("b", side=3, line=1, adj=0, cex=1.1)
```

Plots are shown in Figure 13.1. The annual number of loss events varies between 0 and 4. There appears to be slight increase in the number events over time. The frequency of extreme losses per event decreases nearly linearly on the log scale (see Murnane and Elsner, 2012). A time series of the amount of loss indicates no trend. The mean value is the expected loss while the standard deviation is the unexpected loss.

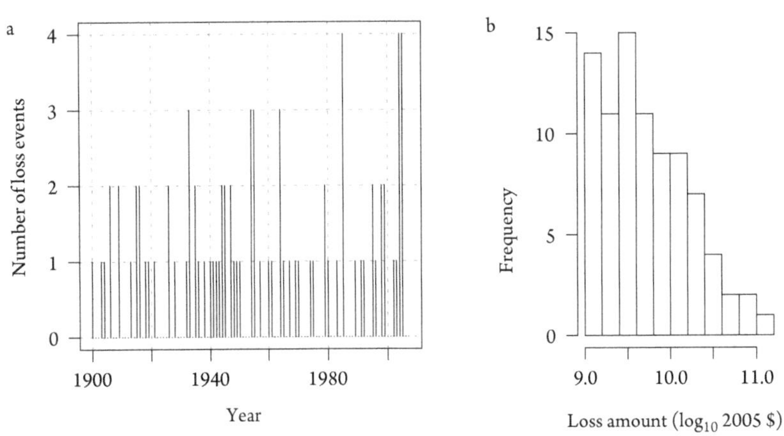

Figure 13.1 Loss events and amounts. (a) Number of events and (b) amount of loss.

Tail values are used to estimate the "value-at-risk" (VaR) in financial instruments used by the insurance industry.

13.1.2 Conditional Losses

You assume that a Poisson distribution adequately quantifies the number of large loss events[2] and that a generalized Pareto distribution (GPD) quantifies the amount of losses above the billion-dollar threshold (see Chapter 8). The frequency of loss events and the amount of losses given an event vary with your covariates including sea-surface temperature (SST), the Southern Oscillation Index (SOI), the North Atlantic Oscillation (NAO), and sunspot number (SSN) as described in Chapter 6.

The model is written in JAGS code available in *JAGSmodel3.txt*. More details about the model are available in Jagger et al. (2011). Posterior samples from the model (see Chapter 12) are available through a graphical user interface (GUI). Start the GUI by typing[3]

```
> source("LossGui.R")
```

Use the slider bars to vary the covariates. The covariates are scaled between ±2 s.d. Select OK to obtain the return-level graph. You can select the graph to appear as a bar or dot plot.

The GUI allows you to easily ask "what if" questions related to future damage losses. As one example, Figure 13.2 shows the posterior predictive distributions of return levels for individual loss events using two climate scenarios. For each return period, the black square shows the median and the red squares show the 0.05, 0.25,

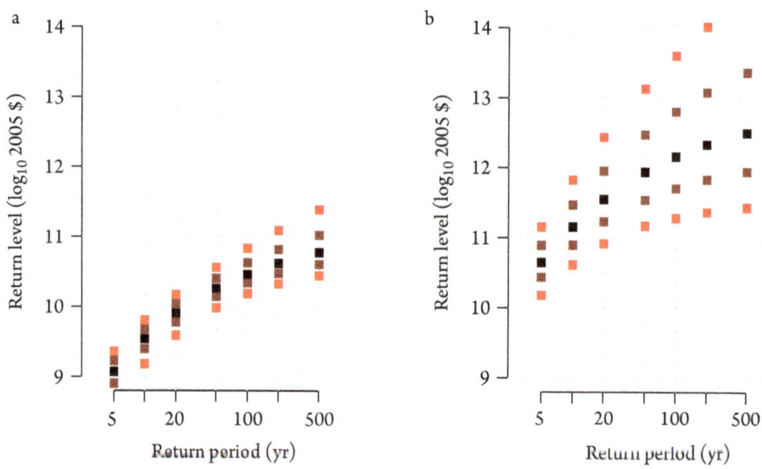

Figure 13.2 Loss return levels. (a) Weak and few versus (b) strong and more.

[2] There is no evidence of loss event clustering.

[3] If you are using a MAC, you will need to download and install the GTK+ framework. You will also probably need the package **gWidgetstcltk**.

0.75, and 0.9 quantile values from the posterior samples. The left panel shows the losses when SST = −0.243°C, NAO = +0.7 s.d., SOI = −1.1 s.d., and SSN = 115. The right panel shows the losses when SST = +0.268°C, NAO = −1.4 s.d., SOI = −1.1 s.d., and SSN = 9. The first scenario is characterized by covariates related to fewer and weaker hurricanes and the second scenario is characterized by covariates related to more and stronger hurricanes. The loss distributions are substantially different between the two scenarios. Under the first scenario, the median return level of a 50-year loss event is $18.2 bn; this compares with a median return level of $869.1 bn under the second scenario.

The results are interpreted as a return level for a return period of 50 years with the covariate values as extreme or more extreme than the ones given (about one s.d. in each case). With four independent covariates and an annual probability of about 16 percent that a covariate is more than one standard deviation from the mean, the chance that all covariates will be this extreme or more in a given year is less than 0.1 percent. The direction of change is consistent with a change in U.S. hurricane activity using the same scenarios and consistent with your understanding of hurricane climate variability.

The model is developed using aggregate loss data for the entire United States susceptible to North Atlantic hurricanes. It is possible to model data representing a subset of losses capturing a particular insurance portfolio. Moreover, since the model uses MCMC sampling, it can be extended to include error estimates on the losses. The model can also accommodate censored data where you know that losses exceeded a certain level, but you do not have information about the actual loss amount.

13.1.3 Industry Loss Models

Hazard risk affects the profits and losses of companies in the insurance industry. Some of this risk is transferred to the performance of securities traded in financial markets. This implies that early reliable information from loss models is useful to investors.

A loss model consists of three components: hurricane frequency and intensity, vulnerability of structures, and loss distributions. A loss model is built using a combination of numerical, empirical, and statistical approaches. The intensity and frequency components rely on "expanding" the historical hurricane data set. This is done by resampling observed hurricane attributes to generate thousands of synthetic cyclones.

The approach is useful for estimating a probable maximum loss (PML), which is the expected value of all losses greater than some high level. The level is chosen based on the quantile of the loss say $L(\tau)$, where $\tau = 1 - 1/\text{RP}$ and where RP is the return period of the event. The PML can be estimated on a local exposure portfolio or a portfolio spread across sections of the coast. The PML defines the amount of money that a reinsurer or primary insurer should have available to cover a given portfolio. In order to estimate the 1-in-100-year PML, a catalogue that contains at least an order of magnitude more synthetic cyclones than contained in the historical data set is required.

The model described in the previous section uses only the set of historical losses so, it allows you to anticipate futures losses on the seasonal and multiyear time scales without relying on a catalogue of synthetic cyclones. The limitation is that the losses must be aggregated over a region large enough to capture a sufficient number of loss events. The aggregation can be geographical, like Florida, or across a large and diverse set of exposures.

The model is further limited by the quality of historical loss data. Although the per hurricane damage losses are adjusted for increases in coastal population, the number of loss events is not. A cyclone making landfall early in the record in a region void of buildings did not generate losses, so there is nothing to adjust. This is not a problem if losses from historical hurricanes are estimated from a hazard model using constant building exposure data available.

13.2 FUTURE WIND DAMAGE

A critical variable in a loss model is hurricane intensity. Here we show you a way to adjust this variable for climate change. The methodology focuses on a statistical model for cyclone intensity trends, the output of which can be used as input to a hazard model. Details are found in Elsner et al. (2011).

13.2.1 Historical Catalogue

You begin by determining the historical hurricanes relevant to your location of interest. This is done with the `get.tracks` function in **getTracks.R** (see Chapter 6). Here your interest is a location on Eglin Air Force Base (EAFB) with a latitude 30.4°N and longitude −86.8°E. You choose a search radius of 100 nmi as a compromise between having enough cyclones to fit a model and having only those that are close enough to cause damage at the base. You also specify a minimum intensity of 33 m s^{-1}.

```
> load("best.use.RData")
> source("getTracks.R")
> lo = -86.8; la = 30.4; r1 = 100
> loc = data.frame(lon=lo, lat=la, R=r1)
> eafb = get.tracks(x=best.use, locations=loc,
+   umin=64, N=200)
```

Tracks meeting the criteria are given in the list object `eafb$tracks` with each component a data frame containing the attributes of individual cyclones from `best.use` (see Chapter 6).

Your interest is restricted further to a segment of each track near the coast. That is, you subset your tracks keeping only the records when the cyclone is near your location. This is done using the `selectTrackSubset` function and specifying a radius for clipping the tracks beyond 300 nmi. You also convert the translation speed (`maguv`) to m s^{-1}.

Figure 13.3 Tracks of hurricanes affecting EAFB.

```
> r2 = 300
> eafb.use = getTrackSubset(tracks=eafb$tracks,
+   lon=lo, lat=la, radius=r2)
> eafb.use$maguv = eafb.use$maguv * .5144
```

The output is a reduced data frame containing cyclone locations and attributes for track segments corresponding to your location. Finally, you remove cyclones that exist for less than 24 hours inside the domain.

```
> x = table(eafb.use$Sid)
> keep = as.integer(names(x[x >= 24]))
> eafb.use = subset(eafb.use, Sid %in% keep)
```

You plot the tracks on a map (Fig. 13.3) reusing your code from Chapter 6. The plot shows a uniform spread of cyclones approaching EAFB from the south with an equal number of hurricanes passing to the west as passing to the east.

Your catalogue of historical hurricanes affecting EAFB contains 47 hurricanes. You summarize various attributes of these hurricanes with plots and summary statistics. Your interest is on attributes as they approach land so you first subset on the land marker (`M`).

```
> sea = subset(eafb.use, !M)
```

A graph showing the distributions of translational speed and approach direction is plotted by typing

```
> par(mfrow=c(1, 2), mar=c(5, 4, 3, 2) + 1, las=1)
> hist(sea$maguv, main="",
+   xlab="Translational Speed (m/s)")
> require(oce)
> u = -sea$maguv * sin(sea$diruv * pi/180)
> v = -sea$maguv * cos(sea$diruv * pi/180)
```

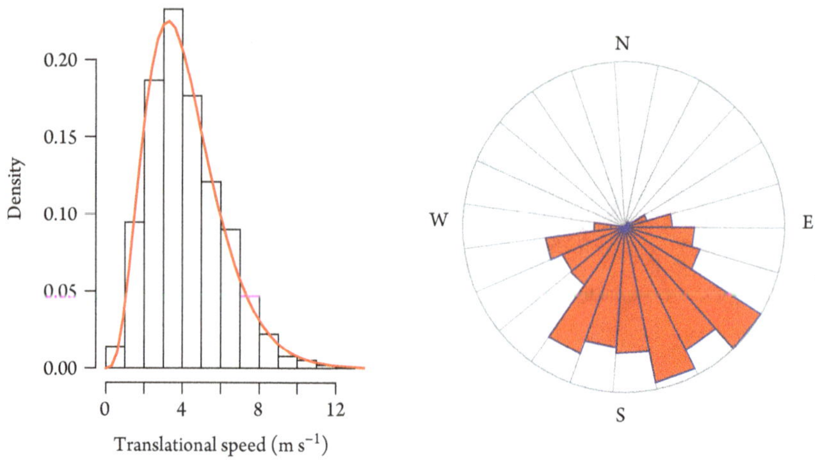

Figure 13.4 Translation speed and direction of hurricanes approaching EAFB.

```
> wr = as.windrose(u, v, dtheta=15)
> plot.windrose(wr, type="count", cex.lab=.5,
+   convention="meteorological")
```

The histograms are shown in Figure 13.4. The smoothed curve on the histogram is a gamma density fit using the `fitdistr` function (**MASS** package). The wind rose function is from the **oce** package (Kelley, 2011). The median forward speed of approaching cyclones is 3.9 m s^{-1} and the most common approach direction is from the southeast.

The correlation between forward speed and cyclone intensity is 0.3. For the subset of approaching cyclones that are intensifying, this relationship tends to get stronger with increasing intensity. The evidence supports the idea of an intensity limit for hurricanes moving too slow due to the feedback caused by the cold ocean wake. This is likely to be even more the case where the ocean mixed layer is shallow.

From a broader perspective, Figure 13.5 shows the relationship between forward speed and cyclone intensity for all hurricanes over the North Atlantic south of 30°N latitude moving slower than 12 m s^{-1}. The plot shows the lifetime maximum intensity as a function of average translation speed, where the averaging is done when cyclone intensity is within 10 m s^{-1} of its lifetime maximum.

The two-dimensional plane of the scatter plot is binned into rectangles with the number of points in each bin shown on a color scale. A local regression line (with a 95% confidence band) is added to the plot showing the conditional mean hurricane intensity as a function of forward speed. The line indicates that, on average, intensity increases with forward speed, especially for slower moving hurricanes (Mei et al., 2012). The relationship changes sign for cyclones moving faster than about 8 m s^{-1}. A linear quantile regression (not shown) indicates that the relationship is stronger for quantiles above the median although forward speed explains only a small portion (about 1 percent) of lifetime maximum intensity.

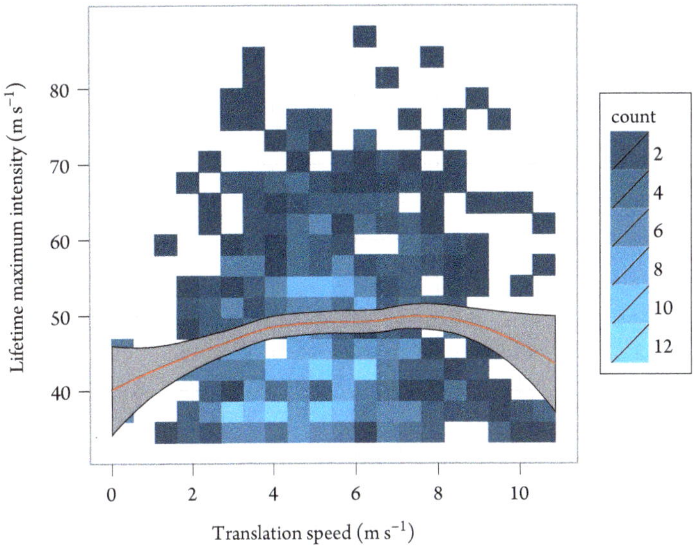

Figure 13.5 Lifetime maximum intensity and translation speed.

13.2.2 Gulf of Mexico Hurricanes and SST

Your historical catalogue of 47 hurricanes is too small to provide an estimate of changes over time. Instead, you examine the set of hurricanes over the entire Gulf of Mexico. Changes to hurricanes over this wider region will likely be relevant to changes at EAFB.

Subset the hurricanes within a latitude by longitude grid covering the region using the period 1900 through 2009, inclusive.

```
> llo = -98; rlo = -80
> bla = 19; tla = 32
> sy = 1900; ey = 2009
> gulf.use = subset(best.use, lon >= llo & lon <= rlo
+    & lat >= bla & lat <= tla & Yr >= sy & Yr <= ey)
```

Next, find the per cyclone maximum intensity for cyclones in the Gulf of Mexico.

```
> source("getmax.R")
> GMI.df = get.max(gulf.use, maxfield="WmaxS")
> GMI.df$WmaxS = GMI.df$WmaxS
```

Use the July SST over the Gulf of Mexico as your covariate for modeling the changing intensity of Gulf hurricanes. The gridded SST data are in *ncdataframe.RData*, where the column names are the year and month concatenated as a character string that includes Y and M. First, create a character vector of the column names.

```
> se = sy:ey
> cNam = paste("Y", formatC(se, 1, flag="0"), "M07",
+   sep="")
```

Then, load the data, create a spatial points data frame, and extract the July values for the North Atlantic.

```
> load("ncdataframe.RData")
> require(sp)
> coordinates(ncdataframe) = c("lon", "lat")
> sstJuly = ncdataframe[cNam]
```

Next, average the SST values in your Gulf of Mexico grid box. First, create a matrix from your vertex points of your grid box. Then, create a spatial polygons object from the matrix and compute the regional average using the `over` function. Next, make a data frame from the resulting vector and a vector of corresponding years. Finally, merge that data frame with your earlier wind data frame.

```
> bb = c(llo, bla, llo, tla, rlo, tla, rlo, bla,
+   llo, bla)
> Gulfbb = matrix(bb, ncol=2, byrow=TRUE)
> Gulf.sp = SpatialPolygons(list(Polygons(list(
+   Polygon(Gulfbb)), ID="Gulfbb")))
> SST = over(x=Gulf.sp, y=sstJuly, fn=mean)
> SST.df = data.frame(Yr=sy:ey, sst=t(SST)[, 1])
> GMI.df = merge(SST.df, GMI.df, by="Yr")
```

The data frame has 451 rows one for each hurricane in the Gulf of Mexico region corresponding to the fastest wind speed while in the domain. The spatial distribution favors locations just off the coast and along the eastern boundary of the domain.

13.2.3 Intensity Changes with SST

Theory, models, and data provide support for estimating changes to cyclone intensity over time. The heat-engine theory argues for an increase in the maximum potential intensity of hurricanes with increases in SST. Climate model projections indicate an increase in average intensity of about 5–10 percent globally by the late twenty-first century with the frequency of the most intense hurricanes likely increasing even more. Statistical models using a set of homogeneous tropical cyclone winds show the strongest hurricanes getting stronger with increases as high as 20 percent per degree celsius for the strongest hurricanes.

Your next step is to fit a model for hurricane intensity change relevant to your catalogue of cyclones affecting EAFB. The correlation between per hurricane maximum intensity and the corresponding July SST is a mere 0.04, but it increases to 0.37 for the set of hurricane above 64 m s^{-1}.

You use a quantile regression model (see Chapter 8) to account for the relationship between intensity and SST. Save the wind speed quantiles and run a quantile regression model of lifetime maximum wind speed (intensity) on SST. Save the trend coefficients and standard errors from the model for each quantile.

```
> tau = seq(.05, .95, .05)
> qW = quantile(GMI.df$WmaxS, probs = tau)
> n = length(qW)
> require(quantreg)
> model = rq(WmaxS ~ sst, data=GMI.df, tau=tau)
> trend = coefficients(model)[2, ]
> coef = summary(model, se="iid")
> ste = numeric()
> for (i in 1:n){
+   ste[i] = coef[[i]]$coefficients[2, 2]
+ }
```

Next use a local polynomial regression to model the trend as a change in intensity per degree celsius and plot the results. The regression fit at intensity w is made using points in the neighborhood of w weighted inversely by their distance to w. The neighborhood size is a constant of 75 percent of the points.

```
> trend = trend/qW * 100
> ste = ste/qW * 100
> model2 = loess(trend ~ qW)
> pp = predict(model2, se=TRUE)
> xx = c(qW, rev(qW))
> yy = c(pp$fit + 2 * pp$se.fit,
+   rev(c(pp$fit - 2 * pp$se.fit)))
> plot(qW, trend, pch=20, ylim=c(-20, 30),
+   xlab="Intensity (m/s)",
+   ylab="Percent Change (per C)")
> polygon(xx, yy, col="gray", border="gray")
> for(i in 1:n) segments(qW[i], trend[i] - ste[i],
+   qW[i], trend[i] + ste[i])
> points(qW, trend, pch=20)
> lines(qW, fitted(model2), col="red")
> abline(h=0, lty=2)
```

Results are shown in Figure 13.6. Points are quantile regression coefficients of per cyclone maximum intensity on SST and the vertical bars are the standard errors. The red line is a local polynomial fit through the points and the gray band is the 95 percent confidence band around the predicted fit. There is little change in intensity for the weaker cyclones but there is a large, and for some quantiles, statistically significant upward trend in intensity for the stronger hurricanes.

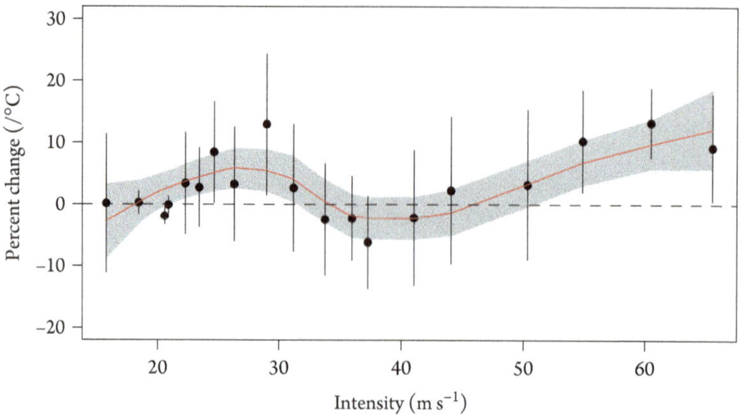

Figure 13.6 Intensity change as a function of SST for Gulf of Mexico hurricanes.

13.2.4 Stronger Hurricanes

Next you quantify the trend in SST over time using linear regression of SST on year.

```
> model3 = lm(sst ~ Yr, data=SST.df)
> model3$coef[2] * 100
   Yr
0.679
```

The upward trend is 0.68°C per century, explaining 36 percent of the variation in July SST over the period of record. The magnitude of warming you find in the Gulf of Mexico is consistent with estimates of between 0.4 and 1°C per century for warming of the tropical oceans (Deser et al., 2010).

Your estimate of the per degree celsius SST increase in hurricane intensities (as a function of quantile intensity) together with your estimate of the SST warming is subsequently used to estimate the increase in wind speeds for each hurricane in your catalogue. You assume that your catalogue is a representative sample of the frequency and intensity of future hurricanes, but that the strongest hurricanes will be stronger due to the additional expected warmth. The approach is similar to that used in Mousavi et al. (2011) to estimate the potential impact of hurricane intensification and sea-level rise on coastal flooding.

The equation for increasing wind speeds w of hurricanes in your catalogue of hurricanes affecting EAFB is given by

$$w_{2110} = [1 + \Delta w(q) \cdot \Delta\text{SST}] \cdot w \tag{13.1}$$

where w_{2110} is the wind speed 100 years from 2010, $\Delta w(q)$ is the fractional change in wind speed per degree change in SST as a function of the quantile speed (red curve), and ΔSST is the per century trend in SST. You certainly do not expect an extrapolation (linear or otherwise) to accurately represent the future, but the method provides an estimate of what Gulf of Mexico hurricanes might be like, on average, during the twenty-second century.

Additional cyclone vitals including radius to maximum winds, a wind decay parameter, and minimum central pressure need to be added to the catalogue of cyclones to make them useful for storm surge and wind-field model (Vickery et al., 2006), like those included in the U.S. Federal Emergency Management Agency (FEMA) HAZUS model. Lacking evidence indicating these vitals will change in the future you would use historical values. Otherwise, you would adopt a method similar to what you used earlier for cyclone intensity. In the end, you have two cyclone catalogues, one representing the contemporary view and the other representing a future view, a view that is consistent with the current evidence and theory of hurricanes intensity and which aligns with the consensus opinion on anthropogenic global warming.

This chapter demonstrated a few ways in which the models and methods described in the earlier chapters are used to answer questions about possible future impacts. In particular, we showed you how data on past insured losses can be used to hedge against future losses conditional on the state of the climate. We also showed you how to model potential changes to local hurricane impacts caused by a changing cyclone climatology.

This concludes your study of hurricane climatology using modern statistical methods. We trust your skills will help you make important new discoveries in the fascinating field of hurricane climate. Good luck.

APPENDIX A

R Functions

Here we list the R functions used in this book. Some of the functions are used "behind the scenes" and are available on the book's Web site.

Table A.1 R functions used in this book.

Function {Package}	*Application*	*Chapters*
`abline` {**graphics**}	add straight line to plot	2 4 5 8–10 12 13
`abs` {**base**}	absolute value	3 12
`acf` {**stats**}	autocorrelation function	12
`add.grid` {**datasupport.R**}	add coastal regions to best.use	6
`all` {**base**}	are all values true?	8
`apply` {**base**}	apply functions over array margins	6
`apropos` {**utils**}	find objects by name	2
`ar` {**stats**}	fit autoregressive (AR) models	10
`arima` {**stats**}	fit ARIMA models	10
`arrows` {**graphics**}	add arrows to plot	5 6
`as` {**methods**}	force object to belong to class	5 9
`as.array` {**base**}	coerce to array	4
`as.character` {**base**}	convert to character vector	5
`as.data.frame` {**base**}	convert to data frame	6
`as.Date` {**base**}	date conversion to/from character	4 5 10
`as.integer` {**base**}	convert to integer	7 9
`as.matrix` {**base**}	convert to matrix	5 10 12
`as.numeric` {**base**}	convert to numeric	4 6 8 10–12
`as.POSIXct` {**base**}	date/time conversion functions	5
`as.ppp` {**spatstat**}	convert to spatial point pattern	11
`as.psp` {**spatstat**}	convert to spatial lines	11
`as.sociomatrix` {**network**}	coerce network to socio matrix	10
`as.vector` {**base**}	convert to vector	5 6 10
`as.windrose` {**oce**}	create wind rose object	13
`attach` {**base**}	attach object to search path	2 10
`attr` {**base**}	object attributes	5 11
`axis` {**graphics**}	add axis to plot	4–8 10 12

Function {Package}	*Application*	*Chapters*
`barplot` {**graphics**}	create bar plot	2 4 7 12
`bbox` {**sp**}	retrieve bounding box	9 12
`beta.select` {**LearnBayes**}	find beta prior given quantiles	4
`betweenness` {**sna**}	centrality scores of graph nodes	10
`bic.glm` {**BMA**}	Bayesian model averaging for GLM	12
`bic.poisson` {**prediction.R**}	Poisson prediction from BMA	12
`blue2red` {**colorRamps**}	create gradient color ramp	9 12
`blue2yellow` {**colorRamps**}	create gradient color ramp	9
`boot` {**boot**}	bootstrap resampling	3
`boot.ci` {**boot**}	bootstrap confidence interval	3
`bootstrap` {**bootstrap**}	bootstrap resampling	12
`box` {**graphics**}	draw box around plot	5 6 11 13
`boxplot` {**graphics**}	create box and whisker plot	5 8 9
`brewer.pal` {**RColorBrewer**}	create color palette	11
`c` {**base**}	create vector or list	2–13
`cat` {**base**}	concatenate and print	2
`cbind` {**base**}	combine objects by columns	4–6 11 12
`ceiling` {**base**}	smallest integer not less than x	5
`cens` {**gamlss.cens**}	fit GAM using censored data	8
`choose` {**base**}	n choose k	3
`chooseCRANmirror` {**utils**}	choose CRAN mirror site	2
`citation` {**utils**}	create citation for package	8 11
`class` {**base**}	find object's class	2 5 6
`coda.samples` {**rjags**}	samples in MCMC list format	4 12
`coefficients` {**stats**}	extract model coefficients	11 13
`colMeans` {**base**}	column means	11
`colnames` {**base**}	column names	6 8–12
`colorRampPalette` {**grDevices**}	color interpolation	5 9 12
`colourmap` {**spatstat**}	color lookup tables	11
`confint` {**stats**}	confidence ints model parameters	7
`contour` {**graphics**}	plot contours	5
`coordinates` {**sp**}	set/get spatial coordinates	5 9 11–13
`coplot` {**graphics**}	create conditioning plot	5
`cor` {**stats**}	correlation coefficient	3 7
`cor.test` {**stats**}	correlation test	3
`cos` {**base**}	cosine function	13
CRS {**sp**}	class coordinate reference system	5 9
`cumsum` {**base**}	cumulative sums	4 5 10
`curve` {**graphics**}	draw function curve	3 4 8 12 13
`cut` {**base**}	divide range into intervals	8
`daisy` {**cluster**}	dissimilarity matrix	11

Table A.1 (cont.)

Function {Package}	*Application*	*Chapters*
`data.frame` {**base**}	create data frames	3 6–13
`dbeta` {**stats**}	density beta distribution	4
`dbinom` {**stats**}	density binomial distribution	3
`dchisq` {**stats**}	density chi-squared distribution	11
`degree` {**sna**}	degree centrality of node	10
`density` {**stats**}	kernel density estimation	4 5 8 11
`densityplot` {**lattice**}	kernel density estimation lattice	4 5
`dev.off` {**grDevices**}	turn off current graphic device	2
`dgamma` {**stats**}	density gamma distribution	12 13
`diameter` {**igraph**}	diameter of graph	10
`diff` {**base**}	lagged differences	3 10
`difftime` {**base**}	time intervals	4–6
`dim` {**base**}	dimensions of object	2 6 7
`dimnames` {**base**}	dimension names	7 12
`discint` {**LearnBayes**}	highest prob int discrete distr	4
`dist` {**stats**}	distance matrix	11
`dnbinom` {**stats**}	density negative binomial dist	12
`dnorm` {**stats**}	density normal distribution	3
`do.call` {**base**}	execute function call	8 10
`dpois` {**stats**}	density Poisson distribution	3 7
`drop1` {**stats**}	drop model term	3
`dweibull` {**stats**}	density Weibull distribution	3
`ecdf` {**stats**}	empirical cumulative distribution	8
`effectiveSize` {**coda**}	effective sample size for mean	12
`ellipse` {**ellipse**}	outline confidence region	3
`envelope` {**spatstat**}	simulation envelope of summary	11
`equal.count` {**lattice**}	create plot shingles	5
`example` {**utils**}	run example from online help	2
`exp` {**base**}	exponential function	4 7–10 12
`expression` {**base**}	unevaluated expressions	3–13
`factor` {**base**}	encode vector as factor	2 10
`factorial` {**base**}	factorial function	3
`filter` {**stats**}	linear filter on time series	10
`findColours` {**classInt**}	assign colors from classInt object	5
`fit.variogram` {**gstat**}	fit model to sample variogram	9
`fitdistr` {**MASS**}	fit univariate distributions	13
`fitted` {**stats**}	extracts fitted values	3 10
`fivenum` {**stats**}	five-number summary	5 7 8 12
formatC {**base**}	c-style formats	6
`gamlss` {**gamlss**}	fit generalized additive models	10
`gelman.plot` {**coda**}	shrink factor	12

Function {Package}	*Application*	*Chapters*
`geocode` {**ggmap**}	geocode map location	5
`get.max` {**get.max.R**}	get cyclone maxima	6 12 13
`get.tracks` {**getTracks.R**}	get cyclone tracks	6
`get.var.ncdf` {**ncdf**}	read from netCDF file	6
`get.visibility` {**get.visibility.R**}	create visibility graph	10
`getwd` {**base**}	retrieve working directory	2
`get_map` {**ggmap**}	get map from map server	5
`ggmap` {**ggmap**}	plot ggmaps	5
`ggplot` {**ggplot2**}	create grammar of graphics plot	6 13
`glm` {**stats**}	fit generalized linear models	7 8 10 12
`gpd.fit` {**ismev**}	fit generalized Pareto distr	8
`graph.edgelist` {**igraph**}	methods for creating graphs	10
`gray` {**grDevices**}	gray-level specification	6
`grid` {**graphics**}	add grid to plot	2 4 7 9–13
`grid.circle` {**grid**}	draw circle	4
`grid.curve` {**grid**}	draw curve	4
`grid.newpage` {**grid**}	move to new page on grid device	5
`grid.text` {**grid**}	add text	4
`gridat` {**sp**}	N–S and E–W grid locations	5
`gridlines` {**sp**}	add N–S and E–W grid lines	5
`gsub` {**base**}	pattern matching and replacement	10
`gwr` {**spgwr**}	geographically weighted regression	9
`gwr.sel` {**spgwr**}	select bandwidth for gwr	9
`head` {**utils**}	return first part of object	2 5–10 12
`help` {**utils**}	open help page	2
`HexPoints2SpatialPolygons` {**sp**}	make polygons from grid object	9 12
`hist` {**graphics**}	compute/plot histogram	4–6 9 12 13
`histogram` {**lattice**}	histogram using lattice	5
`ifelse` {**base**}	conditional element selection	9
`image` {**graphics**}	display color image	5
`image.plot` {**fields**}	image plot with legend	12
`imageplot.bma` {**BMA**}	image plot of models in BMA	12
`imageplot.bma2` {**imageplot.bma2.R**}	modification of imageplot.bma	12
`imageplot.bma3` {**imageplot.bma3.R**}	modification of imageplot.bma	12
`import.grid` {**datasupport.R**}	import grid boundaries	6
`install.packages` {**utils**}	install packages from repository	2
`is.na` {**base**}	which elements are missing?	6
`ISOdate` {**base**}	date/time conversion	4 6
`ISOdatetime` {**base**}	date/time conversion	6
`jags.model` {**rjags**}	create JAGS model object	4 12
`Kest` {**spatstat**}	second-moment spatial function	11

Table A.1 (cont.)

Function {*Package*}	*Application*	*Chapters*
`Kinhom` {**spatstat**}	inhomogeneous K function	11
`kmeans` {**stats**}	perform k-means clustering	11
`krige` {**gstats**}	perform kriging	9
`lag.listw` {**spdep**}	compute spatial lag	9 12
`lag.plot` {**stats**}	time series lag plots	5
`lapply` {**base**}	apply function over list	6 10 12
`layout` {**graphics**}	plot arrangement	4 7 13
`leap_year` {**lubridate**}	is leap year?	5
`legend` {**graphics**}	add legend to plot	3–6 8 10 12
`length` {**base**}	length of object	2–13
`library` {**base**}	load package to workspace	2
`lines` {**graphics**}	add connected lines to plot	3–13
`list` {**base**}	construct and check for lists	4 7 9 11 12
`lm` {**stats**}	fit linear models	3 5 7 9 13
`load` {**base**}	reload saved data sets	4 6 7–13
`loess` {**stats**}	fit local polynomial regressions	13
`loess.smooth` {**stats**}	scatter plot with loess smooth	9
`log` {**base**}	logarithmic function	2
`map` {**maps**}	draw maps	5 6 9 11–13
`map2SpatialLines` {**maptools**}	convert map object to spatial line	5 9 11 12
`marks` {**spatstat**}	get/set marks of point pattern	11
`matrix` {**base**}	create a matrix	5 7 9 13
`max` {**base**}	sample maximum	2 3 8 11
`mean` {**base**}	sample mean	2 3 6–9 11
`median` {**stats**}	sample median	3 9 10 12
`melt` {**reshape**}	reshape object for easy casting	10
`merge` {**base**}	merge data frames	8 13
`min` {**base**}	sample minimum	3 8
`minimum.spanning.tree` {**igraph**}	minimum spanning tree of graph	10
`moran` {**spdep**}	compute moran's I	9 12
`moran.test` {**spdep**}	test spatial autocorrelation	9
`mrl.plot` {**ismev**}	mean residual life plot	8
`mrl.plot2` {**mrl.plot2.R**}	revised mean residual life plot	8
`mtext` {**graphics**}	add margin text	4–8 10–13
`mvrnorm` {**MASS**}	samples from multivariate normal	3
`mycontour` {LearnBayes}	contour bivariate density	4
`names` {**base**}	get/set object names	2 5 6 9 11 12
`nb2listw` {**spdep**}	spatial weights for neighbor list	9 12
`nb2WB` {**spdep**}	output spatial weights for WinBUGS	12
`ncol` {**base**}	number of columns in array	12
`network` {**network**}	make/coerce to network object	10

Function {Package}	*Application*	*Chapters*
`now` {**lubridate**}	get current date	5
`nrow` {**base**}	number of rows in array	9 11
`numeric` {**base**}	create numeric vector	3 4 7 8 10 12 13
`object.size` {**utils**}	space allocation for object	6
`objects` {**base**}	list objects in working directory	2
`open.ncdf` {**ncdf**}	open netCDF file	6
`optim` {**stats**}	general purpose optimizer	12
`options` {**base**}	set/get options	2
`order` {**base**}	return permutation in ascending	10 11
`outer` {**base**}	outer product of arrays	7 12
`over` {**sp**}	overlay points grids and polygons	9 12 13
`pairs` {**graphics**}	matrix of scatter plots	3
`par` {**graphics**}	set graph parameters	2–13
`parse` {**base**}	parse expressions	5
`paste` {**base**}	concatenate character strings	3–13
`pbeta` {**stats**}	beta distribution function	4
`pbetap` {**LearnBayes**}	predict dist binom w/ beta prior	4
`pchisq` {**stats**}	chi-squared distribution function	7
`pgamma` {**stats**}	gamma distribution function	12
`pixellate` {**spatstat**}	convert object to pixel image	11
`plot` {**graphics**}	generic x–y plotting	3–13
`plot.im` {**spatstat**}	plot pixel image	11
`plot.rq.process` {**quantreg**}	plot quantile regression process	8
`plot.windrose2` {**plot.windrose2.R**}	modified plot wind rose diagram	13
`plotfits` {**correlationfuns.R**}	annual count vs cluster rate	11
`plotmo` {**plotmo**}	plot model response	7
`pnbinom` {**stats**}	negative binomial distribution	12
`pnorm` {**stats**}	normal distribution function	3
`points` {**graphics**}	add points to plot	4–8 10–13
`poly2nb` {**spdep**}	neighborhood from polygon list	9 12
`polygon` {**graphics**}	draw polygons with given vertices	4 5 7 8 10 11 13
`Polygon` {**sp**}	create spatial polygons object	9 13
`Polygons` {**sp**}	polygons object	9
`ppois` {**stats**}	Poisson distribution function	3 7
`predict` {**stats**}	generic function for predictions	3 5 7 12 13
`print` {**base**}	print objects	3 5–7 9 10 12
`prod` {**base**}	product of values in object	3
`proj4string` {**sp**}	projection attributes for sp data	9 11 12
`projInfo` {**rgdal**}	list proj.4 tag information	5
`prop.test` {**stats**}	test of equal or given proportions	4
`pt` {**stats**}	Student's t-distribution function	3

Table A.1 (cont.)

Function {*Package*}	*Application*	*Chapters*
`pushViewport` {**grid**}	navigate grid viewport tree	5
`q` {**base**}	terminate session	2
`qbeta` {**stats**}	quantile beta distribution	4
`qnorm` {**stats**}	quantile normal distribution	3 8
`qplot` {**ggplot2**}	quick plot wrapper for ggplot	5 10
`qqline` {**stats**}	plot line on qqplot/qqnorm	5
`qqnorm` {**stats**}	quantile quantile plot	5
`qr` {**base**}	qr decomposition of a matrix	12
`qr.Q` {**base**}	recover Q matrix from qr object	12
`qr.R` {**base**}	recover R matrix from qr object	12
`quadratcount` {**spatstat**}	quadrat counts for point pattern	11
`quantile` {**stats**}	sample quantiles	2–4 8 10 12 13
`randomForest` {**randomForest**}	random forest algorithm	7
`range` {**base**}	range of values	5
`rank` {**base**}	sample ranks	3 8
`rbeta` {**stats**}	random numbers beta distribution	4
`rbind` {**base**}	combine objects by rows	11
`rbinom` {**stats**}	random numbers binomial	4 7
`read.bugs` {**R2WinBUGS**}	read output files in coda format	4
`read.csv` {**utils**}	read comma-separated values file	4 6 7 11
`read.table` {**utils**}	read space-separated values file	2–12
`readCov` {**datasupport.R**}	read environmental covariates	6
`readShapeSpatial` {**maptools**}	read shape files	5
`regionTable` {**correlationfuns.R**}	table counts by region	11
`rep` {**base**}	replicate elements	2 6 8
`require` {**base**}	load package to workspace	2–13
`resid` {**stats**}	extract model residuals	3
`rev` {**base**}	reverse elements	2 4 6–9 11–13
`rgamma` {**stats**}	random numbers gamma distribution	4
`rle` {**base**}	run length encoding	3
`rlWeibPois` {**CountyWinds.R**}	return-level Weibull Poisson model	8
`rm` {**utils**}	remove objects from workspace	2 6 8
`rMatClust` {**spatstat**}	simulate Matern cluster process	11
`rMaternI` {**spatstat**}	simulate Matern inhibition process	11
`rnorm` {**stats**}	random numbers normal distribution	3
`roc` {**roc.R**}	plot roc curve	7
`round` {**base**}	round the number	3–10 12
`rowMeans` {**base**}	row averages	11
`rownames` {**base**}	get/set row names in data frame	6 7 12
`rpois` {**stats**}	random count Poisson distribution	3 7 11
`rq` {**quantreg**}	fit quantile regression models	8 13

Function {*Package*}	*Application*	*Chapters*
`rug` {**graphics**}	add rug to plot	4–6 8 10
`runif` {**stats**}	random number uniform distribution	4 11
`rweibull` {**stats**}	random number Weibull distribution	3
`sample` {**base**}	random samples and permutations	3 7 8 12
`sampleParameters` {**CountWinds.R**}	sample return levels	8
`sapply` {**base**}	wrapper for lapply	8 10–12
`save` {**base**}	save objects	6
`savgol.best` {**savgol.R**}	filter best-track data	6
`scale` {**base**}	scale and center object	11
`scatterhist` {**scatterplothist.R**}	scatter plot with histogram	5
`sd` {**stats**}	sample standard deviation	2 3
`seq` {**base**}	sequence generator	2–5 7–9 12 13
`set.seed` {**base**}	set seed value for random numbers	3 7
`signif` {**base**}	round the number	3
`simcontour` {**LearnBayes**}	random draws bivariate density	4
`sin` {**base**}	sine function	2 13
`slot` {**methods**}	list slots in object	9 12
`sort` {**base**}	sort elements of a vector	5 8 10
`source` {**base**}	input code from file	4–8 10 11 13
`SpatialPointsDataFrame` {**sp**}	create spatial points data frame	5
`SpatialPolygons` {**sp**}	create spatial polygons	9
`SpatialPolygonsDataFrame` {**sp**}	create spatial polygons data frame	9 12
`split` {**base**}	divide into groups	11 12 13
`spplot` {**sp**}	plot method for spatial data	5 9 12
`spsample` {**sp**}	sample locations in spatial object	9
`spTransform` {**rgdal**}	map projections and transforms	5 9 11
`sqrt` {**base**}	square root function	2 3 7 8 12
`step` {**stats**}	choose model stepwise	3
`stl` {**stats**}	seasonal decomposition of series	10
`str` {**utils**}	display object structure	5 6 9 10
`strptime` {**base**}	date/time conversion	5
`strsplit` {**base**}	split elements of character vector	6 12
`subset` {**base**}	subset data objects	3 6 10 11 13
`substring` {**base**}	substring of character vector	12
`sum` {**base**}	sum of vector elements	2–4 6 7 9–12
`summary` {**base**}	summarize objects	2–5 7–9 11–13
`switch` {**base**}	select one from list of choices	12
`Sys.time` {**base**}	get current date and time	6
`Szero` {**spdep**}	give constant for spatial weights	9 12
`t` {**base**}	matrix transpose	11 12
`t.test` {**stats**}	perform Student's t-test	3

Table A.1 (cont.)

Function {Package}	*Application*	*Chapters*
`table` {**base**}	cross tabulations	2–4 7 10 13
`tail` {**utils**}	return last part of object	3 5
`terrain.colors` {**grDevices**}	color palette	11
`testfits` {**correlationsfuns.R**}	test for time clustering	11
`text` {**graphics**}	add text to plot	4 5 9 11
`time` {**stats**}	create vector of times	5
`title` {**graphics**}	add title to plot	12
`toBibtex` {**util**}	convert to bibtex/latex	8
`topo.colors` {**grDevices**}	color palette	9
`traceplot` {**code**}	successive iterations of mcmc	12
`transitivity` {**igraph**}	prob adjacent vertices connected	10
`tree` {**tree**}	regression/classification tree	7
`trellis.par.get` {**lattice**}	get parameters of trellis display	12
`trellis.par.set` {**lattice**}	set parameters of trellis display	12
`try` {**base**}	try expression	2
`ts` {**stats**}	create time series object	5 10
`unique` {**base**}	remove duplicate elements	12
`unlist` {**base**}	flatten lists	6 11
`unmark` {**spatstat**}	remove marks from spatial points	11
`update` {**stats**}	refit model	4 5 9 12
`var` {**stats**}	sample variance	2 3 7
`var.test` {**stats**}	test comparing two variances	3
`variogram` {**gstat**}	sample variogram	9
`viewport` {**grid**}	create grid viewport	5
`wday` {**lubridate**}	get day of the week	5
`week` {**lubridate**}	get week of the year	5
`which` {**base**}	which indices are true?	2 10 12
`which.max` {**base**}	where is the maximum?	3
`wilcox.test` {**stat**}	rank/sign test difference in mean	3
`with` {**base**}	evaluate expression in data enviro	5 6
`with_tz` {**lubridate**}	get date–time in diff time zone	5
`write.table` {**utils**}	output data frame	6
`writeDataFileR` {**writedatafileR.R**}	output WinBUGS data	12
`xtable` {**xtable**}	create export table	3 7 8 10–12
`year` {**lubridate**}	get/set year of date–time object	5
`ymd` {**lubridate**}	parse dates to specified formats	6
`zeroinfl` {**gamlss**}	fit zero-inflated Poisson model	7

APPENDIX B

R Packages

Here we list the packages by chapter with the version number in parentheses that were used to create the book. Before you begin using the code, make sure the listed packages are installed on your computer. Many of the packages depend on other packages that are not listed here. The dependent packages are usually installed automatically.

- Chapter 1: None
- Chapter 2: UsingR(0.1-17)
- Chapter 3: ellipse(0.3-5), xtable(1.6-0)
- Chapter 4: rjags(3-5), R2WinBUGS(2.1-18), lattice(0.20-0), LearnBayes(2.12)
- Chapter 5: lubridate(0.2.6), maps(2.2-5), classInt(0.1-17), maptools(0.8-14), rgdal(0.7-8), mapdata(2.2-1), ggplot2(0.9.0), ggmap(1.2)
- Chapter 6: lubridate(0.2.6), maps(2.2-5), survival(2.36-10), ncdf(1.6.6)
- Chapter 7: pscl(1.04.1), xtable(1.6-0), plotmo(1.3-1), earth(3.2-1)
- Chapter 8: quantreg(4.76), xtable(1.6-0), ismev(1.37), gamlss(4.1-1), gamlss.cens(4.0.4)
- Chapter 9: gstat(1.0-10), maptools(0.8-14), sp(0.9-94), rgdal(0.7-8), maps(2.2-5), colorRamps(2.3), spdep(0.5-43), spgwr(0.6-13)
- Chapter 10: gamlss(4.1-1), xtable(1.6-0), chron(2.3-42), ggplot2(0.8.9), network(1.7), sna(2.2-0), igraph(0.5.5-4)
- Chapter 11: maps(2.2-5), spatstat(1.25-3), sp(0.9-94), rgdal(0.7-8), maptools(0.8-14), cluster(1.14.1), xtable(1.6-0)
- Chapter 12: rjags(3-5), fields(6.6.3), BMA(3.15), sp(0.9-94), rgdal(0.7-8), maps(2.2-5), maptools(0.8-14), colorRamps(2.3), spdep(0.5-43), coda(0.14-6), bootstrap(1.0-22), xtable(1.6-0)
- Chapter 13: maps(2.2-5), ggplot2(0.9.0), oce(0.8-4), sp(0.9-94), quantreg(4.76)

APPENDIX C

Data Sets

Here we list the data sets used by chapters. Most of the data sets are explained in Chapter 6. The data are available by chapter on the book's Web site.

- Chapter 1: None
- Chapter 2: *US.txt, NAO.txt*
- Chapter 3: *H.txt, NAO.txt, SST.txt*
- Chapter 4: *ATL.txt, H.txt, hurart.txt, modelupdates.RData*
- Chapter 5: *SOI.txt, NAO.txt, SST.txt, ATL.txt, H.txt, Ivan.txt, LMI.txt, FLPop.txt, JulySST2005.txt*
- Chapter 6: *BTflat.csv, gridboxes.txt, HS.csv, landGrid.RData, best.RData, best.use.RData, sstna.nc*
- Chapter 7: *US.txt, annual.RData, bi.csv*
- Chapter 8: *LMI.txt, SST.txt, SOI.txt, catwinds.RData, catcounts.RData*
- Chapter 9: *best.use.RData, JulySST2005.txt, FayRain.txt*
- Chapter 10: *annual.RData, best.use.RData, SST.txt*
- Chapter 11: *gridboxes.txt, bi.csv, annual.RData, best.use.RData*
- Chapter 12: *US.txt, annual.RData, best.use.RData, ncdataframe.RData, Chain1.txt, Chain2.txt, Index.txt, SSTChain.txt, SSTIndex.txt, SOIChain.txt, SOIIndex.txt*
- Chapter 13: *Losses.txt, best.use.RData, ncdataframe.RData*

BIBLIOGRAPHY

Albert J (2009) Bayesian Computation with R, 2nd edn. Springer, ISBN 978-0-387-71384-7

Albert J (2011) LearnBayes: Functions for Learning Bayesian Inference. URL `http://CRAN.R-project.org/package=LearnBayes`, r package version 2.12

Anselin L, Syabri I, Kho Y (2006) Geoda: An introduction to spatial data analysis. Geographical Analysis 38(1):5–22

Baddeley A, Turner R (2005) Spatstat: An R package for analyzing spatial point patterns. Journal of Statistical Software 12(6):1–42

Baddeley AJ, Moller J, Waagepetersen R (2000) Non- and semi-parametric estimation of interaction in inhomogeneous point patterns. Statistica Neerlandica 54(3):329–350

Barrodale I, Roberts FDK (1974) Solution of an overdetermined system of equations in the l1 norm [f4] (algorithm 478). Communications of the ACM 17(6):319–320

Besag J, York J, Mollie A (1991) Bayesian image restoration, with two applications in spatial statistics. Annals of the Institute of Statistical Mathematics 43(1):1–59, DOI 10.1007/BF00116466

de Beurs KM, Henebry GM (2008) Northern annular mode effects on the land surface phenologies of northern Eurasia. Journal of Climate 21(17):4257–4279, DOI 10.1175/2008JCLI2074.1

Bivand R, Pebesma E, Gomez-Rubio V (2008) Applied Spatial Data Analysis with R. Springer, New York, NY

Bivand R, contributions by Micah Altman, Anselin L, Assunção R, Berke O, Bernat A, Blanchet G, Blankmeyer E, Carvalho M, Christensen B, Chun Y, Dormann C, Dray S, Halbersma R, Krainski E, Legendre P, Lewin-Koh N, Li H, Ma J, Millo G, Mueller W, Ono H, Peres-Neto P, Piras G, Reder M, Tiefelsdorf M, Yu D (2011a) spdep: Spatial Dependence: Weighting Schemes, Statistics and Models. URL `http://CRAN.R-project.org/package=spdep`, r package version 0.5-40

Bivand R, Yu D, contributions by Tomoki Nakaya, tricube function based on a contribution by Miquel-Angel Garcia-Lopez (2011b) spgwr: Geographically Weighted Regression. URL `http://CRAN.R-project.org/package=spgwr`, r package version 0.6-10

Breiman L (2001) Random forests. Machine Learning 45:5–32

Brewer CA, Hatchard GW, Harrower MA (2003) ColorBrewer in print: A catalog of color schemes for maps. Cartography and Geographic Information Science 30(1):5–32

Brunsdon C, Fotheringham S, Charlton M (1998) Geographically weighted regression - modelling spatial non-stationarity. Journal of the Royal Statistical Society Series D-The Statistician 47(Part 3):431–443

Butts CT (2010) sna: Tools for Social Network Analysis. URL `http://CRAN.R-project.org/package=sna`, r package version 2.2-0

Butts CT, Handcock MS, Hunter DR (2011) network: Classes for Relational Data. Irvine, CA, URL `http://statnet.org/`, r package version 1.7

Chambers JM, Cleveland WS, Kleiner B, Tukey PA (1983) Graphical Methods for Data Analysis. Chapman and Hall, New York, NY

Cleveland WS (1985) The Elements of Graphing Data. Wadsworth, Monterey, CA

Coles S (2001) An Introduction to Statistical Modeling of Extreme Values. Springer–Verlag, London, UK

Coles S, Stephenson A (2011) ismev: An Introduction to Statistical Modeling of Extreme Values. URL `http://CRAN.R-project.org/package=ismev`, r package version 1.36

Cressie N (1993) Statistics for Spatial Data. John Wiley and Sons, New York, NY

Cressie N, Wikle C (2011) Statistics for Spatio-Temporal Data. Wiley Series in Probability and Statistics, John Wiley & Sons, New York, NY

Csardi G, Nepusz T (2006) The igraph software package for complex network research. InterJournal Complex Systems:1695, URL `http://igraph.sf.net`

Dalgaard P (2002) Introductory Statistics with R. Springer, New York, URL `http://www.biostat.ku.dk/~pd/ISwR.html`, ISBN 0-387-95475-9

Deser C, Phillips AS, Alexander MA (2010) Twentieth century tropical sea surface temperature trends revisited. Geophysical Research Letters 37:L10,701, DOI 10.1029/2010GL043321

Diggle P (2003) Statistical Analysis of Spatial Point Patterns. A Hodder Arnold Publication, Arnold, URL `http://books.google.com/books?id=fnFhQgAACAAJ`

Efron B, Tibshirani R (1986) Bootstrap methods for standard errors, confidence intervals, and other measures of statistical accuracy. Statistical Science 1(1):54–75

Eilers PHC, Marx BD (1996) Flexible smoothing with b-splines and penalties. Statistical Sciences 11(2):89–121, DOI 10.1214/ss/1038425655

Elsner JB, Bossak BH (2001) Bayesian analysis of U.S. hurricane climate. Journal of Climate 14(23):4341–4350

Elsner JB, Jagger TH, Hodges RE (2010) Daily tropical cyclone intensity response to solar ultraviolet radiation. Geophysical Research Letters 37:L09,701, DOI 10.1029/2010GL043091

Elsner JB, Kocher B (2000) Global tropical cyclone activity: Link to the North Atlantic Oscillation. Geophysical Research Letters 27(1):129–132

Elsner JB, Kara AB (1999) Hurricanes of the North Atlantic: Climate and Society. Oxford University Press, NY.

Elsner JB, Jagger TH, Dickinson M, Rowe D (2008a) Improving multiseason forecasts of North Atlantic hurricane activity. Journal of Climate 21(6):1209–1219, DOI 10.1175/2007JCLI1731.1

Elsner JB, Kossin JP, Jagger TH (2008b) The increasing intensity of the strongest tropical cyclones. Nature 455(7209):92–95, DOI 10.1038/nature07234

Elsner JB, Jagger TH, Fogarty EA (2009) Visibility network of United States hurricanes. Geophysical Research Letters 36:L16,702, DOI 10.1029/2009GL039129

Elsner JB, Jagger TH (2006) Prediction models for annual U.S. hurricane counts. Journal of Climate 19(12):2935–2952

Elsner JB, Jagger TH (2008) United States and Caribbean tropical cyclone activity related to the solar cycle. Geophysical Research Letters 35(18):L18,705, DOI 10.1029/2008GL034431

Elsner JB, Lehmiller GS, Kimberlain TB (1996) Objective classification of Atlantic hurricanes. Journal of Climate 9(11):2880–2889

Elsner JB, Lewers SW, Malmstadt JC, Jagger TH (2011) Estimating contemporary and future wind-damage losses from hurricanes affecting Eglin Air Force base, Florida. Journal of Applied Meteorology and Climatology 50(7):1514–1526, DOI 10.1175/2011JAMC2658.1

Elsner JB, Hodges RE, Jagger TH (2012) Spatial grids for hurricane climate research. Climate Dynamics 39:21–36.

Emanuel KA (1988) The maximum intensity of hurricanes. Journal of the Atmospheric Sciences 45(7):1143–1155

Embrechts P, Klüppelberg C, Mikosch T (1997) Modelling Extremal Events for Insurance and Finance. Springer-Verlag, Berlin.

Enfield DB, Mestas-Nunez AM, Trimble PJ (2001) The Atlantic multidecadal oscillation and its relation to rainfall and river flows in the continental U.S. Geophysical Research Letters 28(10):2077–2080

Epstein ES (1985) Meteorological monographs. No. v. 20, no. 42 in Meteorological Monographs, American Meteorological Society, URL `http://books.google.com/books?id=P9oWAQAAMAAJ`

Fotheringham AS, Brunsdon C, Charlton M (2000) Quantitative Geography: Perspectives on Spatial Data Analysis. Sage Publications, London.

Friedman JH (1991) Multivariate adaptive regression splines. The Annals of Statistics 19(1): 1–67, URL `http://dx.doi.org/10.2307/2241837`

Gelman A, Rubin DB (1992) Inference from iterative simulation using multiple sequences. Statistical Science 7(4):457–472, DOI 10.2307/2246093, URL `http://dx.doi.org/10.2307/2246093`

Gower JC (1971) A general coefficient of similarity and some of its properties. Biometrics 27:857–874

Grolemund G, Wickham H (2011) Dates and times made easy with lubridate. Journal of Statistical Software 40(3):1–25, URL `http://www.jstatsoft.org/v40/i03/`

Hartigan JA, Wong MA (1979) A k-means clustering algorithm. Applied Statistics 28:100–108

Heckert N, Simiu E, Whalen T (1998) Estimates of hurricane wind speeds by 'peaks over threshold' method. Journal of Structural Engineering ASCE 124:445–449

Hodges RE, Elsner JB, Jagger TH (2012) Predictive models for time-to-acceptance: An example using 'hurricane' articles in AMS journals. Bulletin of the American Meteorological Society 93:879–882.

Hoeting JA, Madigan D, Raftery AE, Volinsky CT (1999) Bayesian model averaging: A tutorial. Statistical Science 15(3):193–195

Jackman S (2009) Bayesian Analysis for the Social Sciences. Probability and Statistics, Wiley, URL `http://books.google.com/books?id=QFqyrNL8yEkC`

Jagger TH, Elsner JB, Saunders MA (2008) Forecasting U.S. insured hurricane losses. In: Murnane RJ, Diaz HF (eds) Climate Extremes and Society, Cambridge University Press, Cambridge, UK, chap 10

Jagger TH, Elsner JB (2006) Climatology models for extreme hurricane winds near the United States. Journal of Climate 19(13):3220–3236

Jagger TH, Elsner JB (2010) A consensus model for seasonal hurricane prediction. Journal of Climate 23(22):6090–6099, DOI 10.1175/2010JCLI3686.1

Jagger TH, Elsner JB, Niu XF (2001) A dynamic probability model of hurricane winds in coastal counties of the United States. Journal of Applied Meteorology 40:853–863

Jagger TH, Elsner JB, Burch R (2011) Climate and solar signals in property damage losses from hurricanes affecting the United States. Natural Hazards 58(1):541–557, DOI 10.1007/s11069-010-9685-4

Jagger TH, Elsner JB (2012) Hurricane clusters in the vicinity of Florida. Journal of Applied Meteorology and Climatology 51:869–877

Jagger TH, Niu XF, Elsner JB (2002) A space-time model for seasonal hurricane prediction. International Journal of Climatology 22(4):451–465

Jarrell J, Hebert P, Mayfield M (1992) Hurricane experience levels of coastal county populations from Texas to Maine. Technical Memo. 46, NOAA NWS NHC

Jarvinen BR, Neumann CJ, Davis MAS (1984) A tropical cyclone data tape for the North Atlantic basin, 1886–1983: Contents, limitations, and uses. Technical Memo. 22, NOAA NWS NHC

Jones PD, Jonsson T, Wheeler D (1997) Extension to the North Atlantic Oscillation using early instrumental pressure observations from Gibraltar and south-west Iceland. International Journal of Climatology 17(13):1433–1450

Kahle D, Wickham H (2012) ggmap: A package for spatial visualization with Google Maps and OpenStreetMap. URL `http://CRAN.R-project.org/package=ggmap`, r package version 1.2

Kamada T, Kawai S (1989) An algorithm for drawing general undirected graphs. Information Processing Letters 31(1):7–15

Kaplan A, Cane M, Kushnir Y, Clement A, Blumenthal M, Rajagopalan B (1998) Analyses of global sea surface temperature 1856-1991. Journal of Geophysical Research 103:18,567–18,589

Keitt T (2009) colorRamps: Builds color tables. URL `http://CRAN.R-project.org/package=colorRamps`, r package version 2.3

Keitt TH, Bivand R, Pebesma E, Rowlingson B (2012) rgdal: Bindings for the Geospatial Data Abstraction Library. URL `http://CRAN.R-project.org/package=rgdal`, r package version 0.7-8

Kelley D (2011) oce: Analysis of Oceanographic data. URL `http://CRAN.R-project.org/package=oce`, r package version 0.8-4

Kimberlain TB, Elsner JB (1998) The 1995 and 1996 North Atlantic hurricane seasons: A return of the tropical-only hurricane. Journal of Climate 11(8):2062–2069

Knuth DE (1992) Literate Programming. Stanford, California: Center for the Study of Language and Information, URL `http://www-cs-faculty.stanford.edu/~knuth/lp.html`

Kodera K (2002) Solar cycle modulation of the North Atlantic Oscillation: Implication in the spatial structure of the NAO. Geophysical Research Letters 29(8):1218, DOI 10.1029/2001GL014557

Koenker R (2005) Quantile Regression. No. 9780521608275 in Cambridge Books, Cambridge University Press, Cambridge URL `http://ideas.repec.org/b/cup/cbooks/9780521608275.html`

Koenker RW, d'Orey V (1987) Computing regression quantiles. Applied Statistics 36:383–393

Lacasa L, Luque B, Ballesteros F, Luque J, Nuno JC (2008) From time series to complex networks: The visibility graph. Proceedings of the National Academy of Sciences of the United States of America 105(13):4972–4975, DOI 10.1073/pnas.0709247105

Leisch F (2003) Sweave and beyond: Computations on text documents. In: Hornik K, Leisch F, Zeileis A (eds) Proceedings of the 3rd International Workshop on Distributed Statistical Computing, Vienna, Austria, URL `http://www.R-project.org/conferences/DSC-2003/Proceedings/`, ISSN 1609-395X

Lunn DJ, Thomas A, Best N, Spiegelhalter D (2000) WinBUGS - A Bayesian modelling framework: Concepts, structure, and extensibility. Statistics and Computing 10(4): 325–337

Madigan D, Raftery AE (1994) Model selection and accounting for model uncertainty in graphical models using Occams window. Journal of the American Statistical Association 89(428):1535–1546

Matheron G (1963) Principles of geostatistics. Economic Geology 58(8):1246–1266

Mei W, Pasquero C, Primeau F (2012) The effect of translation speed upon the intensity of tropical cyclones over the tropical ocean. Geophysical Research Letters 39, L07801, DOI: 10.1029/2011 GL050765

Milborrow S (2011a) earth: Multivariate Adaptive Regression Spline Models. URL `http://CRAN.R-project.org/package=earth`, r package version 3.2-1

Milborrow S (2011b) plotmo: Plot a model's response while varying the values of the predictors. URL `http://CRAN.R-project.org/package=plotmo`, r package version 1.3-1

Moran PAP (1950) Notes on continuous stochastic phenomena. Biometrika 37:17–33

Mousavi ME, Irish JL, Frey AE, Olivera F, Edge BL (2011) Global warming and hurricanes: The potential impact of hurricane intensification and sea level rise on coastal flooding. Climatic Change 104:575–597, DOI 10.1007/s10584-009-9790-0

Murnane RJ, Elsner JB (2012) Maximum Wind Speeds and US Hurricane Losses. Geophysical Research Letters, 39, L16707, DOI:10.1029/2012GL052740

Murrell P (2006) R Graphics. Chapman & Hall/CRC, Boca Raton, FL, ISBN 1-584-88486-X

Neuwirth E (2011) RColorBrewer: ColorBrewer palettes. URL `http://CRAN.R-project.org/package=RColorBrewer`, r package version 1.0-5

Newman M (2010) Networks: An Introduction. Oxford University Press

Ogi M, Yamazaki K, Tachibana Y (2003) Solar cycle modulation of the seasonal linkage of the North Atlantic Oscillation (NAO). Geophysical Research Letters 30(22):2170, DOI 10.1029/2003GL018545

Pebesma EJ (2004) Multivariable geostatistics in S: the gstat package. Computers & Geosciences 30:683–691

Pielke RA, Gratz J, Landsea CW, Collins D, Saunders MA, Musulin R (2008) Normalized hurricane damage in the United States: 1900-2005. Natural Hazards Review 9(1):29–42

Pierce D (2011) ncdf: Interface to Unidata netCDF data files. URL `http://CRAN.R-project.org/package=ncdf`, r package version 1.6.6

Plummer M (2011) rjags: Bayesian graphical models using MCMC. URL `http://CRAN.R-project.org/package=rjags`, r package version 2.2.0-4

Plummer M, Best N, Cowles K, Vines K (2010) coda: Output analysis and diagnostics for MCMC. URL `http://CRAN.R-project.org/package=coda`, r package version 0.14-2

Portnoy S, Koenker R (1997) The Gaussian Hare and the Laplacian Tortoise: Computability of squared-error versus absolute-error estimators, with discussion. Statistical Science 12: 279–300

Raftery AE (1996) Approximate Bayes factors and accounting for model uncertainty in generalised linear models. Biometrika 83(2):251–266

Raftery AE, Hoeting J, Volinsky C, Painter I, Yeung KY (2009) BMA: Bayesian Model Averaging. URL `http://CRAN.R-project.org/package=BMA`, r package version 3.12

Raftery AE, Zheng YY (2003) Discussion: Performance of Bayesian model averaging. Journal of the American Statistical Association 98(464):931–938

Rigby RA, Stasinopoulos DM (2005) Generalized additive models for location, scale and shape, (with discussion). Applied Statistics 54:507–554

Raftery AE, Gneiting T, Balabdaoui F, Polakowski M (2005) Using Bayesian model averaging to calibrate forecast ensembles. Monthly Weather Review 133(5):1155–1174

Ripley BD (1981) Spatial Statistics. Wiley, New York, NY

Ripley BD (1991) Statistical Inference for Spatial Processes. Cambridge University Press, Cambridge

Ripley BD (2011) tree: Classification and regression trees. URL `http://CRAN.R-project.org/package=tree`, r package version 1.0-29

Rupp J, Lander M (1996) A technique for estimating recurrence intervals of tropical cyclone-related high winds in the tropics: Results for Guam. Journal of Applied Meteorology 35(5):627–637

Sarkar D (2008) Lattice: Multivariate Data Visualization with R. Springer, New York, URL `http://lmdvr.r-forge.r-project.org`, ISBN 978-0-387-75968-5

Savitzky A, Golay MJE (1964) Smoothing and differentiation of data by simplified least squares procedures. Analytical Chemistry 36(8):1627–1639.

Scheitlin KN, Elsner JB, Malmstadt JC, Hodges RE, Jagger TH (2010) Toward increased utilization of historical hurricane chronologies. Journal of Geophysical Research-Atmospheres 115:D03,108, DOI 10.1029/2009JD012424

Schlather M (2011) RandomFields: Simulation and Analysis of Random Fields. URL `http://CRAN.R-project.org/package=RandomFields`, r package version 2.0.53

Spector P (2008) Data Manipulation with R. Springer, New York, NY, ISBN 978-0-387-74730-9

Stasinopoulos DM, Rigby RA (2007) Generalized Additive Models for Location Scale and Shape (GAMLSS) in R. Journal of Statistical Software 23(7):1–46

Teetor P (2011) R Cookbook. O'Reilly Cookbooks, O'Reilly Media Beijing

Therneau T (2012) A Package for Survival Analysis in S. R package version 2, 36–14.

Tibshirani R, Leisch F (2007) bootstrap: Functions for the Book "An Introduction to the Bootstrap". R package version 1.0-22

Trenberth KE (1984) Signal versus noise in the Southern Oscillation. Monthly Weather Review 112(2):326–332

Tsonis AA, Roebber PJ (2004) The architecture of the climate network. Physica A-Statistical Mechanics and Its Applications 333:497–504

Tsonis AA, Swanson KL, Roebber PJ (2006) What do networks have to do with climate? Bulletin of the American Meteorological Society [illegible]

Tufte E (1997) Visual Explanations: Images and Quantities, Evidence and Narrative. Graphics Press, Cheshire, CT.

Vickery PJ, Lin J, Skerlj PF, Twisdale LA, Huang K (2006) HAZUS-MH hurricane model methodology. I: Hurricane hazard, terrain, and wind load modeling. Natural Hazards Review 7:82–93, DOI {10.1061/(ASCE)1527-6988(2006)7:2(82)}

Walshaw D (2000) Modelling extreme wind speeds in regions prone to hurricanes. Journal of The Royal Statistical Society Series C (Applied Statistics) 49(Part 1):51–62

Wickham H (2007) Reshaping data with the reshape package. Journal of Statistical Software 21(12), URL `http://www.jstatsoft.org/v21/i12/paper`

Wickham H (2009) ggplot2: Elegant Graphics for Data Analysis. Springer, New York, NY URL `http://had.co.nz/ggplot2/book`

Wilkinson L (2005) The Grammar of Graphics (Statistics and Computing). Springer-Verlag, New York, Inc., Secaucus, NJ, USA

Wilks D (2006) Statistical Methods in the Atmospheric Sciences. International Geophysics Series, Academic Press, Oxford.

Winkler RL (2003) An Introduction to Bayesian Inference and Decision, 2nd Edition. Probabilistic Publishing, ISBN 0-9647938-4-9

Zeileis A, Kleiber C, Jackman S (2008) Regression models for count data in R. Journal of Statistical Software 27(8), URL `http://www.jstatsoft.org/v27/i08/`

INDEX

www.ingramcontent.com/pod-product-compliance
Lightning Source LLC
LaVergne TN
LVHW011835200926
848484LV00002B/359